Introduction to
TECHNOLOGY

About the Authors

Dr. Alan J. Pierce is a Professor in the Department of Technology at Elizabeth City State University. He has been an educator for 33 years at the elementary, junior high, and college levels. During that time he has authored numerous articles in the fields of industrial arts, technology, supervision, and computers. He co-authored the NYC Careers 4 Children (C4C) Curriculum and wrote the computer and technology sections for the ACAST program. This experimental teaching project placed technology and computer instruction as the core for teaching academics to at-risk students. In 1997 Dr. Pierce was the technical consultant for the children's book *Discover How Things Work*.

Dr. Pierce was the recipient of a national research award from the American Vocational Association and was chosen by New York State as their 1990-91 Ambassador of Occupational Education. In 1997 he received a Departmental Teaching Excellence Award from Elizabeth City State University.

He created the "Technology Today" column in *Tech Directions* magazine and has written that column since August 1995. He is an active member in ITEA and serves on their awards and CTTE accreditation committees.

He received his Bachelor of Science and Master of Science degrees in education from The City College of New York. He received his doctorate in Vocational-Technical Education from Rutgers University.

Dennis Karwatka is a Professor in the Department of Industrial Education and Technology at Morehead State University in Morehead, Kentucky. He is a full-time teacher who received his university's Distinguished Faculty Award in 1988. The courses he has taught include basic electricity, introduction to power, quality control, and structural design.

He also taught photography and woodworking for many years in the high school Upward Bound Program and has been a Scout leader for more than 20 years. He has written two middle/high school textbooks, a technical history book, and numerous technical articles.

He created the "Technology's Past" column in *Tech Directions* magazine and has written that column since 1980. His industrial background includes engineering positions with Boeing Company and Pratt & Whitney Aircraft.

Introduction to

TECHNOLOGY

Alan J. Pierce, Ed.D.
Professor of
Technology Education
Industrial Technology
Elizabeth City State
University
Elizabeth City,
North Carolina

Dennis Karwatka
Professor of
Industrial Education
Morehead State University
Morehead, Kentucky

Glencoe
McGraw-Hill

New York, New York   Columbus, Ohio   Woodland Hills, California   Peoria, Illinois

Cover Image: Gregory McNicol

Glencoe/McGraw-Hill

A Division of The **McGraw·Hill** *Companies*

Send all inquiries to:
Glencoe/McGraw-Hill
3008 W. Willow Knolls Drive
Peoria, IL 61614

ISBN 0-02-831275-9 (Student Edition)

Printed in the United States of America

3 4 5 6 7 8 9 10   003/043 02 01 00

Dedication

This book is dedicated to the students and teachers who find learning at its best when it is coupled with real or, at least, simulated life experiences. This book is dedicated to my wife, Shelley, and my children, Stacey and Seth, for without their sacrifices this project could not have been completed.

Alan Pierce

This book is dedicated to my wife and best friend, Carole, who has been with me through fair weather and foul. Also, to my son Alex and daughter Jill who have honored me by being good at what they do. Last but not least, this book is dedicated to all the youngsters who have experienced the thrill of confronting a problem and solving it. May your technical successes be as rewarding for you, as mine have been for me.

Dennis Karwatka

Contents in Brief

■ Table of Contents ■

■■■ **Section II** ■■■ **Communication** **131**

—Chapter 5—
Understanding Communication
Technology 133

■ Section III ■ Production　279

■ Section IV ■ Transportation   383

■■ Section V ■■   Special Topics   457

—Chapter 15— Biotechnology   459

—Chapter 16— Tomorrow's Technology   489

■ Putting Knowledge to Work ■

■ Credits ■

■ PREFACE ■

Technology is *amazing* in the way that it has dramatically changed and improved people's lives over the past 150 years. Technology is *exciting* when one thinks about the technical innovations that are coming—the technical advances that will radically change the way in which people will lead their lives on this planet, and possibly on other planets, in the years to come. And perhaps most importantly, technology is *thrilling* when one actually participates in technology, identifying and solving problems, creating something entirely new and unique with one's own hands.

The major goal of *Introduction to Technology* is to create in students the same amazement, excitement, and thrill that is shared by the authors and others who understand and work with technology on a daily basis. The textbook was designed and written to be fun and interesting for students, so that it might thereby spark an enthusiasm that will hopefully lead to further study and a lifetime appreciation of technology.

Here are a few of the special text features that attempt to generate student enthusiasm for technology:

- **Chapter introductions**, which capture student interest and motivate students to read and study the chapter.
- **Illustrations**, which number over 400 and are as large and colorful as possible, to appeal to today's visually oriented students.
- **Tech Talks**, which are brief discussions of technology related terms, such as "telecommunication," "schematic," "lasers," and "photon surgery," often containing little known but extremely interesting bits of information about technology.
- **Putting Knowledge to Work**, which are high-interest features about important people and innovations in technology, ranging from Ben Franklin's experiments with electricity to "Building an Anti-Gravity Machine."
- **Applying What You've Learned**, which provide a variety of activities within the chapters so students can *do* technology. Examples are constructing Lego traffic signals, metal castings, and model bridges.
- **Problem-Solving Activities**, which are additional, more fully developed, hands-on activities for students at the end of the chapters.

In addition to its main goal of generating student interest and enthusiasm for technology, *Introduction to Technology* strives to teach students the most important concepts and skills of technology. Students are not expected to become computer technicians or highly skilled woodworkers by studying the text, but they should be expected to learn and understand how technology has developed, the major types of technology, its resources and methods, and many of its key principles.

Realizing that middle and junior high students often lack the discipline and know-how to fully comprehend and retain what they read, *Introduction to Technology* has been designed to maximize student learning. A major component of this design is the reading and study technique known as the **SQ3R method**, which has been acknowledged by educators for years as one of the most effective ways to improve student comprehension and retention. Following is a description of SQ3R and the ways in which it has been incorporated into *Introduction to Technology*. SQ3R stands for the five study steps of *survey, question, read, recite,* and *review*.

S = **Survey**—Students will get an overall idea of what a chapter is about by skimming through the chapter before reading it. Using the "Introducing the Chapter" annotation, encourage students to skim the Chapter Outline, read the introduction, and look at the opening photograph for an idea of what the chapter is about. Also have students focus on chapter headings, pictures and illustration captions throughout the chapter, and the summary at the end . . . before they begin reading the chapter. This step alone will improve comprehension.

Q = **Question**—Too often students fail to focus on the key concepts of the passages they are reading. Turning each topic heading into one or more questions will help them focus on the most important concepts. Many of the headings in this text have already been converted into questions for them. Simply direct your students' attention to the questions under the main headings. These questions should help stimulate the students' interest and increase their comprehension. These questions should also help students get into the habit of turning headings into questions on their own.

R1 = **Read**—The first R in SQ3R refers to *read*. Students should read short sections, from one topic heading to the next, trying to answer the questions under the headings. The abundant use of headings and the *Learning Time* boxes serve to divide the reading into manageable sections for the students.

R2 = **Recite**—The second R stands for *recite*. When students have finished reading a section, they should silently *recite* the answer to the question. If they can't answer the question, they should reread the section until they are able to answer the question. There's no point in going on until students understand and remember what they've already read. The "question-read-recite" cycle should be repeated for each heading throughout the chapter. The *Learning Time* questions at the end of each major section can also be used for "reciting" information learned in the chapter.

R3 = **Review**—When students have finished reading they should skim over the chapter again, questioning and quizzing themselves and reciting answers as they go. Use the *Summary* and *Key Word Definitions* at the end of the chapter to further help your students review.

Studies have shown that SQ3R improves both reading comprehension and grades. The special study features of this text, such as the questions under the headings, the *Learning Time* boxes, and the *Summary* at the end of the chapter will help you help your students make the most effective use possible of the SQ3R method.

Introduction to Technology was designed and developed not only for students, but also for teachers. The intention was to produce a useful tool, like any other in your classroom/laboratory—one that you can count on and use effectively in a variety of situations. The textbook itself, with its high student appeal, flexible organization, and effective SQ3R study plan should enable you to obtain maximum involvement and results from your students.

To help minimize the amount of time you need to spend in preparing to teach with *Introduction to Technology*, we have developed the following supplements:

- **Teacher's Annotated Edition**—contains a variety of annotations including ideas to provoke students' interest, material relating to the students' personal lives, critical thinking and problem-solving activities, and ideas for math problems students can solve. A Teacher's Manual is also included in the *Annotated Edition*. Included here are chapter-by-chapter teaching aids and a number of articles on various topics to assist the teacher in teaching this class.
- **Teacher's Resource Binder**—includes daily lesson plans and transparency masters, and a wealth of reproducible activity sheets on topics such as technology careers, technology consumerism, how things work, and the effects of economics on technology. The binder also includes reteaching and interdisciplinary activities. Also included in this second edition is a test section with pretests, quizzes, chapter tests, and a final examination.
- **Student Activity Manual**—contains step-by-step projects and problem-solving activities in the areas of communication, production, transportation, and biotechnology.

ACKNOWLEDGMENTS

Without the assistance of many special people, this project could not have been undertaken or completed. The authors wish to express their special thanks to the following people: Alan J. Jones, a publisher with Caddo Gap Press, who originally recommended Alan Pierce as an author for this book; Shelley Pierce, who assisted in the research and editing of the manuscript and galleys; Fanny Ryngel, a reading specialist in New York City, who helped the authors gain expertise in the CLOZE method of teaching reading comprehension; *Tech Directions* magazine for permission to use material from our "Technology's Past" and "Technology Today" columns; Ron McQueen, a developer at Pitsco, for assistance with the update on the LEGO DACTA activity; Chuck Twiddy, CommTech Computers, and Diane Patterson, University Graphics at ECSU, who kept us on the cutting edge of computer, electronics, and graphic arts technology; Bob Cassel, Glenda Samples, Lynda Kessler, and Mario Rodriguez who helped breathe life and form into the original manuscript; and Trudy Muller, Editorial Director, and John Gorman, Senior Editor, of Glencoe/McGraw-Hill for breathing new life and form into the revised edition.

The authors and publisher would also like to thank the following reviewers/consultants and writers for their valuable suggestions and comments throughout the development of this textbook:

John Allen
Dan McCarthy Middle School
Fort Pierce, Florida

Bob Bauer
Spring Hill Middle School
Akron, Ohio

Chris Chamuris
Lawrence Middle School
Lawrence, New Jersey

Richard Christie
Randolph Intermediate School
Randolph, New Jersey

Sandra Gingrich
J.W. Leary Junior High School
Massena, New York

Marie Hoepfl
Appalachian State University

Wanda Jones
Horace Mann Middle School
Amarillo, Texas

Robert McFarlin
Gunn Junior High School
Arlington, Texas

David Sharp
Sam Rayburn Middle School
San Antonio, Texas

Brian Stemmerman
Lincoln Junior High School
Ypsilanti, Michigan

Clay Taylor
Henderson Junior High School
Little Rock, Arkansas

Brian Webb
Shaker Junior High School
Latham, New York

Barry Wilson
John F. Kennedy Junior High School
Utica, New York

William Youngfert
Herricks Middle School
Albertson, New York

A World of Technology

Did you surf the 'Net today? Ride in a car? Open a door? Eat a bowl of cereal? You may be surprised to learn that *all* of these activities involve technology. This section introduces technology: what it is and how it works. As you read the chapters in this section, ask yourself, "How is technology affecting *my* world?"

Learning About Technology

Chapter 1

Introduction to Technology

Key Words

In this chapter you will learn and study the meanings of the following important technology terms.

TECHNOLOGY
TECHNICIAN
TECHNOLOGIST
SCIENCE
ENGINEERING
COMMUNICATION
PRODUCTION
TRANSPORTATION
ADAPTING

What is **technology**? Let's answer that question backward by asking, What *isn't* technology? Look out a window. Perhaps you see trees, plants, rocks, or other natural items. That's what technology isn't: anything created by nature. Sand doesn't involve technology, and neither does water.

People take those natural materials and turn them into useful products. Trees are turned into lumber to make houses and furniture. Plants such as asparagus and Brussels sprouts are put into cans and sold as food. Special rocks like iron ore are turned into steel, and sand can be changed into glass. Water can be used to make soft drinks.

Technology involves turning natural items into useful products. It is involved in all the products, inventions, and discoveries made by people. Technology can be summed up in one simple sentence. *It is the practical use of human knowledge.* Technology provides us with everything we use in our society. How much do you know about technology? Do you know how inventors invent? Do you know what makes your stereo or bike work the way it works? Do you understand how technology can hurt your environment?

In this book you are going to learn the answers to these questions and many more *about* technology. You will also learn to *do* technology. And best of all, you will probably learn to *enjoy* technology.

Understanding the Word *Technology*

technologist
a person who has special training and skill in technology

technician
a person whose occupation requires specialized training in a certain subject area or who is known for having skill in a particular area

Where do you think the word technology *comes from?*

How many different ways have you heard people use the word *technology*? Maybe you've heard that we live in the *age of technology*. Some folks talk about *high-technology*, or "high-tech," careers. Others mention *industrial technology*, *medical technology*, or *agricultural technology*. See Figure 1.1. There seems to be no end to the ways people use the word *technology*.

The word *technology* comes to us from the ancient Greek word *tecknos*, which means "art." You might think that art means only paintings, and maybe sculpture. To the Greeks, it meant much more. They felt it took a real artist to make useful products from natural materials. They never thought that technology abused nature. They believed instead that technology combined nature with the human spirit. Many centuries later, Walter Chrysler, the man who started the car company, expressed the same idea. He said, "Someday I'd like to show a poet how it feels to design and build a railroad locomotive."

Early workers in technology called themselves *artisans*. It's still a name we hear from time to time. It referred to a highly skilled worker or craftsperson.

During the 1800s, the unusual name of *mechanician* started to be used. From this came our word *mechanic*. The word *mechanician* also came from a Greek word—*mechanikos*, meaning "machine." Two modern names for people who work in technology are **technician** and **technologist**.

Figure 1-1 Checking electronic circuits requires well-educated technologists who use modern up-to-date measuring equipment and their own professional judgment.

Strictly speaking, there is a difference between the two words. A technician is a person skilled in any subject. Writers, artists, and musicians are also technicians. The name *technologist* is more specific. A technologist is a person skilled in some area of technology. However, it is customary to use either word to describe a person who works in technology.

Technicians and technologists usually use equipment as part of their job. Some operate X-ray equipment in hospitals. Others work with industrial robots in factories. Technicians install and operate electronic equipment such as home security alarms. Aircraft technologists inspect and repair airplanes. You can find many different jobs listed under the *Careers* heading in your school's encyclopedia.

Science, Technology, and Engineering

How are the areas of science, technology, and engineering different?

You may notice that the words *science* and *technology* are often used together. A classmate might say, "I want to study science and technology in school." The National Aeronautics and Space

Putting Knowledge To Work

The Brake Fixer

After school one day, Joshua Lewis and David Zewinski were pedaling their ten-speed bikes. As they went down a long hill, David noticed that Josh's bike was speeding up. At the bottom of the hill, Josh ran off the road into Mrs. Jelnik's yard, and was stopped by her bushes. Fortunately, the bike, the bushes, and the rider were not hurt.

"My brakes didn't work! I tried to squeeze the handles as hard as I could, but they wouldn't stop the bike!" Josh was really upset. David calmed him down and then looked at the brakes. He noticed they were way out of adjustment. Without a word, David went to the small tool kit under his bike's seat. He returned with a small wrench and a screwdriver. While explaining what he was doing, David adjusted Josh's brakes.

"I didn't know you could do that," said Josh. David answered, "Oh, I can do lots of stuff. I can adjust the chain, check tire pressure, oil the wheel bearings, and figure out gear ratios. My whole class spent some time fixing bikes and learning how they work." "You mean at school?" asked Josh. "Sure do," said David, "in my technology class."

Administration (NASA) frequently makes comments about "American scientific and technological growth." And your dictionary no doubt uses the word *science* in its very definition of the word *technology*. Although science and technology are related, they're not the same. **Science** explains *why* things happen. Technology *makes* things happen.

For example, the German scientist Wilhelm Roentgen discovered X rays in 1895. As best he could, he tried to explain why X rays were produced. His work helped people develop practical and safe X-ray equipment for hospitals. Today, technicians use that equipment to make X-ray photographs. Such photographs help locate broken bones. Roentgen's *scientific* discoveries led to other people developing new, *technological* equipment and processes.

Scientific effort created electronic chips. Technology used those chips to make automatic cameras, videocassette recorders, tape decks, and many other electronic devices. Scientists discovered nuclear power before technologists built nuclear power plants for generating electricity.

We don't always need science to do technology, however. People made objects out of bronze 5,000 years ago, long before there was a branch of science known as *metallurgy*. Before the sciences of *chemistry* and *biology* were identified, people used certain plants to help cure injuries and illnesses.

Another profession important to technology is **engineering**. It fits between science and technology. Of all the world's professions, only military leadership and religious leadership are older than engineering.

Engineers decide how to make things. For example, chemical engineers design machines that produce plastics and other materials. Civil engineers design bridges and skyscrapers. Electronic engineers design computers and other electronic devices. Mechanical engineers design engines and airplanes. There are also chemical, civil, electronic, and mechanical technologists. They build the products engineers design. See Figure 1.2.

A simple way to remember how each of these three professions is different is to think of the sentence, "**S**ome **W**ild **E**arthquakes **H**ave **T**remendous **M**otion."

Science: Why something happens
Engineering: How to turn the science into something useful
Technology: Making something useful

If you're confused about the differences between science, engineering, and technology, don't feel alone. Many people are. To add to the confusion, three of our country's best *engineering* colleges are the California Institute of Technology, the Georgia Institute of Technology, and the Massachusetts Institute of Technology. Those schools offer college degrees in engineering and science.

Figure 1-2 Technologists take engineering ideas from a computer monitor and convert them to drawings that can be used to make a product.

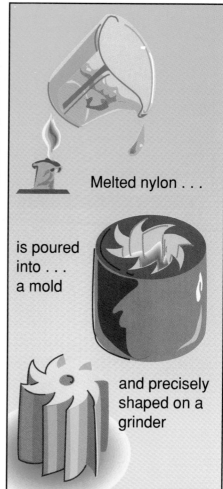

Melted nylon . . .

is poured into . . . a mold

and precisely shaped on a grinder

Figure 1-3 Nylon is turned into useful products by technologists who operate melting, molding, and shaping equipment.

Carothers
(kə-rŭ*th*′ərz)

Let's use the 1935 discovery of nylon to see how the three professions fit together. Nylon was the first plastic strong enough to replace metal. It is regarded as one of the most important discoveries of twentieth-century chemistry.

- *Science* - Wallace Carothers was a chemist for E. I. du Pont de Nemours & Co. in Delaware. The company had hired him to establish a scientific laboratory in Wilmington. One of his experiments was combining two chemicals with unusual names—hexamethylenediamine and adipic acid. Carothers heated the carefully mixed solution in a container much like a pressure cooker and made the first batch of nylon.

- *Engineering* - Engineers tested the hardened nylon to see how strong it was. They found that nylon was weaker than metal, but it was naturally slippery. Nylon didn't need to be oiled. The engineers wanted to use it for gears in small motors and for sliding surfaces under doors. They designed gears by deciding how big they could be and still work well. Nylon was fine for small parts like clock gears, but it wasn't strong enough for automobile or truck gears.

- *Technology* - The final step was to make the gears and be sure all the parts fit together. Technologists made molds to shape the molten nylon. They operated machines that cut the gears to the precise shape. See Figure 1.3. Technologists assembled all the parts and made sure the finished product worked properly. They were the first ones to see the completed product. It provided a strong sense of satisfaction. Some people say that this alone makes a job in technology more rewarding than any other.

7

Why We Study Technology

Why do you think the average person needs to understand technology?

Why study technology? That's an easy question to answer. It's fun, rewarding, exciting, and important to our national security. Technology will also help you develop your problem-solving skills. You will learn to identify a problem and come up with a solution.

Technology is fun because you get to work with tools and materials. Instead of only reading about bridges, you might build a model of one. Instead of only reading about electricity, you might make a small electric motor. Instead of only reading about wood, you might make something useful out of it. It's always fun to work with your hands. See Figure 1.4. Technology is always an up-to-date subject. You'll learn how to make and repair many things that can help you—not only in the future, but also right now.

Technology is rewarding because you can see the results of your work. People who put space shuttles together have a direct

■ **Figure 1-4** ■ People like to work with wood because it's a natural material that can be turned into many different items.

connection to the quality of the rocket. They are the ones who tighten the bolts, assemble the electronic equipment, and test the controls. A successful flight frequently brings tears of joy to many of the workers. The rocket means more to them than just a piece of equipment.

Technology is exciting because each day brings new ideas and new challenges. Some days you and your class might work on computer projects and other days on engine projects. It will be that way if you become a full-time technologist. No two days will be the same. Think how dull your life would be if you always did the same thing. That simply doesn't happen in the field of technology.

Studying technology is important to our national security because it helps keep America technically strong. National defense relies heavily on properly trained and dedicated technologists. In a world with many different forms of government, modern military equipment must be carefully manufactured to help maintain our freedom. See Figure 1.5.

Technology is important in solving problems related to many other national concerns. For example, technology is working to provide pure water to an ever-increasing population. Technology also

Figure 1-5 The F-117 Stealth fighter airplane is almost invisible to radar because of its unusual shape and other technical advancements.

Putting Knowledge To Work

Rome or Bust

Mary Goldstein's history teacher had worked for weeks to organize a field trip to the art and history museum in a large nearby city. The teacher wanted her students to learn more about the ancient Romans. The day they left, everyone rode school buses with "Rome or Bust" written on signs displayed in the windows.

Mr. Maki, the art teacher, led Mary's group of twelve through the large museum. He spoke expertly about the sculptures and the models of ancient buildings. He had a little trouble at one display of Roman road construction. Mary quietly said, "I think I know what that's about, Mr. Maki." He asked her to share her knowledge with the group.

Mary immediately began to explain how the Romans had built 50,000 miles of highway. She said that the Romans were the first to make roads higher at the center so that water would run into the ditches they dug on both sides. She told the group that modern roads are based on designs developed by the Romans. Many of the students asked questions. Mary answered them all.

Mr. Maki thanked her for helping out and casually asked, "Where did you learn all that interesting information?" Mary smiled and said, "In my technology class."

lets us maintain communication links with fire departments and hospitals. Efficient automobiles will extend our limited fuel supplies. New construction techniques can quickly replace older buildings. The list is almost endless.

You will also receive many immediate benefits from studying technology in school. You might learn how to repair your bicycle, make a birdhouse, or launch model rockets. Your friends will come to you when they have questions about how electric motors work, or why a two-by-four isn't two inches by four inches, or what the gear ratios mean on a ten-speed bicycle. You'll know all that "stuff" and much more.

— Learning Time —

Recalling the Facts

1. What is industrial technology?
2. Is sand an example of technology? Explain.
3. What is the meaning of the Greek word *tecknos*?
4. What is the difference between science and technology?
5. Who decides how big a gear should be: a scientist, an engineer, or a technologist?
6. Who makes the gear: a scientist, an engineer, or a technologist?
7. What are four reasons for studying technology?

Thinking for Yourself

1. Does wood involve technology? Explain.
2. The Greeks thought that technology combined nature with the human spirit. What do you think that means?
3. Why is the word *technologist* more specific than the word *technician*? Explain.
4. Give two examples of technological advances that did not first require science. Don't use any example from this section of the book.
5. Give two examples of technological advances that did first require science. Don't use any examples from this section of the book.

Applying What You've Learned

Convert some natural materials into useful products:
1. Tree branches into small wooden items
2. Locally obtained rocks into jewelry or other decorative items
3. Dyed sand into paintings, as done by native Americans
4. Grass or straw into woven hot pads

Technology Careers

What do technologists do?

It's never too early to think about a career. Many world-class dancers, gymnasts, and athletes started training before they entered first grade. When you were a youngster, you may have wanted to be a fire fighter or a nurse or a teacher. Those are all excellent career choices. Now that you're a young adult, it's possible you may have changed your mind and are looking at other careers.

If you like technology, you might want to be a scientist, engineer, or technologist. If you like animals, you might want to be a farmer or a veterinarian. If you like the field of medicine, you might want to be a physician, or a pharmacist. Now is a good time to give some thought to what you want to be. See Figure 1-6. You don't have to decide what job you want, but you can be thinking about it in general terms.

This book will give you an idea of what technology is like. Three main sections of the book deal with **communication, production,** and **transportation.** People have found that practically all forms of industrial technology fit into one of those three general areas.

What do I want to be?

Figure 1-6 You can be whatever you want to be: builder of skyscrapers, surgeon, farmer, or anything else.

communication
the process of exchanging information between individuals. This can be done through behavior or by use of symbols or signs.

production
making products available for use

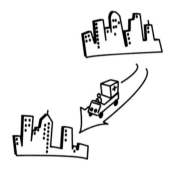

transportation
moving people and products from one place to another

- *Communication* - This refers to communication of all different types. It can mean information shown on a computer monitor, or reading the plans for a new building, or printing your school's newspaper. See Figure 1.7. Of course, there are other types of communication too. Can you think of what some might be?
- *Production* - This refers to making products. That includes manufacturing clothing, building a new restaurant, and any other activity that results in a finished product. Can you think of other types of production?
- *Transportation* - This refers to moving people and products from one place to another. We use cars, trucks, airplanes, trains, ships, and other vehicles for transportation. See Figure 1.8. Can you think of other examples of transportation?
- *Combinations* - Few technologies exist by themselves. It is more common to find them overlapping each other. For example, maybe you like electronics. Electronics fits into the communication field because of television sets and telephones, for example. It fits into production because companies use computer-controlled machines like industrial robots. It fits into transportation because all modern cars have electronic engine control.

Perhaps you like photography and printing. Those two certainly fit into communication, since people look at photographs and read magazines. Production companies use photography in their reports, and they use printing to reproduce information for all employees. Transportation uses printing because all cars, trucks, airplanes, and other vehicles have operation manuals, repair manuals, and parts manuals.

■ **Figure 1-7** ■ Word processing and desktop publishing programs help students put a school publication together.

■ **Figure 1-8** ■ Modern highways provide an efficient way to move large numbers of people to many different locations. Trains, another form of transportation, move materials from one location to another.

——— Technology's Progress ———

≡ *How does technology change to meet people's needs?*

In everything we do, we build on the efforts of people who came before us. Your teachers teach you a bit more or a bit better than they were taught. See Figure 1.9. Our government built onto the U. S. Constitution by adding the Bill of Rights. We continue to build onto the English language. We often add brand new words such as *fax* (to send a facsimile of a written message) and *rapper* (a person who talks to the beat of rap music).

■ **Figure 1-9** ■ Not too many years ago, American youngsters began their education in crude one-room schools staffed by dedicated teachers. Today's classrooms are more like the one on the right.

Edison's cylindrical records adapted to . . .

Flat records that adapted to . . .

Cassette tapes that adapted to . . .

Compact disk recordings

Figure 1-10 Because new technologists always have fresh ideas, every aspect of technology continually changes.

Isaac Newton was a famous British scientist during the 1700s who investigated the motion of the planets. He once said, "If I have seen further than others, it's because I've stood on the shoulders of giants." Newton didn't really stand on anyone's shoulders. He meant that his accomplishments were based on the work of previous people. That's something you share with Newton. Everyone builds on what came before.

In technology, we build on the accomplishments of early artisans, mechanicians, technicians, engineers, and scientists. Even Abraham Lincoln helped out technology. He's the only president to receive a *patent*. A patent is a legal document issued by the government. It gives a person sole rights to any invention they develop. Lincoln's patent No. 6469 was an unsuccessful attempt to float barges over shallow river sections.

Strange as it may sound, failures are a type of success. Thomas Edison had some trouble finding a material for the filament in his incandescent lamp. At one point he said, "Now I know 2,000 materials that won't work." Many of the successes in technology have been made possible by "failures" like Edison's.

Today's technology might seem to be far advanced over the technology of earlier eras. However, the telegraph was as important to people 100 years ago as telephones are today. Factories powered by flowing streams were as modern in 1800 as any factory now in operation. The steam engines of 1850 were as advanced as jet engines are today. All these examples were part of the evolutionary process of progress.

New Technologies

What are some examples of brand new technologies?

Thank goodness that technology is continually growing. Growth keeps the profession fresh and interesting and allows it to change as new products or processes enter our lives. Thomas Edison invented phonograph records, but he knew nothing about tape recordings. Tapes are now making room for compact disk recordings. This evolutionary process is normal and never stops. See Figure 1.10.

Boeing 707 airplanes were designed by people who used slide rules for calculating. Today's airplanes are designed by people who use pocket calculators and desktop computers. Technology continually moves forward by *adapting*, so that each upward step is an improvement over existing products.

Some technologies, however, are completely new. *Biotechnology* is an example. Biotechnology deals with health, plants, waste disposal, and other biological-type topics. Many people are finding exciting and rewarding careers in this new area of technology. You will learn more about biotechnology in Chapter 15.

Perhaps sometime in the future, we will see hydrotechnology (water) or econotechnology (economics) or geotechnology (geography). That's okay. There's enough room in technology's house for everyone.

Teenage Technology

How have teenagers contributed to the development of technology?

Believe it or not, technological progress was sometimes started by teenagers. Have you ever heard of George Westinghouse? He started a company that still uses his name. Westinghouse is known for inventing improved brakes for trains. However, when he was only nineteen, he patented a new type of steam engine. See Figure 1.11. It wasn't a successful invention, but at least he tried. Perhaps his experience with that steam engine encouraged him to invent the train brakes.

Westinghouse wasn't the only technically minded young adult. George Washington, our first president, was a self-taught surveyor long before he became a military leader. He surveyed 5 million acres for the largest landowner in Virginia and helped lay out the city of Alexandria. He was appointed official surveyor of Culpeper County, Virginia, when he was only seventeen.

■ **Figure 1-11** ■ George Westinghouse started his train brake company when he was 25. This photograph was taken many years later.

■■■ Putting Knowledge To Work ■■■

The Video Game

One rainy Saturday, Sarah Orteig was at Ann Laflamme's house to play a video game. One of the joysticks wouldn't move the screen figure to the right. "Phooey," said Ann, "I wanted to learn how to play this new game."

Sarah suggested they take the joystick apart to see if they could tell what was wrong. Ann wasn't so sure, and Sarah had to talk her into it. Ann went to the tool drawer in the utility room. She returned with a small screwdriver. She gave it to Sarah, who used it to remove four small screws from the bottom of the joystick.

"We have to be very gentle with electrical and electronic parts," said Sarah. She carefully separated the two halves of the joystick base and looked inside. Some metal pieces had a thin sticky coating. After a bit of detective work, they decided that Ann's brother may have accidentally spilled soda pop inside.

After unplugging the game, Sarah asked Ann for some cotton and rubbing alcohol. She cleaned the electrical contacts by wiping them with a small amount of alcohol on the cotton. Sarah gently reassembled the joystick.

They tried it out, and the joystick worked perfectly. Ann couldn't believe that her best friend had just repaired such a complicated electronic part. She asked Sarah where she learned to do it. Sarah answered, "In my technology class."

Elmer Sperry was a poor youth raised in hilly central New York State. At fourteen, he invented a swiveling head lamp for locomotives so the engineer could see around curves. Although the head lamp wasn't successful, the gyrocompass he later invented was one of the world's most remarkable inventions. Sperry's gyrocompass is used in all ship and airplane guidance systems.

— Learning Time —

Recalling the Facts

1. Give three examples of how communication fits into technology.
2. Give two examples of how production fits into technology.
3. Give five examples of how transportation fits into technology.
4. What does the word *adapting* mean?
5. Give two examples of how technology adapted to new methods.

Thinking for Yourself

1. Give three examples, not given in this section of the book, of how communication fits into technology.
2. Give three examples, not given in this section of the book, of how production fits into technology.
3. Give three examples, not given in this section of the book, of how transportation fits into technology.

Applying What You've Learned

Investigate a favorite career in technology. Make a display board or diorama about it.

—— Superstars of Technology ——

Who were some important technologists of the past, and what were their accomplishments?

The professional lives of our early technologists make fascinating reading. It is interesting to learn how these people wouldn't give up until they had solved the problems facing them. Their hard work and problem-solving abilities have made life easier and more enjoyable for all of us.

Here are a few lesser-known stories of some important people in technology. You will be reading about many more technologists as you read through this book. Some are famous—you will have learned about them in your social studies classes. Others are not so famous. Still, we like to think of all of them as the "superstars of technology."

Tech Talk

Instead of a magnet, Sperry's *gyrocompass* used a high-speed motor to always point north. A rapidly rotating disk will always point in one direction. That's the *gyroscopic effect* principle. Sperry's gyrocompass, which he named "Metal Mike," was first used on a ship in 1911.

Communication Pioneers

Which technologists developed fresh ideas in communication?

Albert Blake Dick (1856–1943). You might notice your teachers using products with the brand name AB Dick on them. That company was founded by the man who invented a method of duplicating called *mimeograph*. He made up the word from the Greek *mimeos*, which means "to draw," and *graphos*, which means "writing." Modern mimeograph machines are used in every school in the United States.

Dick worked for a lumber company near Chicago, and he sometimes had to send fifty of the same letters. Each had to be written individually. At that time, there was no simple way to make many copies.

One day, he ate a piece of candy that had been wrapped in waxed paper. Dick casually placed the waxed paper over a fingernail file. He gently dragged a metal nail over the paper. Holding the paper up to the light, he saw many holes along the line he made with the nail. He thought, "Maybe ink could be forced through the holes to form an image of the line." It could. The idea turned out to be the principle of the mimeograph duplicator.

Dick discovered that Thomas Edison had patented a vibrating pen that also made holes in waxed paper. Edison's invention wasn't practical, but his waxed paper was better than Dick's. The two formed a partnership and started manufacturing mimeograph materials. They eventually became such close friends that Dick named one of his sons Edison.

Chester Carlson (1906–1968). What we sometimes call *Xerox* copies are really *xerographic* copies. The word comes from the Greek *xeros,* which means "dry," and *graphos*, which means "writing." Before xerographic images were made, only photographic techniques could make exact duplicates. Those copies required wet chemicals and a darkroom.

While working for an electronics firm, Carlson noticed that there never seemed to be enough copies of the electronic circuits. He thought about ways to make dry copies. Carlson spent his evenings in a small laboratory he set up in a rented room in Astoria, New York. After a great deal of experimenting, he made the world's first xerographic image in 1938. He had written simply, "10-22-38 ASTORIA."

In today's world, we are used to instant copies and take them for granted. We might forget that people didn't always feel the need for so much duplication. In 1938, no one was interested in the process because they could see no need for it. Carlson was turned down by twenty companies. See Figure 1.12.

Finally, a research organization began to work with Carlson. Companies eventually became interested and the first dry office

■ **Figure 1-12** ■ Using light, chemicals, heat, and waxed paper, Chester Carlson produced the world's first xerographic image in 1938.

17

copying machine was made available to the public in 1959. It was the Xerox 914, manufactured by the Xerox Corporation.

Steven Jobs (1955–) and Stephen Wozniak (1950–). This team doesn't go back as far as other pioneers, but their work was certainly pioneering. Together they designed and assembled the Apple II computer in 1977. It was the world's first widely accepted personal computer.

Steven Jobs and Stephen Wozniak met as teenagers at the Homebrew Computer Club in Cupertino, California. The club is considered the first computer users group. They found a common interest in electronics. Together, the two Steves designed and assembled an early version of their computer.

To get money to build and market a computer for the public, Jobs sold his Volkswagen bus and Wozniak sold his Hewlett-Packard calculator. Their first Apple computers were made in Jobs's garage and became an immediate success. The name Apple was chosen because it was a simple word and because Jobs once spent a summer picking fruit in Oregon. Their Macintosh computer came out in 1984. It was named after Wozniak's favorite type of apple, the McIntosh. See Figure 1.13.

They both left the company in 1985 to pursue other opportunities. Wozniak took the opportunity to go back to college and earn his degree in computer science. Jobs has since started a new computer company called NEXT.

Production Pioneers

Which technologists helped improve factories?

Samuel Slater (1768–1835). It's not unusual for people to remember firsts. For example, Christopher Columbus was the first European to reach North America. George Washington was the first president of the United States. And the first U. S. factory was built by Samuel Slater in 1790.

Slater was born in England, where the factory system of manufacture began. The idea of bringing many people together to make products in one building was brand new. The British didn't want other countries to know how it was done. They would not permit any information about factories to leave the country. Slater knew that and used his remarkable memory to memorize every detail before he sailed to the United States.

A textile mill owner from Rhode Island was looking for a manager, and Slater applied for the position. He was hired, but found that the mill at Pawtucket wasn't set up well for factory production. Slater worked for a year to modify the mill. Using power from a waterwheel, the factory produced its first cotton in 1790.

Slater didn't invent or improve anything he brought to America. However, he did establish the factory system of manufacture. He is frequently called the founder of the U.S. cotton industry.

Figure 1-13 Stephen Wozniak (with beard) and Steven Jobs were teenaged electronic wizards who started Apple Computer, Inc.

Norbert Rillieux (1806–1894). Some people are so intelligent, they get upset when others can't understand technical detail. That's the way it was with Norbert Rillieux a brilliant, inventive, and fiery-tempered African American. His invention of a vacuum evaporator is little known to most people because it's not a consumer product. You can't buy one at a discount store.

To people who work in factories that make liquid products, however, his invention has proved to be irreplaceable. He found a way to quickly evaporate large quantities of liquids without using much energy. It's now used by companies that make soap, glue, paint, soup, sugar, paper, evaporated milk, and countless other items. See Figure1.14.

Rillieux first used his evaporator to make white granulated sugar from cane juice. Near his hometown of New Orleans, he partly filled a large container with the liquid. He knew it would boil at a temperature lower than 212°F if some of the air were removed from the sealed container. Before his vacuum evaporator, almost all sugar was brown. It carmelized under the high temperature necessary to evaporate water from the cane juice. Rillieux's sugar was white because the water was boiled away at a much lower temperature.

His method laid the foundation for all modern industrial evaporation. His name is so little known that the National Geographic Society called Rillieux "one of the most neglected of major American inventors."

Alice Hamilton (1869–1970). It is impossible to remove all dangers from factory work, just as it is impossible to remove all dangers from just about everything we do. However, factory work

Rillieux
(rĭl′ē-ū)

Figure 1-14 One of the least known of all major American technologists was Norbert Rillieux who invented an evaporator used by companies that process liquids.

became much safer because of Alice Hamilton, a production pioneer in industrial medicine.

After earning a medical degree, she went to work for a laboratory in Chicago. She lived in a neighborhood with uneducated immigrants. Many of them worked in unsafe factories. Her medical training allowed her to see and understand the health effects of improper factory handling of hazardous and toxic materials. She became involved with state agencies that could help the workers.

By 1916, Hamilton was the leading American authority on lead poisoning. Her work resulted in Illinois passing laws requiring industrial safety measures, medical examinations, and workers' compensation. Other states soon followed.

Hamilton generally shunned publicity. With quiet dignity, she became familiar with a company's processes and was completely honest and persistent. She was well respected by both workers and company owners alike.

She regularly associated with Margaret Mead, the American who studied South Pacific native cultures, and Florence Sabin, the American tuberculosis researcher. Hamilton also knew Marie Sklodowska Curie, Polish-born discoverer of radium and the first person to win two Nobel prizes. Harvard University made Hamilton their first-ever woman professor in 1925.

Sabin
('sā-bin)

Sklodowska Curie
(sklǝ-'dȯv-skǝ) (kyu̇-'rē)

Transportation Pioneers

≡ *Who were some of the technologists inventing better transportation devices?*

Elisha Graves Otis (1811–1861). When you think of transporting people, you might forget that one of the safest ways to travel is by elevator. That wasn't the case before Elisha Graves Otis invented the safety elevator in 1852.

The safety of early elevators relied completely on the strength of the rope. If it broke, the result could be serious injury, or worse. That's why early elevators rarely carried people.

Otis worked for a New York City company that was expanding. They asked him to design an elevator to move products from one floor to another. What he constructed was something the world had never seen before.

Strong wooden guide rails on each side had *ratchets* cut into them. The ratchets looked like large saw teeth. A piece of curved steel was at the top of the elevator. A rope passed through a hole in the steel and up to the lifting equipment. The weight of the elevator exerted enough tension on the steel to pull its ends out of the ratchets. If the rope broke, tension was released and the steel ends jammed into the ratchets. The elevator was locked in place. It couldn't move up or down. See Figure 1.15.

Figure 1-15 Elisha Otis invented an elevator that would lock in place if the hoisting rope broke. Modern elevators use carefully designed safety mechanisms.

Some people feel that this invention made the skyscraper possible. Whether or not that's true, the French certainly liked it. An Otis elevator was the only non-French item allowed on the 1889 Eiffel Tower.

Ole Evinrude (1877–1934). Ole Evinrude was a Norwegian-born transportation pioneer whose company made outboard motors for small boats. He decided to build the motor because he disappointed his future wife during a picnic on an island in a Wisconsin lake. She wanted him to get some ice cream. Although he was large and strong, Evinrude was unable to row back to the island before the ice cream melted. The next day, he started working in his spare time on a motor for a rowboat.

Employed as a machinist in Milwaukee during the day, it took Evinrude a year of night work to get an engine working. He called it his coffee grinder and used it for personal enjoyment. It developed about 1.5 horsepower. Two years later, after additional work on the engine, he lent it to a friend. The friend was excited with how well it performed and returned with an order for ten more. Another order for twenty-five soon followed and Evinrude formed his own company to produce the outboard motors. See Figure 1.16.

Evinrude's engine design was so remarkable that it was dedicated as a national historic mechanical engineering landmark. It was the first consumer product to be so recognized. There's even an early Evinrude on display at the Science Museum in London, England.

Figure 1-16 Ole Evinrude's first outboard motor was affectionately called a "coffee grinder" by his wife Bess.

Edwin Albert Link (1904-1981). Computer *simulation* games allow people to fly airplanes, operate submarines, and race automobiles in ways that resemble real life. Simulation is a type of imitation. Simulators are used to train ship captains, astronauts, and airplane pilots. The age of simulation began in 1929 when Edwin Link built the first mechanical trainer to provide ground instruction for pilot trainees.

His father owned a piano and organ company, and Link worked there for a while. Much of what he learned in the factory he would use to build his future trainers. While taking flying lessons himself, Link wondered if it might be possible to learn basic flight techniques in a nonflying model.

Link's trainer took him eighteen months to build and he made it look like a stubby airplane. There was just enough room inside for one person. The pilot trainee moved controls like those in a real airplane. Air pressure caused the box to move. It gave the same feel as a real airplane. Simulated instruments inside showed air speed and altitude. See Figure 1.17.

The government was interested in the new method of ground instruction and purchased several trainers. Link started the Link Aviation Company to build them. His trainers were usually painted a bright blue. That's why they were also called Blue Boxes. Blue Boxes were used to train over half a million World War II pilots. The company is still making trainers today in Binghamton, New York. One difference is that they are now called simulators.

■ **Figure 1-17** ■ Simulators train pilots and astronauts by giving them the sensation of flight without leaving the ground.

— Learning Time —

Recalling the Facts

From the nine short stories you just read, choose the person best described by each of the following phrases.

1. Invented safety elevator.
2. A file, waxed paper, and nail led to this person's invention.
3. Spent a year rebuilding a factory.
4. Invention was first consumer product to be a national historic mechanical engineering landmark.
5. Made first widely accepted personal computer.
6. Personally knew the individual who was the first to win two Nobel Prizes.
7. Invention was usually painted bright blue.
8. Laboratory was in Astoria.
9. First product was white sugar.
10. Invented the mimeograph.
11. Had a photographic memory.
12. Born in Norway.
13. Invented the vacuum evaporator.
14. First product was made in a garage.

(Continued on next page)

Continued

15. Invented airplane flight simulator.
16. Partner was Thomas Edison.
17. Lived to be over 100 years old.
18. Important part of invention was piece of curved steel.
19. Built first U.S. factory.
20. Invented xerographic copying.
21. Invention used air pressure to move a box.

22. One of the most neglected of major American inventors.
23. Pioneer in industrial medicine.
24. Invented only non-French item on Eiffel Tower.
25. Product named after a favorite fruit.
26. Invented outboard motor.
27. Name of first product included the number 914.

Applying What You've Learned

Investigate the lives of other important early American technologists. Here are some suggestions.

Technologists Important to Communication
1. Granville Woods (1856–1910): African American inventor of telegraph method to communicate between moving trains
2. Lee DeForest (1873–1961): inventor of vacuum tube for early radios and televisions
3. Margaret Bourke-White (1904–1971): industrial photographer, whose magazine pictures introduced factories to the public

Technologists Important to Production
1. Erastus Bigelow (1814–1879): inventor of power loom for weaving carpets
2. Elijah McCoy (1843–1929): Canadian-born black American inventor of lubricating equipment for factory machines
3. Lillian Gilbreth (1878–1972): developer of techniques to improve factory efficiency

Technologists Important to Transportation
1. Augustus Fruehauf (1868–1930): inventor of the semitrailer truck, now sometimes called an "18-wheeler."
2. Robert Goddard (1882–1945): inventor of liquid-fueled rockets, necessary to land people on the moon
3. Igor Sikorsky (1899–1972): Russian-born American inventor of the first practical helicopter

People and Technology

What are some ways in which technology affects people?

More than anything else, technology involves people. People make things happen. People do the doing. We use technology to improve our lives. We live in more comfortable houses than any eighteenth-century king or queen. We eat better food, wear better clothes, and enjoy life more. Technology has provided all of these benefits.

The United States is a country many people want to come to. The reason is not just because we are technically strong, but also because we recognize the importance of the individual person. Technology has helped each American maintain independence and dignity. Here are just a few examples of how technology helps individuals.

- **Education** - Your school probably provides bus service to transport students. No individual is denied an education because of distance from school or physical or mental handicaps. You also have the benefit of learning with computers, videos, laser disks— all examples of the latest technology.
- **Current Events** - For very little money you can purchase a small black and white television set to connect you to the entire world. America owns communication satellites to keep people informed.
- **Housing and Food** - You live in a dry, warm, and comfortable house, apartment, or mobile home. You take running water and electricity for granted. Conveniently located stores sell food at reasonable prices.
- **Transportation** - Most Americans can afford to purchase their own automobile, and we have excellent roads to travel. Bus, airplane, and train schedules are as convenient for the individual as possible.
- **Health** - Our hospitals and doctors' offices are equipped with high-powered, sensitive machines and instruments that enable health care workers to keep us alive and healthy longer than ever before.

People have created technology for people. People will continue to create new technology so that all people can live more comfortable, satisfying lives.

——— Thinking About Safety ———

Why is safety instruction important?

When the umpire yells, "safe," the baseball player is no longer in danger of being tagged by the ball. In the technology laboratory being safe means you won't be "tagged" by a machine, chemical, slippery floor, or sharp tool. Since you are getting ready to work on your first "hands-on" activity, this is the perfect time for you to stop and **think seriously about safety.**

Equipment, tools, materials, and activities determine the dangers of a particular situation. That is why the safety rules around a swimming pool, gymnasium, and technology laboratory are different. Even the rules from one technology lab to the next are different. That is why your teacher will provide you with a set of safety rules and safety instructions specifically geared to your technology lab.

In general, you should know that recognizing hazards is the best way to avoid danger. Accidents usually occur because people are not aware of the dangers that exist around them. You can avoid accidents just by paying careful attention to what you are doing.

The Do's and Don'ts of Shop Safety

What dangers should you avoid to prevent injury?

Using common sense and following some simple safety rules will make your experience in the shop a very enjoyable one. Here are a few basic rules:

- Protect your eyes by wearing proper eye protection.
- Wear a shop apron and roll up your sleeves.
- Never use equipment, tools, and materials unless your teacher has approved their use.
- Never attach an electrical device without your teacher's permission.

■■■ Putting Knowledge To Work ■■■

People Helping People

Aaron Rostok's grandfather was recuperating at home from a heart attack. Mr. Rostok was confined to his bed and was very depressed. He was tired of reading books and watching television. He needed to do something with his hands.

Wanting to help his grandfather, Aaron discussed some possibilities with his teachers. His technology teacher offered to help Aaron build a simple three-string guitar-like musical instrument called a dulcimer. The class was already working on wood projects. When the other class members found out about the project, they eagerly offered to help. Some even occasionally stayed after school for an hour or so.

It took about a week to complete the dulcimer. Although it looked good from a distance, a closer look was something else. Some of the glued joints didn't fit well, and the sanding could have been better. However, it made music, and Aaron thought it sounded wonderful.

On his way home after school, Aaron stopped at his grandfather's house. He gave Mr. Rostok the dulcimer and a library book on how to play it. His grandfather couldn't believe his eyes. It was the last thing he expected to receive, and his spirits soared. He asked Aaron where he built such a beautiful stringed instrument. Aaron beamed a huge smile and answered, "In my technology class."

Safety First

■ **Figure 1-18** ■ Proper safety equipment such as goggles, shop coats, and gloves help prevent injuries.

- Inform your teacher if you are cut or injured.
- Inform your teacher if you find any broken, dull, or damaged tools or equipment.

Most injuries occur because people don't think about what they are doing. Instead, they

- touch a material that could be hot without wearing heat resistant, non-asbestos gloves.
- touch spinning rollers, which causes their fingers to be pulled into a machine.
- rest their fingers in areas where they can be pinched.
- use chemicals without wearing the proper eye and clothing protection. (Figure 1.18)
- wear loose clothing and jewelry around machines that can grab them.
- use electric tools even though their wires contain broken insulation.
- use tools that should be plugged into a three-prong plug in non-grounded wall outlets or extension cords.

Safe Use of Machines

What safety precautions should you take when using machines?

The machines in your technology lab are designed to process materials. If you play with these machines as if they were toys, they can "process" you as if you were a material. Remember—those machines are designed to cut, bend, and reshape what goes inside them. You don't want to have your hand bent, smashed, or cut.

To avoid any chance of injury when working with machines, keep the following rules in mind at all times:

- Stay out of the safety area that surrounds a machine.
- Never use any machine until after your teacher has shown you how to operate it.
- On a session-by-session basis, never use a machine until after your teacher has granted you permission. This will protect you from using a machine that has been damaged since the last time you entered the room.
- Make certain that other people are clear of the machine area before you start any machine.
- Work alone.
- Wear safety goggles.
- Watch what you are doing, don't rush, avoid talking, and concentrate.
- Shut down the machine and request your teacher's assistance if you are having any difficulties.
- Never walk away from a machine that is running.

Safe Use of Hand Tools

⊤ *What causes most hand tool accidents?*

Your teacher will show you the correct way to use hand tools as they are needed to complete the activities in this book. Here are some important reminders for all hand tools:

- Never use a tool to perform a job unless the tool was designed to do that job.
- Always cut *away* from yourself. Most accidents with hand tools happen because people cut *toward* themselves rather than away from themselves.
- Use sharp tools. A dull tool is more dangerous than a sharp one because you often need to force the dull tool, which makes it more likely to slip.
- Never use broken tools or tools without proper handles.

Summary

Directions
The following sentences contain blanks. For each numbered blank, pick the answer that makes the most sense in the entire passage. Write your answer on a separate sheet of paper.

Technology is the practical use of human knowledge. It involves turning natural items into useful products.

The word *technology* comes from the ancient Greek word *tecknos*, which means ___1___ . This is why early workers in technology called themselves *artisans*. Later, in the 1800s, workers in technology called themselves *mechanicians*, which is very similar to our modern-day word ___2___ .

The areas of science, technology, and engineering are related but different. ___3___ try to explain *why* things happen; they conduct experiments in laboratories. Engineers figure out *how* to make things; they ___4___ parts for machines and structures. Technologists *make* things; they do this by operating machines and assembling parts. All of these people work together to create and produce the ___5___ we need.

There are many reasons to study technology. It's fun, rewarding, and exciting. It's also important for many reasons, such as

1.
   a. machine
   b. art
   c. tacks
   d. school

2.
   a. technologist
   b. mechanic
   c. uses
   d. products

3.
   a. Engineers
   b. Technologists
   c. Teachers
   d. Scientists

4.
   a. design
   b. build
   c. make
   d. sell

5.
   a. machines
   b. paintings
   c. products
   d. food

(Continued on next page)

(Continued)

maintaining our freedom through national __6__.

Many types of careers are available in technology. If you're interested in communication technology, you might enjoy working with computers or __7__ newspapers. In production technology, you'd be making products that range from clothes to buildings. Transportation careers involve moving __8__ and products from one place to another. And of course, many careers involve combinations of all the areas of technology. For example, you might like working with __9__. Electronics fits into communication, production, and transportation in many different ways.

In technology, we build on the accomplishments of earlier scientists, engineers, and __10__. Even failures are important in the progress of technology. Thomas Edison once said, "I now know 2,000 materials that won't work." Edison had to discover what __11__ work before he knew what would work.

Some pioneers in communication technology were Albert Dick, Chester Carlson, Steve Jobs, and Steve Wozniak. Dick invented the mimeograph machine. This __12__ could duplicate letters and other documents. Carlson invented a dry copier. Before Carlson's process was invented, making copies required __13__ chemicals and a darkroom. Jobs and Wozniak assembled the Apple II computer in 1977. It was the world's first widely accepted personal __14__.

Production technology pioneers included Samuel Slater, Norbert Rillieux, and Alice Hamilton. Slater built the first factory in the United States. Before coming to the United States, he memorized every detail about English __15__. Rillieux invented a vacuum evaporator that is crucial to the manufacture of __16__ products. Alice Hamilton studied the health effects of improper handling of hazardous

6.
a. emblem
b. deficit
c. goals
d. defense

7.
a. delivering
b. printing
c. reading
d. recycling

8.
a. animals
b. people
c. machines
d. textbooks

9.
a. watercolors
b. plastics
c. electronics
d. adults

10.
a. mechanics
b. teachers
c. doctors
d. technologists

11.
a. couldn't
b. shouldn't
c. wouldn't
d. might

12.
a. machine
b. laser
c. typewriter
d. engine

13.
a. combustible
b. rare
c. wet
d. modern

14.
a. work station
b. laptop
c. video recorder
d. computer

15.
a. ships
b. factories
c. laws
d. schools

16.
a. plastic
b. metal
c. liquid
d. vapor

(Continued on next page)

Summary

(Continued)

and toxic materials. Her work resulted in _____17_____ working conditions for factory workers.

Transportation pioneers included Elisha Otis, Ole Evinrude, and Edwin Link. Otis invented the safety _____18_____ in 1852. Some people feel that this invention made tall buildings, such as skyscrapers, possible. Evinrude invented a motor for small boats, and Edwin Link built the first mechanical trainer to provide simulated ground instruction for pilot trainees. Today these types of machines are called _____19_____ .

Technology, above all else, involves _____20_____ . It involves people making things for other people.

17.
a. safer
b. dangerous
c. tighter
d. warmer

18.
a. furnace
b. door
c. window
d. elevator

19.
a. airplanes
b. simulators
c. space capsules
d. incubators

20.
a. scientists
b. nature
c. people
d. physics

Key Word Definitions

Directions

The column on the left contains the key words from this chapter. The column on the right contains a scrambled list of phrases that describe what these words mean. Match the correct meaning with each word. Write your answer on a separate sheet of paper.

Key Word	*Description*
1. **Technology**	a) Information on a computer monitor
2. **Technician**	b) Making each step an improvement
3. **Technologist**	c) Making products
4. **Science**	d) Practical use of human knowledge
5. **Engineering**	e) Person skilled in technology
6. **Communication**	f) Turning science into useful items
7. **Production**	g) Person skilled in some subject
8. **Transportation**	h) Explains why something happens
9. **Adapting**	i) Moving people and products

High-Technology Paper Airplane System

Did You Know?

You probably already know how to fold a simple paper airplane. The ordinary pointed-nose style has been around since the 1920s. However, you may not have had the opportunity to fold a high-technology paper airplane (HTPA). An HTPA takes careful planning and folding. It's not seen very often.

Several years ago, *Scientific American* held their 1st International Paper Airplane Competition. There were almost 12,000 entries from 28 countries. The plans in this activity are based on those used in the competition.

Resources You Will Need

- Paper, pencil, and ruler
- Colored markers
- Yardstick, meterstick, or tape measure
- Stopwatch

Problems To Solve

Your task is to make two different HTPAs and see which one flies farther and stays up longer. The ordinary pointed-nose style usually flies about 15 feet and can stay up for about 4 seconds. Of course, many other distances and times are possible. A breeze from an open window or a heating vent, for example, could help or hurt a flight.

A.

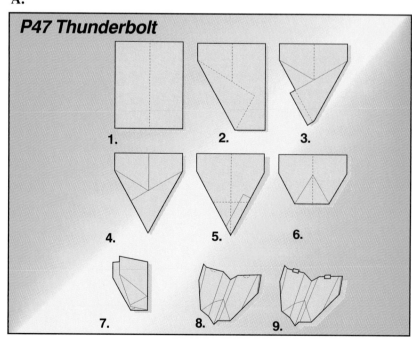

P47 Thunderbolt

1.   2.   3.

4.   5.   6.

7.   8.   9.

During this activity, you will use a simple *systems technique*. A systems technique breaks down a complex project into basic elements. Here are the basic elements for each HTPA construction:

1. Select an HTPA design.
2. Draw the plans.
3. Construct the HTPA.
4. Collect flight information.
5. Evaluate the flight information.

What Did You Learn?

1. Which HTPA flew farther and stayed up longer? Why do you think one was better than the other?
2. Which HTPA was more fun to fly? Why do you think one was more fun than the other?

A Procedure To Follow

1. Select a paper airplane design from the plans below. Figures A, B, and C.
2. Carefully draw the plans for your design on a sheet of ordinary notebook paper.
3. Fold the airplane according to the plans. Be sure your folds make sharp edges. The key to a good paper airplane is to have straight, sharp folds.
4. Decorate your airplane with colored markers so that you can easily identify your plane in case it becomes mixed with others.
5. Fly your HTPA several times.
6. Use the stopwatch to measure the time of the flight. One winning time in the *Scientific American* competition was 10.2 seconds.
7. Use the tape measure to determine the straight-line distance from the point where the HTPA was launched to the point where it stopped on the floor. One winning distance in the *Scientific American* competition was 91 feet.
8. If your plane doesn't fly well at first, try placing a very small weight at the nose. A piece of tape or small paper clip sometimes helps.
9. Repeat the preceding steps with another paper airplane design.

B.

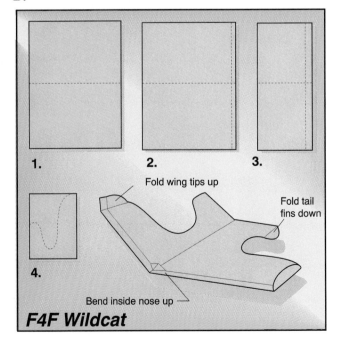

Fold wing tips up

Fold tail fins down

Bend inside nose up

F4F Wildcat

C.

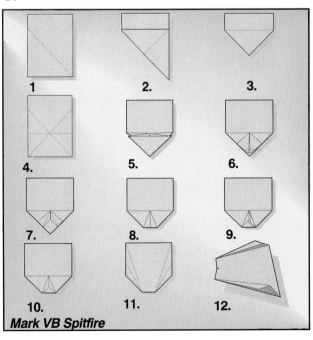

Mark VB Spitfire

The Resources and Methods of Technology

— **Key Words** —

In this chapter you will learn and study the meanings of the following important technology terms.

RESOURCE
TOOL
MACHINE
RAW MATERIAL
PROCESSED MATERIAL
MANUFACTURED
 MATERIAL
SYNTHETIC MATERIAL
SYSTEM
SUBSYSTEM
FEEDBACK
PROBLEM SOLVING
BRAINSTORMING

If a time machine could take you back to prehistoric times, would you be smarter, stronger, or faster than the people of that time period? Before you answer, you should know that there is a limitation to this fictional time machine. It requires you to go back through time just like Arnold Schwarzenegger in the movie *The Terminator,* without any of the things that have been created by technology.

Would you know how to survive in the wilderness? You would have to do a lot of improvising and use your knowledge to survive in this ancient environment. To survive against the meat-eating animals, you would need to do a lot of inventing or hiding. You wouldn't be able to run as fast, see as far, or hear as small a sound as the predators that would view you as a possible dinner.

What resources and methods did our earliest ancestors use to create the tools and weapons they needed to protect themselves and keep themselves alive? What resources and methods do we use today to create communication-age tools and devices to make our lives easier and more enjoyable? In this chapter you will learn the answers to these questions.

The Technology Recipe

≡ *What ingredients do you need to create technology?*

Technology comes from the knowledge and skill that is needed to use raw materials, tools, and energy to create the products and services that we want. The same resources used in ancient days are still being used today to develop new technology.

Our early ancestors knew very little compared to what people know today. But our ancestors were still able to use their limited knowledge to create Stone Age tools. They used their hands to grasp, shape and form natural materials into useful tools. See Figure 2.1. Their tools were simple and crude by our standards. Our tools will seem simple and crude to people in the twenty-second century.

To create new technology today, people use the same ingredients that were used by our earliest ancestors. These ingredients are called **resources** and they include all of the following:

resource
something that supplies help or aid; could be a source of information or expertise, wealth or revenue, supply or support

- People
- Knowledge
- Creativity
- Skill
- Tools and machines
- Capital
- Time
- Materials
- Energy

People

≡ *Why can't we create technology without people?*

Any list of ingredients needed to create technology must start with people. People are the creators of technology.

If we subtract the human element from our list of resources, what do we have left? The resources time, energy, and materials existed for millions of years before people. But technology didn't advance at all during these millions of years. It wasn't until people came along that technology began to advance.

Many animals use sticks and stones to get food, but most have never learned to improve these tools that came from their natural surroundings. People, on the other hand, not only learned to create new tools, but also passed their inventions on to future generations. Each generation has the opportunity to benefit from the accomplishments of the past.

Besides being the creators of technology, people are also the end users of the products that their technology has built. Between the designer and the end user are many jobs that must be done by people. People build the tools and machines, set up the factories, run the machines, and finally package and ship the end products.

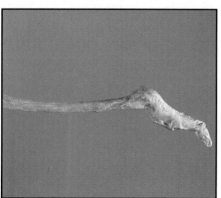

■ **Figure 2-1** ■ Ancient people using crude tools were still able to create beautiful carvings on the ends of their spears. These spears were carved out of animal bone during the Stone Age.

Other people are in the service area of our technology. They sell, install, and repair these products. People use technology in health care facilities, on farms, in businesses, and in schools.

Remember, it is our capacity to create technology that led to our powerful position on our planet. Our early weapons turned us into hunters rather than the hunted. Our telescopes and microscopes gave us the eyes to see the invisible. Our telephones and satellites gave us the ability to hear whispers over fantastic distances. The computer gave us the ability to recall the smallest of details and solve problems in seconds that people wondered about for centuries.

Knowledge

Why do we need knowledge to create new technology?

Knowledge is wisdom, information, learning, scholarship, and understanding. If an animal uses a stick to push food into its reach, is knowledge or technology being used?

Our definition of technology calls for the use of knowledge, skill and natural resources to meet our needs and wants. If a chimp takes a branch (natural resource) and moves an object into its reach (skill), it is using technology to get food (needs). A bird's nest is a complex construction project that uses bird technology.

The big difference between humans and other species is that we learn from our past experiences. When our early ancestors used sticks to gather food, they used elementary technology similar to that used by a chimp. This basic tool was refined by each generation and passed down to us. People learned (gained knowledge) that a stone attached to the stick improved its performance. Others learned that the reaching stick could be thrown.

Creativity

Why do we need creativity to develop new technology?

Knowledge is a key ingredient of our technology recipe. However, the possession and understanding of facts alone will not bring about the development of new technology. Knowledge gives you the power, but creativity gives you direction. Creativity is your ability to use your imagination to develop new approaches. You use your creative energies to develop experimental solutions to solve problems.

Have you ever been confronted by problems that you couldn't solve? All people have the ability to solve problems. However, some people are better at it than others. Creativity is hard work. To quote Thomas Edison, "Genius is 1 percent inspiration and 99 percent perspiration." You have to work hard to come up with new ideas. Techniques that you can use to improve your problem-solving skills will be detailed later on in this chapter.

Tech Talk

There are two basic kinds of telescopes: *refracting* and *reflecting*. Galileo Galilei invented the refracting telescope in 1609. It has a large lens at one end and a smaller lens for your eye at the other end. The reflecting telescope was invented by Isaac Newton in 1668. It uses a curved mirror in place of a large front lens.

Celebrating Multicultural Technology

The achievements of our multicultural nation have been accomplished through the efforts of people of every nationality. But for a long time, the history of American technology talked only of our European male ancestors. It ignored contributions made by women and non-European men. Therefore, we will now take a special look at some of the technological developments that the historians of the past forgot to celebrate.

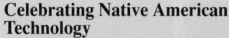

Celebrating Native American Technology

Many centuries before Europeans set foot in the New World, Native Americans made some major breakthroughs. In agriculture they domesticated and selectively bred potatoes, chocolate, peanuts, cotton, and tobacco. In medicine they developed an ointment, now called *petroleum jelly*, that they applied to wounds, and they converted the bark of poplar trees into a liquid, similar to aspirin, used to cure headaches. The North American Indians used asphalt to waterproof their cloth and baskets and to seal the seams of their boats.

More recently, in the mid 1980s, Andrea H. Proudfoot, a Native American, designed a baby carrier in which the infant is carried in a pouch on the parent's back. The *Baby Pack* was patented and is made in a small factory.

Perhaps the greatest technological achievement of the Native Americans was their ability to take from the land only what they needed for survival without threatening the future of their environment.

Celebrating Hispanic American Technology

The ancestors of Hispanic Americans were responsible for the development of many everyday technologies.

In the sixteenth century, Mexicans produced a scarlet red dye that was used to color the fabric used to make British Army uniforms called "redcoats."

Using the stone *obsidian*, the Aztecs made scalpels that were sharper than our modern steel scalpels. South American Indians used rubber to make items such as balls, bottles, raincoats, rubber-soled shoes, and even ropes.

The Incas devised a method of freeze-drying that allowed them to store potatoes for five or six years without spoiling. They also devised a system of drying and preserving meat that the Europeans called "jerky."

More recently, Severo Ochoa, a biochemist, won the Nobel Prize for his discovery of the enzyme that produces RNA; Roberto Gonzalez Barrera developed a time-saving process for making corn tortillas; and Manuel A. Villafana was instrumental in the design, testing and production of a mechanical artificial heart valve that was presented at a 1991 heart surgeons' convention.

Celebrating African American Technology

Let's look at some of the technology developed by African Americans.

In 1821 Thomas L. Jennings patented a process for cleaning clothing called "dry cleaning." Lewis Howard Latimer, an electrical engineer, invented both the bathroom toilet for railway cars and the first electric light with a carbon filament.

Before Andrew Beard's invention of an automatic train coupler, train cars were locked together by a railroad yardman. The job was dangerous and if the yardman's timing was off, he lost a hand, a foot, or possibly his life.

How many times have you used a pencil sharpener? The pencil sharpener was invented by J.L. Love in 1897.

Richard B. Spikes invented the automatic gear shift and the automatic directional signal for cars, and a multiple-barrel machine gun for the military.

Elijah McCoy invented automatic oilers that allowed machines to lubricate themselves while they were still running.

A number of very familiar items were patented by African American inventors: the lawn mower, by J.A. Burr; the golf tee, by G.F. Grant; the guitar, by R.F. Flemming, Jr; the kitchen gas burner, by B.F. Jackson; the air conditioner, by F.M. Jones; the fire extinguisher, by T.J. Marshall; the lawn sprinkler, by E.J. McCoy; the clothes dryer, by G.T. Sampson; and the refrigerator, by J. Standard.

Celebrating the Technology of American Women

As we explore some of the accomplishments of women, you will notice that their inventive genius was for the most part directed by their place in society, which in earlier times was the home and the fields.

Mrs. Samuel Slater was the first woman to receive an American patent for one of her inventions—a sewing thread made out of cotton.

In the early 1920s, Kate Gleason developed the idea of mass-producing homes at reasonable prices by using prefabricated wood forms to cast the concrete for the homes.

Have you ever used correction fluid to cover a typing error? "Mistake Out" (you know it as Liquid Paper) was developed by Bette Nesmith Graham.

Some very familiar items were patented by American women: the Bissell carpet cleaner by Anna Bissell; six models of the automatic washing machine by Margaret Colvin; drip coffee by Melitta Bentz; the disposable diaper by Marion Donovan; the safety auto seat by Gertrude Muller; modern cosmetics by Helena Rubinstein and Elizabeth Arden; and pantyhose by actress Julie Newmar.

Celebrating Asian American Technology

Let's look at some of the contributions that Asian Americans have made to the development of technology.

In the late 1800s, Dr. Jokichi Takamine began isolating the hormone adrenaline. Today doctors use adrenaline to restore blood pressure and to treat severe cases of asthma.

Dr. Chien-Shiung Wu developed an improved Geiger counter, which is a machine that measures radiation.

When the United States was doing underground nuclear testing, it used a trigger developed by Dr. Marguerite Shue-Wen Chang. Dr. Chang also developed some new and highly classified explosives and missile propellants for the U.S. Navy.

In 1991, Li Fu Chen developed a new process that turns agricultural waste into rayon. Chen's process will produce the same quality rayon that is used today at a fraction of today's prices.

Young-Kai Chen and Ming Wu, researchers for AT&T, have developed the world's fastest pulsed laser for use in long distance fiber-optic communications.

Unmentioned Heroes of Technology

We have mentioned only a few of the hundreds of men and women from all nationalities who have contributed to the technological advancement of our country and, ultimately, the world. You may want to visit your school or public library to learn more about these and other heroes of technology.

Skill

Why do we need skill to develop new technology?

Can you define the term *skill*? Are there activities in which you have already developed a high level of skill?

Skill is the combination of knowledge and practice that allows you to perform activities well. If your definition included the ability to do something well, you have a basic understanding of this term.

You need knowledge and skill to make an axe out of a tree limb and a stone. You use your knowledge to figure out the correct specifications. Then you use your skill to chip the stone and cut the tree limb to meet those specifications.

To learn how to use any new tool, you will have to practice using it until you have mastered the technique. You do the same thing when you practice to improve your ability at video games, sports, dancing, reading, and writing.

— Learning Time —

Recalling the Facts

1. What ingredients do we need to create technology?
2. Why do we need knowledge to create technology?
3. Why do we need creativity to develop technology?
4. Why do we need skill to develop technology?

Thinking for Yourself

1. If you traveled back in time, how different would you be from our earliest ancestors?
2. Why can't we create technology without people?
3. Rate the level of importance of each ingredient of technology from one to nine. Indicate your reason for placing the ingredients in this order.

Applying What You've Learned

1. Working in teams of three or four, build a paper bridge that spans 9 inches. Each group must build a bridge using a maximum of five sheets of paper. The paper can be cut, bent, or folded, but no staples, glue, or tape can be used. Determine the best bridge by adding weights to the center of each span to see which one holds the greatest weight.
2. Pick an invention, and research what resources were needed by the original inventor to design and construct the invention. Your final project can be a written report, chart, video, or model of the invention.

Tools and Machines

Why are tools and machines needed to develop new technology?

Did your teacher ever complain that you came to school "without the tools of learning"? Your teacher considered pencils, pens, and books as tools of learning. **Tools** increase our ability to do work. If learning is work, then pens, pencils, and books are tools.

Today all people use tools to help them perform their jobs better. Can you think of an occupation that is performed without the use of any tools?

The first tools invented were all hand held and muscle powered. These early tools were used to construct things that met early human wants. These tools were also used to make other tools. Without the development of these *primary tools*, more complex technology would never have developed. The original (primary) tools increased a person's ability to hold, cut, drill, bend, and hammer materials.

People consider all devices that help them perform their job as tools of their trade. **Machines** are often referred to as tools. A tool becomes a mechanical machine when a power system takes advantage of certain scientific laws and makes the tool work better. You will learn more about power and power systems later in this book.

All mechanical power systems use one of the following machine principles to change the direction, speed, or force that is powering the tool. Very often machines will use a number of these principles in combination:

- *The wheel.* The best-known power system is the *wheel*. It is round and connected to an *axle*, which is the central shaft. See Figure 2.2 on the next page.
- *The gear.* The *gear* is a wheel with teeth running around its circumference. The teeth allow gears to mesh (run) together without any chance of slipping. Your bicycle has a gear that you turn by rotating your legs with your feet on the pedals. This gear is meshing with the chain that is meshing with the gear that drives your rear wheel. See Figure 2.3 on page 41.
- *The lever.* The *lever* is another power system that you are probably quite familiar with. When you were younger, you probably played on a seesaw. This playground toy consisted of a long board that was raised off the ground and fastened securely at its middle so that each end could swing up and down. If you placed heavy people closer to the middle of the board, lighter people could easily lift them. See Figure 2.4 on page 41.
- *The inclined plane.* The *inclined plane* is an angled ramp that makes it easy to raise things by rolling them uphill. See Figure 2.5 on page 41. Inclined planes are often used to move cars into parking garages.

tool
something such as an instrument or an apparatus that increases one's ability to do work

machine
tool with a power system that takes advantage of certain scientific laws and makes the tool work better

Tech Talk

The greatest scientist of ancient times was Archimedes. He worked out the principle of the lever. Archimedes showed that a small weight far from the lever's balance point could support a large weight near the balance point. A famous statement of his was "Give me a place to stand and I can move the world."

Figure 2-2 The wheel made it possible for heavy objects to be rolled rather than dragged.

- *The wedge.* The *wedge* is actually a small inclined plane that is used to spread things apart. See Figure 2.6 on page 41. Its shape converts downward movement into a force that separates things. The axe is a wedge on a stick. The scissor is two wedges joined together. The plow is one of the most important wedge-shaped tools ever invented.

- *The screw.* The *screw* is actually an inclined plane that is running around a metal rod. Figure 2.10c on page 42 shows an outdoor parking garage that is entered by driving around the circular ramp. Notice how the ramp looks like a giant screw.

- *The cam.* The *cam* uses the principle of the wheel in combination with the principle of the inclined plane. Most cams look like wheels that aren't perfectly round. See Figure 2.8 on page 41.

- *The pulley.* The *pulley* uses the principle of the wheel in combination with a rope or chain to lift heavy objects. See Figure 2.9 on page 41. In a one-pulley system, the full weight of the object must

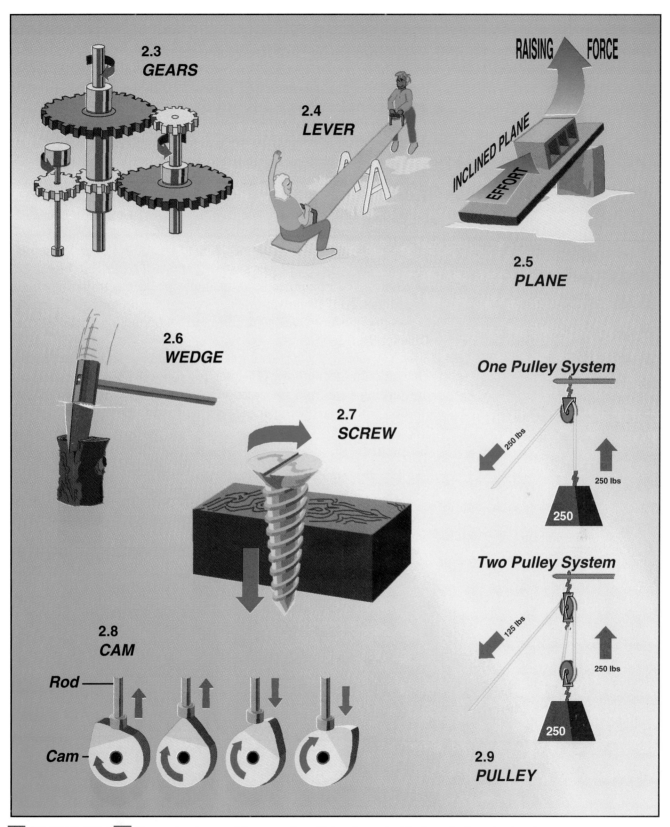

2.3 GEARS

2.4 LEVER

RAISING FORCE

INCLINED PLANE

EFFORT

2.5 PLANE

2.6 WEDGE

2.7 SCREW

One Pulley System

250 lbs

250 lbs

250

Two Pulley System

125 lbs

250 lbs

250

2.8 CAM

Rod

Cam

2.9 PULLEY

Figures 2-3/2-9 These simple machines are the driving force behind most of our technological devices.

be lifted by pulling the rope. In a two-pulley system, the person pulling the rope will have mechanical advantage and feel as if the object weighs one-half its actual weight.

These power systems are responsible for the operation of almost all mechanical machines. Figure 2.10 shows how these simple principles make familiar devices work.

Nanotechnology is a new area of technology that involves the design and development of very small machines. Nanomachines will someday be injected into the human body or placed inside other objects to repair things in places that people can't reach. Today motors are being built that have gears so small that you need a magnifying glass to see them. They are used to run mechanical fingers and special medical tools.

Not all machines have mechanical power systems. Some machines use electronic power systems, and still others are biological in nature. The computer is an example of an electronic machine. See Figure 2.11. Its power transfer system has no moving parts, but works by pushing *electrons* through a conducting material (see Chapter 9).

You are a perfect example of a biological machine. In the chapter on biotechnology (Chapter 15) you will learn how scientists are turning cells into microscopic machines that can manufacture needed biological chemicals.

Capital

≡ *Why do we need capital (money) to develop new technology?*

Capital is barter, money, credit, or property—in other words, accumulated wealth. At the dawn of technology, inventors created

■ **Figure 2-10 a, b, c** ■ Here are some photographic examples of our simple machines at work. Gears move the hands on non-digital watches. Inclined planes are used to lift heavy objects. Many parking garages move cars around a building that is designed like a giant screw.

their Stone Age tools without the help of others. Our earliest ancestors didn't need capital.

Later inventors couldn't get all the necessary tools and materials without the assistance of others. Once they had to purchase or trade for tools, materials, and labor, they were using some form of capital. Capital's importance grew tremendously.

Using investment capital, people can now buy all the ingredients needed to create new technology. See Figure 2.12 on the next page. Today a team approach is used to develop most new ideas. Under the direction of large corporations, experts are hired. The company buys materials, tools, knowledge, skill, and creative people. These companies construct the needed laboratories, mix all the ingredients together, and then wait for team effort to deliver useful products.

Inventors once worked alone and needed little or no financing. You can still find the independent, underfinanced inventor at work in America. Can the basement or garage inventor still succeed? Spending great sums of money won't guarantee success. A person working part-time as an inventor might create the next invention that will spur the growth of a multibillion-dollar business.

Time

⯇ *How does time affect the development of new technology?*

Can you name something that takes place instantly without the passing of time? Everything takes time, even if the time is measured in millionths of a second.

People are paid for the time that they work. Products where

Tech Talk

Capital has many different meanings:
1. A capital crime is punishable by death.
2. Capital letters are upper-case letters.
3. Capital is money, property, buildings, inventory, machines, and tools.

Figure 2-12 Capital can be used to purchase all the necessary ingredients needed to create new technology.

people rather than machines supply most of the labor are usually more expensive than products made by machines that are controlled by other machines.

All recipes call for measured amounts of different ingredients. Once the ingredients are put together, you must stir, mix, heat, or freeze the contents for a specific amount of time.

Too much or too little of any ingredient, including time, could ruin your results. Whether you are making a cake, building a car, or designing a new product, your results will take shape over time.

— Learning Time —

Recalling the Facts

1. Why are tools and machines needed to develop new technology?
2. Explain how the wheel, gear, lever, inclined plane, screw, cam and pulley work.
3. How can capital help someone develop new technology?
4. How does time affect the development of new technology?

Thinking for Yourself

1. Name and describe occupations where people don't use any tools at all.
2. You are a scientist of the future, and you have been asked to design the ideal smart creature. Describe the physical and mental qualities of your creature.

(Continued on next page)

Applying What You've Learned

1. Construct a wall plaque that shows the development of a hand tool.
2. Construct a marble roll (pinball machine concept) where the ball moves along an inclined plane and activates wheels, levers, gears, and other basic machine devices.

Materials

Why do we need materials to develop new technology?

Technology developed because people learned how to use and process various materials into useful objects. People have also learned to create new materials by combining or refining natural resources in ways not done by nature. Material resources can be classified by how they were formed.

Materials that are found in a natural state are called **raw materials.** Raw materials are found on or in our land, sea, and air. Raw materials include rocks, metal ores, crude oil, coal, sand, soil, clay, animals, plants, and trees.

Processed materials are natural resources that have been changed by technology into a more useful form. They include lumber from trees, leather from animals, and crushed or slabs of stone from rock quarries.

Manufactured materials are created when natural resources are altered by processes that do more than merely change the size or shape of the material. Examples include gasoline, kerosene, concrete, and metals. When you look at processed material, you can usually recognize where it came from. Crushed rock resembles the stone it came from. Manufactured materials are so changed in form that you can't recognize where they came from. Metals don't resemble the ores that they were made from.

Synthetic materials are created artificially - they are *not* natural materials. Synthetic materials are made by scientifically combining chemicals and elements into rare natural materials or other materials not found in nature. Industrial diamonds, human-made rubber, and plastics are all produced synthetically.

raw material
natural materials that can be converted into new and useful products

processed material
raw material changed by technology into a more useful form that still resembles the raw material

manufactured material
raw materials that have been so altered by processes that they are unrecognizable

Energy

Why must we use energy to create new technology?

After a hard workout in a gym, did you ever feel like you ran out of energy? Your muscles use a great deal of energy to perform the tasks that you do daily. Even when you are at rest, you use energy to breathe, think, and pump blood throughout your body.

Energy is the source of power that runs all of our technological systems, too. There are many different sources of energy. These sources may be natural or synthetic, and in limited or unlimited supply. You will learn more about this important resource of technology in Chapter 4.

——— The Systems Approach ———

What are systems and subsystems?

system
an organized or established way of doing something through objects or ideas that work together to complete a task

When we talk about a **system,** we are talking about a way of doing something. A system is made up of objects or ideas that work together to complete a task.

Your body digests food through the combined efforts of the organs that make up your digestive system. Many of your parents get to work using a highway or public transportation system. See Figure 2.13.

■ **Figure 2-13** ■ A ferry is a subsystem of our transportation system. It allows cars and people to cross waterways.

In mathematics you are taught how to use different systems to solve addition, subtraction, division, and word problems. In science you are taught the scientific method, which is a system used to solve science problems.

In social studies you study the American system of government. Our country has formed many governmental, military, legal, and educational subsystems to guarantee the basic ideas of our American system of freedom and democracy.

Subsystems are minisystems that exist within larger systems. A system is a subsystem when it can't exist without its surroundings. For example, the digestive system is a subsystem of any living organism. The jet engine is one of many subsystems of the airplane. The airplane is a complete system, but it is also a subsystem of the transportation system.

Systems turn ideas, facts, and principles into the things that we want. This is done through the skilled use of people, capital, tools, machines, materials, energy, and time.

subsystem
a minisystem that cannot exist outside of the larger system it is a part of

Diagraming Systems

Why do people diagram their plans?

Football coaches often diagram plays to help team members understand what they are going to do. Technology has adopted a method of diagraming that helps people understand how any system operates. This same diagram can also help people organize their plans when they want to develop new ideas. In fact, this method of diagraming a system was originally developed by engineers.

Open-Loop Systems

What is an open-loop system?

When a system has no way of measuring or controlling its product, it is called an *open-loop system*. A bathtub, stove, kerosene heater, and old-fashioned traffic light are all examples of open-loop systems. What makes them open-loop systems? What can't they control?

Each of these devices cannot shut themselves down at the appropriate time. The bathtub overflows. The stove burns the food. The kerosene heater overheats and consumes the oxygen in the room. The traffic light prevents the main flow of traffic while it gives a green light to empty lanes in the intersection. Can you think of other open-loop systems?

Figure 2.14 on the next page shows the three parts of the open-loop system diagram: input, process, and output.

In the movie *Short Circuit*, a robot comes to life and constantly asks for *input*. It wants input (information) to determine appropriate behavior.

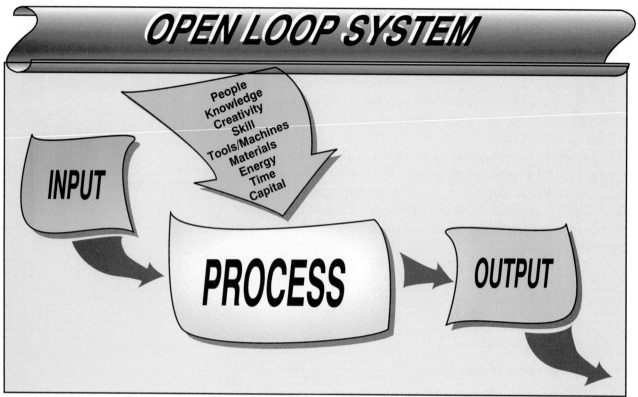

Figure 2-14 ■ Technology that is based on the open-loop system has no way of controlling what it produces.

In our system diagram, input is the information, ideas, and activities that we need to provide to determine what we need to accomplish. For example, you want to run for school president. You and your friends discuss how to run your campaign. You decide to make posters and buttons that will display your picture and a slogan. All the steps that lead up to the idea of creating these posters and buttons are part of input. What the buttons and posters will look like and how they will be made are part of the next part of our system diagram.

Process is the conversion of ideas or activities through the use of machines, resources, and labor into useful products. The production of your posters and buttons would take place in the process part of the system diagram. In the movie *Short Circuit*, process includes all the actions the robot takes to prevent people from disassembling it.

Output is simply what the system produces. Your posters and buttons would be the outcome (output) of your election planning. Survival would be the outcome of the robot's action.

Simply stated, the three parts of a system diagram contain an idea (input), which leads to an action (process), which leads to an outcome (output).

Can an open-loop system measure effectiveness? Will your buttons and posters accurately convey your message to the other students at school?

How would you add a controlling device to regulate a bathtub, stove, kerosene heater, old-fashioned traffic light, and the effectiveness of your buttons and posters? By adding a controlling device, you create what is called a closed-loop system.

Closed-Loop Systems

What is a closed-loop system?

When quality control is added to the open-loop system you get information back about your end product (output). If you knew that your slogan was turning students off, what would you do? You would change your slogan to correct the problem.

Feedback is the name given to the part of a system that measures and controls the outcomes of the system. By adding feedback to an open-loop system, it becomes a closed-loop system. See Figure 2.15.

Can you name examples of closed-loop systems? The heater in a fish tank warms the water in the tank and shuts off when the water reaches the desired temperature. If it didn't shut off, you would have cooked fish.

Many other technological devices use feedback loops. A traffic light that has metal detectors in the intersection automatically stays green unless cars are waiting in the less-used lane of the intersection.

feedback
the part of a system that measures and controls the outcome of that system

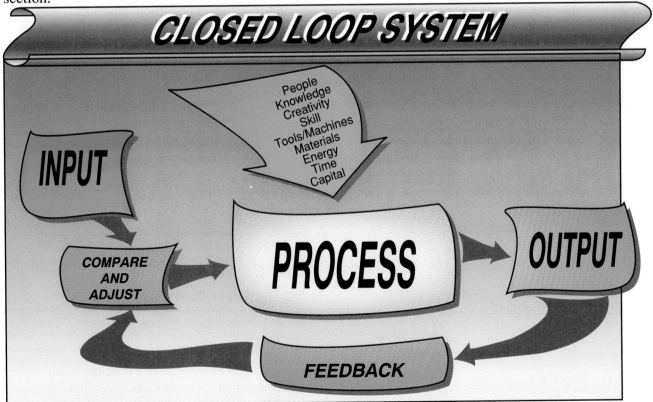

CLOSED LOOP SYSTEM

INPUT

People
Knowledge
Creativity
Skill
Tools/Machines
Materials
Energy
Time
Capital

COMPARE
AND
ADJUST

PROCESS

OUTPUT

FEEDBACK

■ **Figure 2-15** ■ A closed-loop system has feedback. Feedback provides a constant flow of information about the quality of what is being produced.

This same inspection system is used to open doors in some public buildings. Two systems are used to open these doors. In one system a special pressure plate is placed in front of the doors. When you step on this plate, your weight acts as the activator of the door. The other system uses a motion detector that is placed just above the door. See Figure 2.16. This detector senses your movement in the same way that police radar detects a speeding car. If your parents have a radar detector in their car, it might beep when the car passes stores with automatic doors or radar alarm systems.

Ovens that have cooking probes are another example of feedback loops in closed-loop systems. When the food is cooked, the probe tells the oven to shut down.

— Learning Time —

Recalling the Facts

1. What are systems and subsystems?
2. Why do people diagram plans?
3. What is an open-loop system?
4. What is a closed-loop system?

Thinking for Yourself

1. What is the smallest operating system that you can think of?
2. When would an open-loop system be superior to a closed-loop system?

Applying What You've Learned

1. Construct two Lego traffic lights or other Lego objects. Make one of your projects an example of an open-loop system. Make the other an example of a closed-loop system.
2. Construct an object using only raw materials; only processed materials; only manufactured materials; or only synthetic materials. Regardless of which materials you use, you can use glue to hold the parts together. Your object can be useful or just an artistic design.
Your teacher will assign the type of materials that you or your group can use.

—— Creative Thinking and —— Problem Solving

What is creative problem solving?

There is no simple formula for creative thinking, which makes it difficult to learn. Once learned, however, creative thinking can be used to create original ideas or solve problems.

Figure 2-16 The little box at the top of this supermarket entrance contains a motion detector. When people move in front of the entrance, it activates the motors to open the doors.

Today all occupations need creative people who can solve problems. People who are creative often work on a problem for a while and then set the problem aside. Their mind continues to work on the problem at an unconscious level. In time they come up with solutions that they can put into action.

Other people use *group dynamics* to bring about creative solutions. Here many people work together to bounce ideas around and get group inspiration.

Creative thinking is like riding a bicycle. Explanations will only give you guidance, because the act of learning to ride demands that you get on the bicycle. Solving problems demands that you understand the rules of **problem solving** and then practice finding creative solutions to problems.

problem solving
finding solutions to problems

⸺ The Problem-Solving Method ⸺

What are the seven steps of the problem-solving method?

When you attack a problem, you must be aware of what you already know and what you need to find out. People tend to ignore information that is given to them that could save them time and energy. The following seven steps will give you a basis of attack when solving problems:

1. Identify the problem.
2. Establish what you want to achieve.
3. Research past solutions to similar problems.
4. Brainstorm possible solutions.
5. Pick the best solution.
6. Build a working model.
7. Adjust your solution if necessary.

Now let's take a closer look at each step. See Figure 2.17.

Identify the problem

What does it mean to identify the problem?

Before you can solve a problem, you must know what the problem is. Sometimes this is easy to do, but even when it's easy, it's important to state the problem. Make sure you know exactly what the problem is that you are trying to solve. Suppose problem 1 is to design a button that will raise people's awareness of environmental issues. Do you understand what the term *environmental issues* means? This problem is asking you to design a button that tells people how important clean air, clean water, and clean streets are.

■■■ Putting Knowledge To Work ■■■

The Thinker-Doer Society

Once upon a time, there was a thinker-doer society. In this society all the tasks were divided between two groups—the thinkers and the doers. The people were very happy doing the tasks that their society had given to them.

The thinkers supplied the ideas and developed the new technology. The doers supplied the labor and built the machines that were designed by the thinkers. Special doers who were more like the thinkers (they thought a little) controlled and repaired the machines that the thinkers had already designed and the doers had already built.

The thinkers thought so hard that they designed labor-saving machines for the doers to build. After these machines were built, the doers had little to do.

"What shall we do?" cried the doers.

"Think!" answered the thinkers.

Fear moved through the schools of this society because no one ever taught doers to think before. In fact, no one ever taught thinkers to think before either. They just learned to think on their own.

The society in this fable isn't very different from our own. It is very important that you learn how to be a creative thinker.

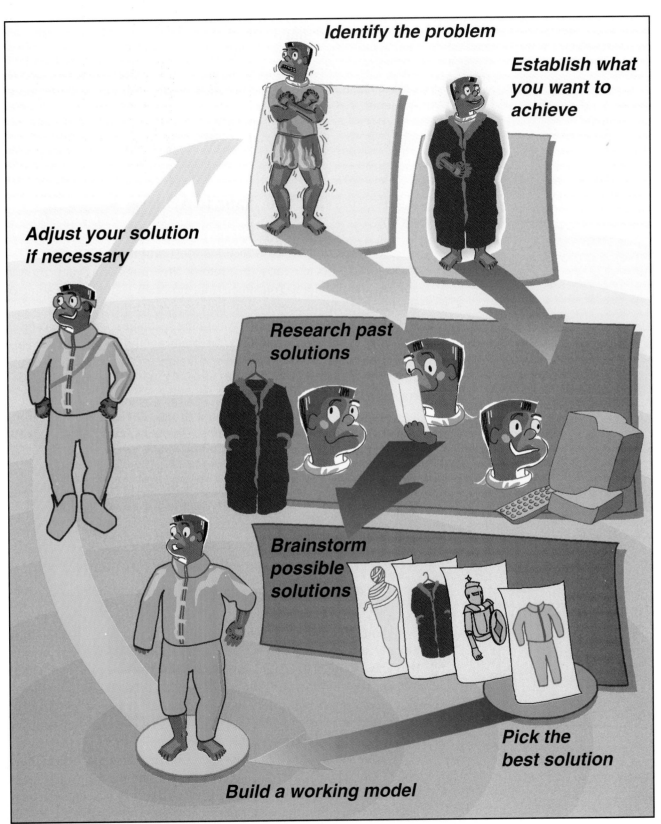

Figure 2-17 Manufacturers, regardless of what they are producing, have to go through these seven steps to produce the products you buy.

53

Problem 2 might be to build a car that contains a seat for an egg and an egg safety restraint system. This system must protect the egg from a high-impact crash. Your egg-mobile must protect the egg from a drive down an 8-foot ramp. The ramp angle is quite steep (60°). You can construct your car and ramp using any materials found in the shop. You are to team up with three or four other students.

Do you need more information before you can start to develop solutions? Do you know what a 60° ramp would look like?

Establish What You Want To Achieve

What do you do in step 2?

There are many ways to solve most problems. Therefore, it's a good idea to decide early on some of the things you want to do in solving the problem. You stated your problem — now state some of the basic guidelines you will follow in solving the problem.

How large a button are you going to make? Are you going to use only words, only pictures, or pictures and words? How easy will it be to read the button from a distance? Are you going to use color? You must make certain that you follow all the rules that were stated in the original problem.

In safety problem 2, you must protect the egg from breaking when it races down the angled ramp. If you can build a car that rolls slowly, there will be less danger to the egg. Can you hard-boil your egg or coat it with nail polish to give it a suit of armor? Does the car need to absorb some of the energy of the crash to prevent the egg from breaking even if it doesn't fly out of its protective seat?

Research Past Solutions To Similar Problems

What are the advantages of finding out about past solutions to problems such as yours?

You don't want to rediscover the wheel. If other people have already solved the same problem, or a similar problem, look at what they have done. This could save you a great deal of time and lead to a better solution to your problem. Research on the environment will give you button ideas. The egg problem has been worked on by many students studying engineering. Their eggs were often dropped from great heights, which caused much greater stress on their eggs.

You won't always have time to research what others have done. And when you do, you won't always find the information that you want. Still, research can be a good source of ideas and can even lead you to a solution to your problem.

Brainstorm Possible Solutions

What is brainstorming?

Brainstorming is a group technique that helps you develop possible solutions to a problem very quickly. Instead of tucking the problem into the back of your mind for unconscious problem solving, you have a number of people freely associate (call out) possible solutions. Your mind will join ideas that are expressed and possibly give you sparks of inspiration.

To use this procedure, your class must break up into small groups. One member of the group is chosen to be the recording secretary, and another member is chosen to be the group leader.

The group leader states the problem, and each member of the group calls out possible solutions. All ideas are recorded, and no one stops to criticize or explain a suggested solution. You try to join on to the last person's solution and spin off new ideas.

In the movie *Big*, a boy wished he was "big" and was turned into an adult, but he still had the mind and experiences of a child. He worked for a toy manufacturer. Sitting in a top-level executive brainstorming session on new toy ideas, he thinks, "What's fun

brainstorming
a group problem-solving technique in which group members call out possible solutions

---

Putting Knowledge To Work

Perpetual Motion Machine

Can a machine be built that will ride forever on its own energy? The name of such a machine is a *perpetual motion* machine. This kind of machine, once set in motion, will continue to move under its own power indefinitely.

President George Washington granted the first patent on July 31, 1790. For over 200 years, people have been taking out patents on all kinds of machines. Many people have tried to build and patent perpetual motion machines, but no one has yet built a machine that works. The reason is a scientific law that states that energy can't be created or destroyed.

Builders of perpetual motion machines know that they can't get around this law of science, but they try to find a back door. What if they can capture the work and heat energy and use it over and over again?

Some designs in the past used waterwheels that were self-propelled and solar panels that powered their own light source. Do you have any ideas that might prove that perpetual motion is possible?

Your public library has books that will show you perpetual motion ideas. Your inventions can be built and tested.

about a toy that turns from a robot into a building?" His inspiration is a toy that turns from a robot into a prehistoric bug. Everyone loves his idea, and he gets a promotion.

For brainstorming to work, you must be relaxed and not afraid that you will say something stupid. You also can't just sit silently waiting for unconscious inspiration.

Pick the Best Solution

What is one factor to keep in mind as you pick the best solution?

If your group came up with ten possible solutions, you now have to decide, as a group, which one is the best. Don't look at your choices without considering your resources. The tools and materials that are available could easily determine which solution is best. The characteristics of the materials that you have to work with can now help you determine the best solution to the problem.

Build a Working Model

What is the best way to test a solution?

Once you've chosen what you think is the best solution, you are ready to test it. The best test of your solution is the working model. If your choice of material was a poor one, your solution might not work. Does this mean that your solution wasn't any good? Not at all. Your solution might have just exceeded your abilities.

Leonardo da Vinci designed many inventions that weren't built during his lifetime. Charles Babbage designed an early computing device that couldn't be constructed because the precision parts that he needed couldn't be built by the technology of his time.

Adjust Your Solution if Necessary

Should you always stick with your chosen solution? Why not?

An expression that is often used is "Back to the drawing board." It means that you should learn from your mistakes. Be on the lookout for better solutions and use new knowledge gained from experimentation to create new and better solutions. Don't feel that you must stick to your first solution. You will probably see ways to improve it as you work on your project.

In business, solving problems affects the survival of the company. Poor performance will show up as poor sales. People buy from companies that they feel have the best products. In a free-market society such as we have in the United States, lack of creative problem solving means going out of business.

— Learning Time —

Recalling the Facts

1. What are the seven steps of the problem-solving method?
2. Explain what goes on during a brainstorming session?
3. If you test and adjust your solution, are you following an open-loop or closed-loop system? Why?
4. When you are solving a technological problem, why is it important to build a model of your solution?

Thinking for Yourself

1. What do you think is the most important step in the problem-solving method? Why?
2. What do you feel is the least important step in the problem-solving method? Why?
3. Why do companies build working models before placing a new invention into production?
4. Your solution to a problem fails to work properly. Assuming your idea is a good one, what might be the reasons for your design failure?

Applying What You've Learned

1. Working as a team, use the problem-solving method to determine the best solution to the egg-mobile. Design, build, and test your team's ideas using materials that are in the shop.
2. Design a button that draws attention to present-day environmental concerns.

Summary

Directions
The following sentences contain blanks. For each numbered blank, pick the answer that makes the most sense in the entire passage. Write your answer on a separate sheet of paper.

"Genius is 1 percent inspiration and 99 percent perspiration." Do you understand what that statement means? Most of our ____1____ developed because special people created ideas that they were able to build into ____2____ devices. The recipe for developing new technology requires nine ingredients: people,

1.
   a. problems
   b. technology
   c. political problems
   d. land

2.
   a. dangerous
   b. strange
   c. useful
   d. non-working

(Continued on next page)

Summary

(Continued)

knowledge, creativity, skill, tools and machines, capital, time, raw materials, and energy.

Today large companies buy all the ___3___ needed to create new technology. They hire ___4___ with the knowledge and skill needed to use tools and machines to convert materials into ___5___ products. The company supplies the energy to run the machines and gives its experts the time that they need to create new ideas.

Will the people hired have the genius needed to create and perfect new ideas? What can be done to help them ___6___? Past experience with problem solving has led to a seven-step problem-solving method. This technique is used by people who ___7___ for a living. It can also be used by you and your friends to solve your problems.

To solve a problem, you ___8___ identify the problem, establish exactly what you want to achieve, determine what others have done to solve similar problems, brainstorm possible solutions, pick the best solution, build a working model, and finally test your solution.

You must limit what you want to achieve or you are bound to ___9___ . Brainstorming is a method that can help you limit and solve your problems. This technique requires your group to bounce ideas off ___10___ . Hearing the ideas of others can spark a creative idea that no one has thought of before.

Technology has developed many systems and subsystems. A system is made up of objects or ideas that work together to complete a task. A system is a subsystem when it can't exist without its surroundings. A ___11___ is a subsystem of the automobile. An automobile is a complete system, but it is also a subsystem of our ___12___ system.

People ___13___ a system diagram to help understand how a system works. In an open-loop system's ___14___ , input is the information, ideas, and activities needed to plan for production. Process is the construction stage,

3.
a. ingredients
b. patents
c. plans
d. products

4.
a. teachers
b. students
c. lawyers
d. experts

5.
a. old
b. strange
c. foreign
d. new

6.
a. find a job
b. get to California
c. identify and solve problems
d. go shopping

7.
a. solve problems
b. drive a bus
c. get in trouble
d. mow lawns

8.
a. can't
b. shouldn't
c. must
d. never

9.
a. succeed
b. get in trouble
c. get a raise
d. fail

10.
a. the wall
b. other members of the group
c. your teacher
d. the television set

11.
a. car engine
b. jet engine
c. television
d. telephone

12.
a. communication
b. manufacturing
c. biological
d. transportation

13.
a. draw
b. score
c. steal
d. destroy

14.
a. machine
b. analysis
c. diagram
d. process

(Continued on next page)

Summary

where machines, labor, and materials ___15___ the product. Output is the final stage of the open-loop system. It is what the system has produced.

When feedback is ___16___ the open-loop system, the system becomes a closed-loop system. Feedback measures and thereby controls the outcome of the system. ___17___ products are sent back to the drawing board. Products that ___18___ pass inspection should lead to changes in the processing stage. If you can't design and produce a quality product, your company is going to go out of business.

15.
a. destroy
b. build
c. conquer
d. lose

16.
a. added to
b. subtracted form
c. multiplied by
d. divided by

17.
a. Superior
b. Foreign
c. Mass-produced
d. Inferior

18.
a. do
b. don't
c. will
d. can

Key Word Definitions

Directions

The column on the left contains the key words from this chapter. The column on the right contains a scrambled list of phrases that indicate what these words mean. Match the correct meaning with each word. Write your answer on a separate sheet of paper.

Key Word

1. **Resource**
2. **Tool**
3. **Machine**
4. **Raw material**
5. **Processed material**
6. **Manufactured material**
7. **Synthetic material**
8. **System**
9. **Subsystem**
10. **Feedback**
11. **Problem solving**
12. **Brainstorming**

Description

a) Material in a natural state

b) Made by scientifically combining chemicals and elements

c) Increases ability to do work

d) Finding solutions

e) Material that no longer resembles the resource from which it was made

f) Can't exist without its surroundings

g) Measures and controls the outcome

h) A tool with a power system

i) People call out possible solutions

j) A way of doing something

k) Raw material in a more useful form

l) Ingredients used to create new technology

PROBLEM-SOLVING ACTIVITY

Resources You Will Need

- Paper
- Markers
- Rulers
- Styrofoam sheets and shapes
- Clay
- Masonite
- Dowels
- Wood
- Glue
- Nails
- Wood screws
- Large steel ball bearings
- Rubber bands
- Electric drill press
- Speed bores
- Scroll saw
- Woodworking vises
- Hand woodworking tools

Designing and Building a Game of Skill

Did You Know?

New products appear in stores every day. These products are manufactured to meet many different needs. These items include toys, electronic gadgets, cars and trucks, home appliances, clothing, sporting goods, and even business supplies.

Manufacturers spend a great deal of money determining what you, the consumer, want. They even hire people to test early versions of their products. The suggestions of these product testers are often included into a product's final design.

The Arcadian Pinball Machine Company has decided to design and market a new game of skill. Their top designers, after months of testing their own ideas, have decided that this problem needs the assistance of a teenage design team. Your group has been asked to design and build a prototype of the game.

Problems To Solve

This game is to be a non-electric game of skill. Figures A and B show two very different solutions that game manufacturers have produced that meet this challenge. The game must consist of a board on which a ball or puck will roll, slide, or drop. The game board cannot be larger than 1 X 2 feet. A scoring system must exist that eventually takes the player's ball or puck out of play. The game should have some obstacles that the player must overcome. Extra points should be given if the player is able to reach a more difficult area of the board.

Remember, a simple ball roll or knock hockey game isn't a satisfactory solution to this

A.

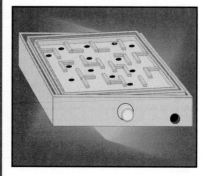

B.

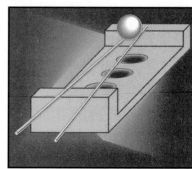

problem. The final score achieved by a game player should reflect the player's level of skill.

Your design team shouldn't have more than four members. You should develop a solution to this problem by following the seven steps in the problem-solving method explained earlier in this chapter. The key to building this game is to find ways of using some of the simple machines explained in this chapter.

A Procedure To Follow

1. Identify the simple machine principles used in the games shown in Figures A and B.
2. Use brainstorming to develop a game board, controllers, obstacles, and method of scoring. Figure C.
3. Develop a rough paper drawing of all promising ideas.
4. Pick the design that you all like best.
5. List all the materials that you will need to construct the game.
6. Call in an outside consultant (your teacher) to determine if your design can be produced with equipment and materials that you have on hand.
7. Select construction materials for your model. This prototype doesn't have to be built out of the same material as the final project. Automobiles are often modeled in clay to refine the design idea.
8. Construct your model and test it. Does it hold your attention? Is it a game of skill?
9. Present your solution to the class.
10. The class picks the best solution or brainstorms how to combine a number of the ideas presented into a super game.
11. Combine the talents of the entire class, and assign which parts you and your classmates will build.
12. Construct enough games for each member of the class.
13. Decide, as a group, if you want to produce extra games for sale or donation to an orphanage.

What Did You Learn?

1. Did you and your design team receive feedback during any phase of this project?
2. What part of this activity was the most fun?
3. What part of this activity was the most frustrating?
4. Did you learn anything that will be of use outside of your technology laboratory? What did you learn?

C.

Chapter 3

Computer Technology

In the "Star Trek" T.V. shows and movies, Mr. Spock is often accused of acting as if he had a computer for a brain. What is it about this Star Trek character that makes people think he acts like a computer? Is it that he has no emotions? That he remembers and instantly recalls information with fantastic accuracy? That he uses his past knowledge to make "logical" decisions?

People are very impressed by the fact that computers can solve all kinds of math problems, store records, act as intelligent typewriters, and run a Nintendo game. Computers can also control the operation of a car engine and record time on the face of a digital watch. See Figure 3.1 on the next page. Computers can do all of these things. But are computers smart?

Although computers can do all these things, they still can't think. They are just electronic devices that can be turned on or off like a radio or television.

But, like radios and T.V.'s, computers have completely changed the way our society works. Computers have increased the rate at which technology can change and improve our lives and the world we live in. Computers are rapidly bringing us closer to the world Spock lives in. Don't you think that's a good reason to study computers and computer technology?

■ **Figure 3-1** ■ This car not only controls its engine through the use of computers, but it also has a computerized navigation system.

The History of Computers

≡ *What is a computer?*

A **computer** is an electronic device that calculates, stores and processes data. The modern computer was built using the knowledge of many inventors. Some of these people lived and designed their machines a very long time ago.

Early Computing Devices

≡ *How would you describe the first "computers"?*

The first computing aid consisted of a number of different-colored stones that were used to help its owner perform business transactions. Here people were just one step up from counting on their fingers. In this ancient system, certain colored stones had more value than others.

In Babylonia, people strung stones together and created the *abacus* about 5,000 years ago. On an abacus you represent numbers by moving the beads from side to side. See Figure 3.2. A person accustomed to using an abacus can solve mathematical problems very quickly. Even by our modern standards, the abacus is a true mechanical computer.

The *slide rule* was developed in 1621 by William Oughtred. For hundreds of years, math students used Oughtred's mechanical computer on a stick, sliding the center bar to solve complex math

computer
a programmable, electronic device that calculates, stores, and processes information

■ **Figure 3-2** ■ An abacus is a mechanical computer that uses beads to represent numbers.

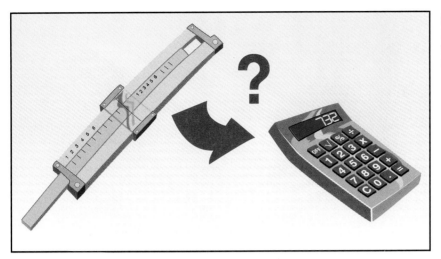

Figure 3-3 The slide rule is a mechanical, hand-held computer which is being replaced by electronic calculators.

problems. See Figure 3.3. You could always tell who the college engineering students were just by the long slide rule case they wore on their belts. Slide rules were replaced by small hand-held calculators in the mid-1970s.

A French mathematician named Blaise Pascal invented the first adding machine in 1642. Pascal's machine actually added and subtracted numbers through the movement of mechanical wheels. See Figure 3.4 on the next page. His calculator worked but was too expensive to build.

In the 1830s an English mathematician by the name of Charles Babbage invented a machine that could perform complicated calculations by following a set of instructions. Babbage's *analytical engine* worked, but couldn't be perfected because technology wasn't advanced enough to build the precision parts that he needed.

In mathematics, you often need to convert word statements into addition, subtraction, multiplication, and division problems. In the mid-1800s, George Boole, an English mathematician, developed a system that took logical statements and processed them in a mathematical (symbolic) way. This form of math was called *Boolean algebra*, and it is this system that makes the modern computer work.

Pascal
(pas-ʹkal)

Tech Talk

Charles Babbage's analytical engine was a computing machine. You will learn in Chapter 10 that an engine uses heat to produce motion. Babbage's machine wasn't really an engine.

An Electric Computer

When was the first electric-powered computer built?

The Mark I computer was built in 1944 by an American Harvard professor named Howard Aiken. See Figure 3.5 on page 67. This machine was built with the financial support of a twenty-year-old company named International Business Machines (IBM).

Aiken
(ʹā-kən)

Figure 3-4 The Pascaline used gears to perform mathematical calculations mechanically.

This computer had many switches that were opened and closed by electricity. The opening and closing of switches formed a type of Morse code that the machine, rather than a person, could follow. The computer used the math system developed by George Boole.

The First Electronic Computer (First-Generation Computers)

The perfection of what device made first-generation computers possible?

In 1946 two Americans, Presper Eckert, Jr. and John Von Mauchly, built the ENIAC electronic computer, which used vacuum tubes instead of the mechanical switches of the Mark I. See Figure 3.6.

The *vacuum tube* was invented around the same time that Thomas Edison invented the light bulb. In fact, the vacuum tube looked very much like a light bulb, but its purpose was to act like an *amplifier* and a *switch*. See Figure 3.7 on page 68. Without any moving parts, vacuum tubes could take a very weak signal and make the signal stronger (amplify it). Vacuum tubes could also stop and start the flow of electricity instantly (switch). These two properties made the ENIAC computer possible. ENIAC stands for Electronic Numerical Integrator and Computer. The ENIAC worked 1,000 times faster than the Mark I. It was able to perform 5,000 math operations per second. That means that ENIAC could do a year's worth of math homework for you in a matter of seconds. This computer was, however, 100 feet long and 10 feet tall, and it contained 18,000 vacuum tubes.

Figure 3-5 The Mark I computer was about 50 feet long and 8 feet tall. It used mechanical switches to open and close its electric circuits. It was the first computer built by IBM and it contained over 500 miles of wire and 750,000 parts.

Figure 3-6 The ENIAC was the first electronic computer. It used vacuum tubes instead of mechanical switches.

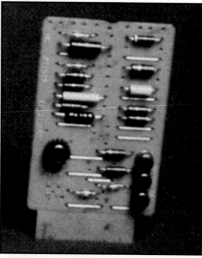

Figure 3-8 Transistors are much smaller than vacuum tubes. They were wired into computer circuit boards which could be plugged into a computer.

Tech Talk

One day a flying bug got into a computer, and its body caused a vacuum tube to burn out. The person that fixed this machine said that there was a "bug" in the system. This expression is still used today to describe computer problems. However, most computer system and programming bugs are no longer caused by insects.

In 1951 the builders of ENIAC constructed UNIVAC I, the first mass-produced computer. Because this machine also used vacuum tubes, it gave off a lot of heat. To control this heat, the computer rooms had to be air-conditioned. When a tube burned out, it had to be replaced. Computers that used vacuum tubes were later known as first-generation computers. Were your parents born before 1951? Are they older or younger than the first commercially sold electronic computer (UNIVAC I)?

Second-Generation Computers

What device led the movement into second-generation computers?

In 1947 a small replacement for the vacuum tube was invented by three scientists working at Bell Labs. Bell Labs was the equipment development company of AT&T, the telephone company founded by Alexander Graham Bell. This device was called a **transistor** and it ran cooler, used less power, and worked faster than the vacuum tube—and it didn't burn out as often. See Figure 3.8.

This 1947 development led to an improved series of computers. The transistor was developed four years before the vacuum tube UNIVAC I was built. But a computer designed around the transistor didn't appear until 1956. By 1960, all computers were using transistors instead of vacuum tubes. These computers were called second-generation computers.

Third-Generation Computers

How were technologists able to develop third-generation computers?

In the 1960s, engineers developed a method that enabled them to deposit dozens of transistors onto the same chip. These chips were called **integrated circuits.** They looked like little plastic caterpillars with a lot of little metal feet running down both sides. See Figure 3.9 on page 70. The machines that used these integrated circuits belonged to the third computer generation. They were capable of carrying out instructions in billionths of a second. The size of these machines dropped to the size of a small file cabinet.

Fourth-Generation Computers

What is a monolithic integrated circuit?

The computer made the jump into the fourth generation with the development of *monolithic integrated circuits.* We can now put the equivalent of millions of transistors onto one integrated circuit chip. See Figure 3.10 on page 70. Have you ever looked inside a computer? What you would see is a series of these monolithic integrated circuits sitting on top of a flat board. Looking inside a new computer, you might see one monolithic integrated circuit that is so complex that it does the job of many circuit boards in older computers.

integrated circuit
a tiny chip that contains dozens of electronic components. It replaced the transistor.

— Learning Time —

Recalling the Facts

1. What is a computer?
2. Name the first electron devices, and explain how they made the electronic computer possible.
3. What device moved computers into their second generation?

Thinking for Yourself

1. Why have computers decreased in size while at the same time they have increased in power?
2. Describe some of the similarities and differences between a modern computer and the Star Trek character Mr. Spock.

Applying What You've Learned

1. Construct an abacus.
2. Make a poster that shows the historical development of the computer.

—— How Computers Work ——

What are the main parts of a computer system?

The *central processing unit,* or **CPU,** of the computer processes your **data** into useful information. It doesn't matter if your com-

CPU (Central Processing Unit)
the part of a computer that processes data into useful information. It is the "brain" of the computer.

data
facts, such as numbers and symbols, that are put into the computer

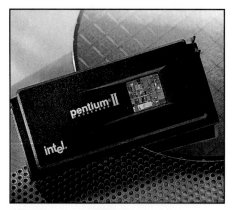

Figure 3-10 The Pentium II processor can perform about 300 million instructions per second. It contains the equivalent of approximately 6 million transistors, twice the number found on the original Pentium processor.

Figure 3-9 An integrated circuit looks like a little plastic caterpillar. This chip contains the equivalent of thousands of transistors.

input device
any control device, such as a keyboard, mouse, game controller, or joystick, that is used to get information into a computer's CPU

output device
equipment such as a printer or monitor that receives information from the CPU

monitor
a computer output device that resembles a television screen

computer hardware
the physical parts of a computer system. Included are such parts as the CPU, monitor, keyboard, disk drives, mouse, CD-ROM drives, and joystick.

puter is a NintendoN64 game machine, Sony PlayStation, or digital watch. It still has to have a central processing unit.

Did you know that a NintendoN64 video game and a digital watch are computers? They certainly don't resemble the computers that you use in school. A review of the parts of a NintendoN64 game machine can help you understand the parts of any computer system.

How do you tell the NintendoN64 game computer (CPU) what moves you want it to make? Any control device that is used to get information into a computer's CPU is called an **input device**. For your game machine, your main input device is a joystick or a controller. For your school computers, home computers, and office computers, the main input device is the keyboard or mouse.

The CPU of your NintendoN64 game converts your moves into electronic signals. It shows you how well you are doing on your T.V. screen. Equipment that receives the information from the CPU is called an **output device**. The most common output device for a computer is the **monitor**, which resembles a T.V. screen. The second most common output device is a printer.

Some computer equipment can receive information from the CPU (output) and record it on magnetic disks or tape. These same machines can send recorded information and programs into the computer's CPU (input). These machines are called input/output devices. *Disk drives* are the most common input/output devices that you will use.

All machine accessories that can be attached to the CPU become parts of the computer system. All of these machine parts are called **computer hardware**. See Figure 3.11 on page 72. Comput-

The Information Superhighway

The network of copper wire, glass fiber, and airwave transmission towers that brings the Internet to you is being compared to the roads of our highway system. Instead of trucks, cars, and buses, the Internet roadway carries information. It is therefore called the Information Superhighway or Infobahn. (In some European countries, highways are called autobahns.) Each month, more than a million people are getting the necessary hardware and software to surf the Net.

Net surfers today are using the Internet to shop at home, play interactive video games, leave messages on computer bulletin boards, join online groups to discuss just about anything, make videophone calls, send electronic messages and files, upload and download software, and receive multimedia information presentations for fun or college credit. One of the newest uses of the Internet allows parents to view their children on their office computer screen while their children are at home or in day-care centers.

To surf the Internet, you need a computer, special cellular phone, or TV Internet hookup (WebTV). The Internet is a free computer network that links nations as well as ordinary people to most of the world's universities, government agencies, research laboratories, and corporations. However, your gateway to cyberspace is usually a pay-for-service provider such as AOL (America Online), AT&T WorldNet, CompuServe, or Prodigy, just to name a few. Your connection to this provider might be through your phone line, cable TV line, wireless phone service provider, or a fiber optic network.

The last piece of the puzzle is a browser software package that makes Net surfing easy and fun. Microsoft Explorer and Netscape are now the most popular browsers. Many computer manufacturers prepackage one or both of these browsers with their computers. Most online services give Netscape, Explorer, or their own browser to customers for free when they sign up for the service.

What you receive over the Internet is constantly growing in sophistication. Play-by-play broadcasts of sporting events allow people from all over our planet to watch their favorite team play in "real time." However, the sports cybercasts will have to compete with 500 channels of interactive video games and Net surfing in 3D. Internet II is now under development, and it promises faster connection times coupled with new technology that will knock your socks off.

ers can be very small or very large. You can no longer judge the power of a computer by its size.

Room-size computers are called *mainframe computers*. Computers that fit on top of a desk are called desktop computers. Computers that are very portable and can be used on your lap are called laptop or notebook computers. Very small computers that can fit in the palm of your hand are called palm-size computers or personal digital assistants (PDAs). Figure 3.12 on page 73 shows what some of these computers look like.

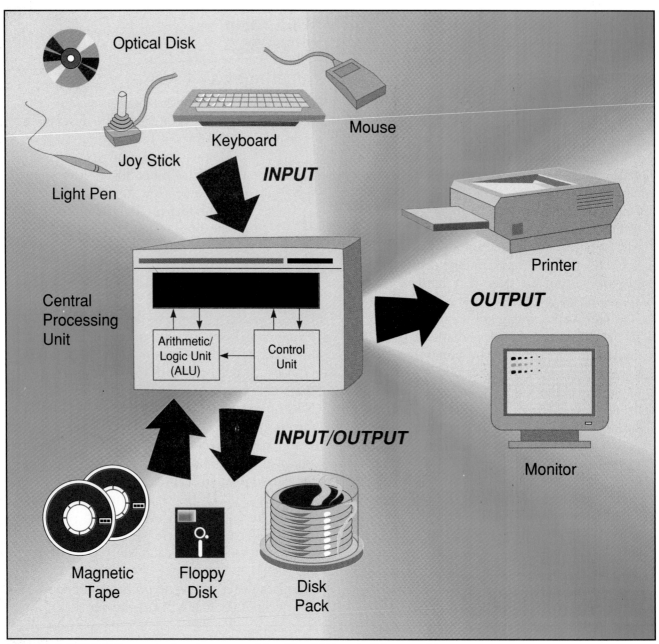

■ **Figure 3-11** ■ Computer hardware consists of the CPU and all the machine accessories that are connected to it.

Computer Programs

≡ *What is the function of a computer program?*

A **computer program** is a set of instructions that the computer follows to do its work. The program controls the computer. It tells the CPU exactly how to handle all the data that are entered into the machine. The program turns your computer into a game machine, word processor, artist's drawing board, or teaching machine.

The program even controls which input and output devices the CPU will recognize. In a game program, the programmer determines new inputs for the letters of the keyboard. While playing the game, you will move animated characters around by pressing certain keys on the keyboard.

All computer programs are called **computer software**, and they are usually stored on **floppy disks** or hard disk drives. Nintendo games store their programs in game cartridges that you plug into the machine.

How does a computer learn to play games or write reports? The computer program provides the directions that tell the computer what to do.

How does a computer know how to run a program or take information from the keyboard or other input device? Many memory chips inside the computer have special **operating system** programs permanently burned into their circuits. Other start-up programs are placed on start-up disks or hard drives. When the computer is turned on, these programs immediately tell the computer how to run its own systems. In a sense, the computer reads an entire instruction book on how to operate a computer each time it is turned on.

Figure 3-12 This tiny, palm-sized computer has the power of a desktop computer.

Putting Knowledge To Work

Crime-Solving Computers

In July of 1987, a former prosecutor from New Jersey was murdered. After three years of investigation, the detectives weren't able to solve the crime. They had a perfect fingerprint from a water glass that was found at the scene of the crime, but they couldn't match it up with a potential suspect. To check the 1 million fingerprints on file would have taken 167 experts one full year. This was a case for Dick Tracy or Sherlock Holmes—or a special IBM computer!

The computer did what the experts didn't have the time to do. Following its program, it examined millions of fingerprint images, looking for a match. It identified a man who was serving time for another crime.

Detectives went to the prison to interrogate the person the computer had identified. He quickly confessed to the crime. The computer that solved this crime didn't exist just a few years ago.

Part of the computer revolution is the constant creation of more powerful machines. Desktop and laptop computers that cost $10,000 today are more powerful than the large mainframe computers that cost $3 million fifteen years ago. The new top-of-the-line mainframes have the power to do image identification of fingerprints and documents. They can solve very complex problems that are beyond the ability of smaller computers. Soon these computers will be able to identify people's faces. Such computers attached to video cameras could scan people walking through an airport terminal and pick out the ones who are wanted by the police.

ROM (read only memory)
one of the two types of memory contained in the CPU. It is permanent memory that cannot be deleted or changed by a computer's instructions.

RAM (random access memory)
one of two types of memory contained in the CPU. This memory is lost when the computer is turned off.

FRAM (ferroelectric random access memory)
flash memory that is saved indefinitely without electricity

binary code
an electronic Morse code based on the binary number system that the computer can understand

For the computer to operate, it must have programs. These programs (software), together with the computer hardware, make up a computer system.

Errors in computer programs sometimes cause computer systems to stop working or do weird things. Computer programmers call this *GIGO*, which stands for "garbage in" (program with wrong directions), "garbage out" (computer does things that have no meaning).

The Central Processing Unit (CPU)

What are the three main parts of the CPU?

The central processing unit (CPU) is the part of a computer system that performs all calculations and follows the step-by-step procedure of the program.

You can compare a CPU to a highway system. Many roads must sometimes be taken to get from one place to another. Information travels the circuits (roads) of the different parts of the CPU to get processed. The CPU's *control unit* guides all information according to the program. The *arithmetic/logic unit* performs math calculations with the data that the control unit sends. The *memory unit* stores the programs and the information before and after processing. These three parts of the CPU work together to accomplish whatever task is assigned to the computer.

The CPU has two types of memory. *Read only memory*, or **ROM**, contains the basic knowledge that the computer needs to run any program or perform any operation. This knowledge is permanently etched onto certain memory chips. Part of this knowledge is the binary code for each letter press of the keyboard. When you shut off the power, this memory is not erased.

The other type of CPU memory is called *random access memory*, or **RAM**. All data that are fed into your computer by your input devices are put into random access memory. The data are temporarily held as binary code in the RAM locations of the computer. When the computer is shut off this information is lost. Some new computing devices have FRAM (flash memory) chips that can retain temporary data after the electricity is shut off.

Binary Code

How does a computer use binary code to process data?

The computer program and all the information that the computer will use must be changed into **binary code**, which is sort of an electrical Morse code that the computer can understand. The computer must also convert data that it receives into a form that it can use. The only thing that a computer can use is the presence or absence of electricity. Each letter, number, punctuation mark, and space is converted into an electronic coded signal that is eight *bits* long. A **computer bit** is the smallest piece of information that a computer can receive. A computer bit is represented by a 1 or a 0.

Bits carry only one message—electricity on, electricity off. You need more information than just a simple on and off to carry a message. By stringing eight of these bits together, you get a **computer byte**—enough space to create little meaningful message units. Each byte carries the individual letter, number, or punctuation mark represented as 1s and 0s. Each byte is like a small car license plate that will whiz around the computer's circuits and be used by the computer to identify a letter, number, or punctuation mark.

Instead of the dots and dashes of Morse code, binary code uses 1s and 0s. In Morse code you translate the dots and dashes into words. In binary code, electricity flows when a 1 is the message, and electricity stops flowing when a 0 is the message. In Morse code the dots and dashes are strung together to represent letters. In the same way, binary code bytes have eight computer bits strung together to represent any number, letter, or punctuation mark.

The computer converts everything that you type into these binary bytes. Each of the eight spaces (bits) in this byte can only contain a 1 or a 0. When the computer reads the binary code 00000100, it is reading the number 4. When the computer reads the binary code 00000000, it is reading the number 0. Figure 3.13 shows you how any number can be represented in binary code.

Why does a computer have to use binary code? Electricity on and electricity off are the only two messages that a computer can

computer bit
the smallest piece of information that a computer can use

computer byte
an information unit made up of eight bits

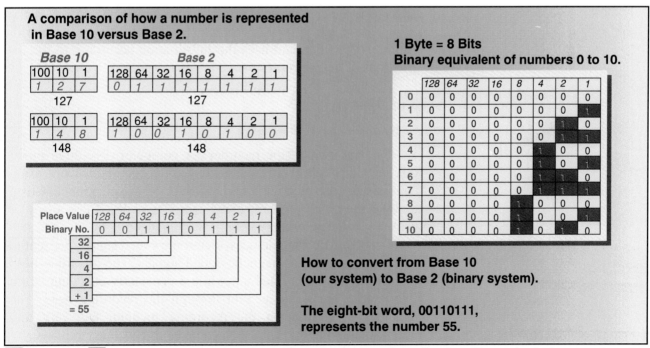

A comparison of how a number is represented in Base 10 versus Base 2.

Base 10

100	10	1
1	2	7

127

Base 2

128	64	32	16	8	4	2	1
0	1	1	1	1	1	1	1

127

100	10	1
1	4	8

148

128	64	32	16	8	4	2	1
1	0	0	1	0	1	0	0

148

Place Value	128	64	32	16	8	4	2	1
Binary No.	0	0	1	1	0	1	1	1

32
16
4
2
+ 1
= 55

1 Byte = 8 Bits
Binary equivalent of numbers 0 to 10.

	128	64	32	16	8	4	2	1
0	0	0	0	0	0	0	0	0
1	0	0	0	0	0	0	0	1
2	0	0	0	0	0	0	1	0
3	0	0	0	0	0	0	1	1
4	0	0	0	0	0	1	0	0
5	0	0	0	0	0	1	0	1
6	0	0	0	0	0	1	1	0
7	0	0	0	0	0	1	1	1
8	0	0	0	0	1	0	0	0
9	0	0	0	0	1	0	0	1
10	0	0	0	0	1	0	1	0

How to convert from Base 10 (our system) to Base 2 (binary system).

The eight-bit word, 00110111, represents the number 55.

Figure 3-13 Computers represent our base 10 numbers using a base 2 system. These charts show you how to convert numbers between these two systems.

sense. The computer does all of its work by having the power quickly turned on and off in parts of its circuits.

It doesn't seem possible that a computer can do all these things just by having electricity on, electricity off. Let's try to understand how a computer adds two numbers using binary code. Remember that the computer doesn't understand addition. It just blindly turns on and off very tiny electronic switches according to the step-by-step procedure of the program.

Look at Figure 3.13 again. It shows a chart of the binary equivalent of numbers 0 to 10. The computer has special parts called *logic gates*. These gates look at the binary codes of two bytes, one bit at a time. See Figure 3.14. It's as if two teams made up of boys and girls are waiting to go through a gate. One member from each team enters the gate together with a member from the other team. In this people model, the gatekeeper looks at the sex of the two who are waiting and decides on the sex of the one person who will be allowed to pass though the gate. In the computer model, the computer's clock pulses electricity and sees if the two bits waiting in the gate are 1s (they carry the electricity) or 0s (they block the electricity). The logic gate then combines the two bits into one bit.

The Keyboard

What is the function of a keyboard?

All inputs to the computer must be converted to binary code. When you press down on a keyboard letter, you cause contacts that are under the key to send the binary code for that letter. This coded message is sent to the screen and into a memory location in the CPU. The screen or monitor shows you what you have just typed, and the CPU holds and uses the information according to the instructions of the program.

The Disk Drive

What does a disk drive do?

Do you own a tape recorder? A computer disk drive might look very different from your recorder, but they both work using magnetism. Your recorder and the disk drive use electromagnetism to write messages onto magnetic tapes or disks. These tapes or disks are called *recording media*.

On both recording devices, electronic coded signals are sent to the recording head as magnetic coded signals. The recording head touches the recording medium while a constantly changing North Pole/South Pole signal is sent through the recording head. The metal oxide surface on the recording medium has very tiny North/South poles that can be magnetized into a coded message.

When the recorder's playback head passes along the recording medium, it picks up this magnetic coded message. This magnetic

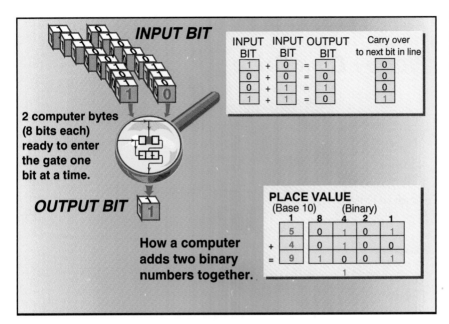

signal, which consists of positive and negative impulses, is then converted back into an electronic signal. Figure 3.15 on page 78 shows just how a floppy disk drive and a hard disk drive operate.

CD-ROM Drives

What is the difference between a CD-ROM drive and a regular disk drive?

Digital information is stored on a CD as microscopic pits or dots that are optically read by a laser. Most current CD-ROM drives are read only devices that can store about 650 MB of data. In 1996-97, the CD-R drive began to gain popularity. It can write only once on its special discs and read them repeatedly. In 1997, the CD-RW (re-recordable drive) entered the market with unlimited read-write potential. At the end of 1997, Toshiba introduced the DVD ROM drive. It can hold 17 gigabytes of data, which is equivalent to 11,500 floppy disks. For more information about the DVD ROM drive, see Chapter 9.

The Joystick, Game Control Unit, and Mouse

How do input devices such as a joystick, game control unit, and mouse change coordinates on a computer screen?

World globes, maps, and computer screens all divide their surfaces into horizontal and vertical coordinates. In social studies you learned that the horizontal and vertical coordinates on a world map are called longitude and latitude.

The horizontal coordinates on a computer screen are called columns, and the vertical coordinates are called rows. The computer uses these coordinates to draw and locate things on the screen.

Tech Talk

Adventurous early airplane pilots thought it was a real *joy* to fly like the birds. Their airplanes were controlled by a stick coming up from the floor. Push it forward and the plane went down. Pull it back and the plane went up, and so on. They called it a *joystick*. Computer joysticks closely resemble those from early airplanes.

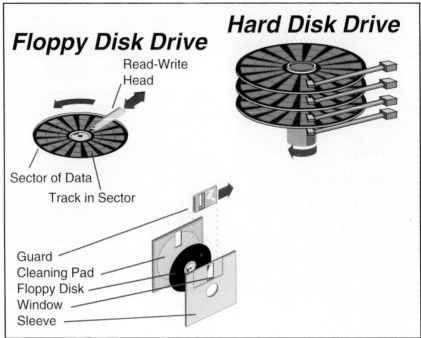

Floppy Disk Drive / Hard Disk Drive

Read-Write Head

Sector of Data
Track in Sector

Guard
Cleaning Pad
Floppy Disk
Window
Sleeve

Figure 3-16 The joystick has a bar that controls movement on the computer screen. The SideWinder joystick provides feedback for your actions. Shoot a gun and you feel its recoil; stall a plane and you feel it shudder. The computer mouse is rolled to select a location on the computer screen. Extra wheels and buttons on some new mice let you scroll pages and perform other tasks.

We can identify any location on the screen by giving its row and column. This can actually be done using a pen-like device directly on the screen or on a special pad.

In most arcade games, you use a joystick or track ball to move your game figure through the coordinates of the screen. The mouse that is used with a desktop computer is nothing more than a track ball that has been turned upside down. How do these input devices actually work? When you move the lever of a joystick or roll the ball of a mouse or track ball, the row and column location is changed, and these changes are fed into the computer. The computer responds to these changes by moving your location on the screen. Figure 3.16 shows a mouse and joystick. They use mechanical and electrical systems to convert your physical movement into electronic signals that the computer can understand.

Printers

What devices are used to make hard copies?

A student says that he did his school report on his home computer but couldn't bring the report into school because the computer was too heavy. What is wrong with this excuse?

The image on your computer screen is called soft copy because it is only temporary. Many different types of printers are attached to computers to make permanent copies of computer-generated material. These permanent copies are called *hard copies*.

A *dot matrix printer* prints pictures and letters by printing small, carefully spaced dots. Take a pencil and just tap the point on a piece of paper. Try to keep hitting the same space over and over again. After a while, the dots will resemble a large dark spot. It will be very difficult to tell that this spot was actually made up of only dots. The CPU tells the printer the exact pattern and how many dots to print. The printing is done by a printer head that has many small wires that are all located in perfect alignment. Each time the printer receives a binary code for a letter, it sends a message that powers the tiny hammers that direct the individual pins against the printer ribbon. See Figure 3.17.

The letters printed by a *laser printer* are actually made up of individual dots. However, these dots are spaced so close together that it is impossible to see their individuality. See Figure 3.18. The signal that determines the letter is the trigger for a laser beam. As the beam shoots, it creates a static electrical charge on a rotating drum. This drum is part of an electrostatic copying machine that produces the final print. See Chapter 7 for a detailed explanation on how electrostatic printers work.

If you want to print in color, the printer that you are most likely to use is an *ink-jet printer*. These machines are often called "Bubble Jets" because they print by squirting small dots of ink onto the paper. Many of the color ink-jet printers use two ink cartridges. The black cartridge is used alone when you are printing only in black. A second cartridge that contains the three primary color inks is used in combination with the black cartridge to let you print pictures that contain every color of the rainbow. See Figure 3.19.

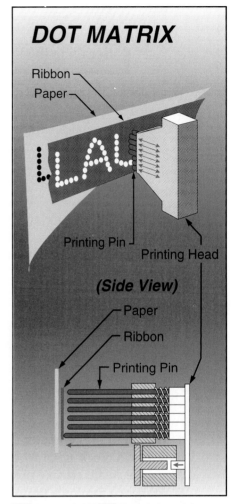

Figure 3-17 The dot matrix printer prints a series of dots to form letters or pictures.

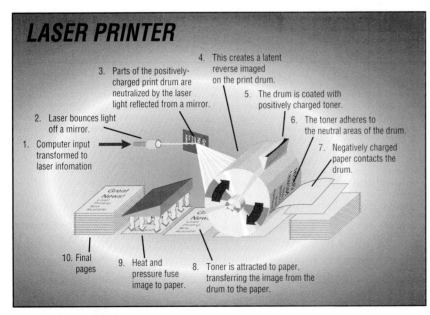

Figure 3-18 The laser printer uses the same technology of the electrostatic copying machine.

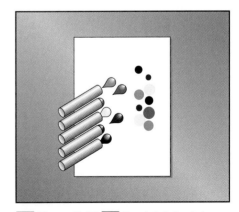

Figure 3-19 The ink-jet printer actually has very tiny spray guns that shoot the ink directly onto the paper.

— Learning Time —

Recalling the Facts

1. What does CPU stand for, and what does a CPU do?
2. Name and describe an input device and an output device that are used with a NintendoN64 game machine.
3. What is the purpose of a computer program?
4. What does GIGO stand for? What does it mean?
5. What is the difference between ROM and RAM?

Thinking for Yourself

1. What are the positive and negative effects of computers on our society?
2. Why is printing quality important to people who use computers?
3. Why are NintendoN64 game machines more popular than the machines made by other companies?
4. Can a computer system work without a CPU?
5. Why aren't all computer data and programs treated in a ROM format?

Applying What You've Learned

1. Construct a human computer using the members of your class as individual bits to add two bytes (numbers) together.
2. Working with two or three of your classmates, design the ideal computer. Complete all of the following tasks.
   a. List all of the computer's features.
   b. Create a mock-up of what the computer would look like.
   c. Create an advertising poster.
   d. Prepare a 5-minute presentation to the class.

Artificial Intelligence

What is artificial intelligence?

By understanding the basic concept of how a computer works, you can see why a computer gets no satisfaction when it solves a problem and why these machines are considered dumb. You must remember that computers can only run programs and manipulate data. In the movie *Short Circuit*, the computer-brained robot starts to think. Thinking computers are, at this time, just science fiction.

Artificial intelligence programs give the impression that a computer can think. The programmer has provided the computer with a number of answers that will be triggered by certain requests.

The program just has a big bag of tricks that the computer automatically uses giving the impression that it is thinking.

Perhaps in time computers will think for themselves. Wouldn't you like to be the computer engineer who designs the first thinking computer?

People often say things that make computers sound human. You now know how computers work, so you know that computer error is really human error. Computer "bugs" don't bite. Computers that are "down" aren't depressed. A computer that "crashes" didn't bang into another computer. And a computer with a "virus" can't get you sick.

▰▰▰ Putting Knowledge To Work ▰▰▰

The Chess Grandmaster Computer

The best example of what many would call artificial intelligence (AI) is the IBM chess champion computer, Deep Blue. In 1997, the chess world was shocked when Deep Blue defeated Garry Kasparov, the world's greatest living chess player. Deep Blue won the match 3½ points to 2½ points. Yet, Deep Blue is not conscious of its surroundings or even the fact that it beat Kasparov.

An understanding of how Deep Blue combines its expert chess program with its many microprocessors should help you understand how today's AI machines appear intelligent.

IBM built Deep Blue by combining the power of many microprocessors. They all work on solving the same problem—moving a chess piece—together. This is called *parallel processing*. In a sense, Deep Blue has many equally powerful microprocessors performing the same tasks that your computer does using one microprocessor. Because many microprocessors work together, speed is greatly enhanced.

Deep Blue's parallel processing can examine 100 million chess moves per second. (Kasparov has reported that he can examine about 3 moves per second.) Each time Deep Blue's opponent makes a move, the computer can quickly examine 100 million possible scenarios that this single move created. Since the players in classical chess have 3 minutes to move a chess piece, Deep Blue plays out billions of possible scenarios before it moves a chess piece on the real game board. Its knowledge of chess comes from the fact that its database contains the chess knowledge of all past chess champions, including Kasparov.

The rules that it uses to apply this knowledge come from the chessmasters and programmers that built the machine.

Not many people will want to buy a chess-playing computer that never lets them win. However, by using what AI designers call *heuristic reasoning,* all kinds of intelligent assistants become possible. In heuristic reasoning, the computer program uses rules that are based on past experiences filed in the program's database. The computer runs test scenarios that help it to make what appear to be intelligent decisions. If the outcomes of its decisions are changed or ignored by the computer user, the program re-writes its rules and also adds its new experiences to its database.

Computer Viruses

computer virus
a destructive computer program that "infects" the computer system and can cause damage to data in the system

How are computer viruses created?

A **computer virus** is the result of a destructive program that someone has written and placed inside a computer program, which unsuspecting people then place in their computer system. Many people get contaminated computer programs by downloading files from the Internet, by opening files that have been sent as part of an e-mail, or by trading programs with other people.

Most of the time, programs that arrive by modem or trade are perfectly safe to use. However, you do stand a chance of getting a program that has been tampered with. If the host computer already has a virus, its copied disks can be contaminated. Here a computer program virus is hiding inside the normal program. Many computer programs that are traded were copied illegally.

When this program enters your computer through your input device, it hides in your computer's memory and starts to duplicate itself like a disease. When you save your data, you also save the virus. Slowly but surely, the virus crowds out your data and causes major system problems.

The virus can't affect the computer's read only memory, but it can affect random access memory and your computer disks and hard drive.

If the virus is on your disk or hard drive, it will return to the computer when you use the program again. If you switch from one program to another, the virus will attach itself to the new program. In this way, it can slowly infect all your programs before you know that it exists. Today millions of dollars are being spent to rid and protect computer systems from these virus programs.

New anti-virus software now exists that can find and destroy any viruses that exist on your computer. These anti-virus programs must be updated frequently to protect against new viruses.

Tech Talk

Can the Y2K bug be exterminated? Experts predicted that after midnight on 12/31/99, many older computer systems would act as if the new year was 1900. If the problem is not corrected, a one-minute phone call could be billed as if you were on the phone for a hundred years.

Computer Careers

What type of computer career might you be interested in?

You will find computers in use in almost every occupational area. *Computer literacy,* which is the ability to use a computer, will be needed by all people entering the labor force in the future.

You don't need to understand the laws of gravity to ride a bicycle. You don't need to understand how an automobile engine works to learn how to drive a car. And you don't need to understand programming and the inner workings of computer hardware to operate a computer.

If you enter any career that uses a computer as an information tool, you will only need to know how to work with specific software and operate certain pieces of computer hardware. However, certain careers involve the designing and construction of computer systems, and these careers demand a thorough understanding of how and why these machines work.

A computer specialist will need a great deal of knowledge of mathematics, computer programming, computer languages, and computer systems. Your studies could be done at a university as part of a computer science, engineering, mathematics, or business degree.

As a computer engineer, you might design new computer systems. A computer technician could work alongside the engineer or repair and service computers.

■■■ Putting Knowledge To Work ■■■

Smart Highway Systems

A new computer navigation system now provides drivers with the ability to avoid traffic jams. These systems receive information from satellites that are placed in a stationary orbit in space. These navigational systems are called global positioning systems (GPS). The owners of these systems will have a computer monitor mounted in the instrument panel of their car. On this screen, an electronic road map would flash problem areas and alternate routes that would take them to a less congested route. Some of these systems provide verbal instructions so the driver doesn't have to watch the GPS map. It is estimated that between $50 billion and $100 billion a year is wasted by motorists who are stuck in traffic.

How does an electronic navigation system operate? Road sensors along the highways, helicopters, or mounted cameras will send data about traffic conditions to a control center. Computers at these centers will use this data to plot alternate routes around the congestion. The automated system would send this information via an earth satellite to the computerized road map in the car.

In 1997, General Motors (GM) introduced OnStar, which is a GPS enhanced navigational system. OnStar combines your hands-free cellular phone with its GPS mapping system to provide auto theft prevention, voice directions, and emergency road service 24 hours a day.

During a family trip, have you and your family ever gotten lost? If your car was equipped with an OnStar system, all you would need to do is press a button. An advisor at GM's 24-hour service center would use GPS to find your location and give you precise directions to your destination. If an accident caused your air bag to open, the system would dispatch emergency help automatically to your car's location. If your car was stolen, the system would call the police and give them precise directions to your vehicle.

GM is offering OnStar as a 1998 option on all Cadillacs and Buicks. The system is also available on some of the other models made by GM's other divisions.

Computer programmers write the software and system programs that make these machines perform all the wonderful things that they do. At this time, programmers have all levels of education. Some are still in their teens and have learned how to program computers through computer clubs, books, and basic courses. Other programmers have completed advanced degrees from universities. Many of these young computer programmers are often called *computer hackers* because they spend enormous blocks of time at their computers, using a system of trial and error until they accomplish what they want.

A *systems analyst* matches computer systems with a company's needs. You would need to know the different computer systems, the needs of different companies, and the ability of the workers who would be using the system that you recommended.

Other workers in the computer field include data entry operators, managers of data processing departments, and computer teachers.

— Learning Time —

Recalling the Facts

1. What is a computer virus, and how does it get into your computer? Can a person catch this virus?
2. Define computer literacy.

Thinking for Yourself

1. How does artificial intelligence differ from human intelligence?
2. What are some possible concerns of having a global positioning system in a vehicle?

Applying What You've Learned

1. Working with two or three classmates, prepare an interview with a person in a computer related occupation. Complete all of the following tasks.
   a. Videotape the interview.
   b. Prepare appropriate captions using a graphics program such as *Print Shop*.
   c. Videotape and edit your interview.
   d. Present the videotape to the class.
2. Using a camera, document the changes that have taken place in our society because of computers.

Summary

Directions

The following sentences contain blanks. For each numbered blank, pick the answer that makes the most sense in the entire passage. Write your answer on a separate sheet of paper.

A computer is an electronic device that calculates, stores, and processes information. Today we find ___1___ in every area of our society. They can be as ___2___ as a digital watch and as ___3___ to use as a NintendoN64 game machine. Your telephone, television, and automobile are now under the ___4___ of computers.

The very first computing device, an abacus, was built by our ancestors about 5,000 years ago. The abacus is a/an ___5___ computer. In 1944 the first electric computer was built by a small, twenty-year-old company named IBM. This ___6___ used mechanical switches that were opened and closed by electricity.

The electronic computer couldn't be ___7___ until a device was ___8___ that could convert electricity into an electronic signal. The invention of the vacuum tube made it possible for two Americans to build the first ___9___ computer, named ENIAC. The ENIAC had 18,000 ___10___ tubes that were the size of light bulbs.

When the transistor was invented, the job performed by the ___11___ could be done by a device that was about the size of a pencil eraser, produced less heat, and worked faster. By 1960 all computers were built using ___12___ .

Engineers soon learned how to deposit

1.
   a. machines
   b. computers
   c. problems
   d. conflict

2.
   a. gigantic
   b. fragile
   c. small
   d. strange

3.
   a. fun
   b. boring
   c. complex
   d. dangerous

4.
   a. gun
   b. control
   c. leadership
   d. heading

5.
   a. electric
   b. electronic
   c. solar
   d. mechanical

6.
   a. toy
   b. place
   c. factory
   d. machine

7.
   a. built
   b. purchased
   c. sold
   d. used

8.
   a. purchased
   b. invented
   c. stolen
   d. used

9.
   a. mechanical
   b. electric
   c. electrical
   d. electronic

10.
   a. hollow
   b. solid
   c. vacuum
   d. open

11.
   a. integrated circuit
   b. vacuum tube
   c. hollow tube
   d. television tube

12.
   a. light bulbs
   b. metal
   c. batteries
   d. transistors

(Continued on next page)

Summary

dozens of transistors onto a chip that they called an integrated circuit. Computers once again ___13___ in physical size. Today the equivalent of millions of transistors are being placed on integrated circuits that are smaller than your fingernail. As a result, very ___14___ computers can now be built.

A computer system consists of hardware and software. The hardware of an average computer system includes the central processing unit (CPU), keyboard, monitor (T.V.), disk drive, joystick, mouse, and printer. The programs that tell the ___15___ what to do are called computer software. ___16___ turn your computer into a game machine, typewriter, artist's drawing board, or teaching machine.

The computer is the central processing unit (CPU). All the other ___17___ provides a window through which information is entered, stored, and retrieved. All computers have ___18___ of memory called ROM and RAM. ROM contains the ___19___ that the computer needs to run programs and operate its own system. RAM is temporary, and it is the place where the computer stores data that are received from the keyboard and other computer input devices. RAM must be saved to your disk or hard drive or it will be ___20___ when the computer is shut down.

The computer monitor shows you what the computer is working on. The printer gives you a permanent copy of your work. Computer programmers are now writing ___21___ that make you think that a computer can think. These programs are called artificial intelligence. Some ___22___ are writing programs containing computer viruses. These programs make your computer act as if it has a memory disease.

13.
a. grew
b. shrunk
c. increased
d. multiplied

14.
a. tiny but powerful
b. big and powerful
c. small and useless
d. strange and colorful

15.
a. transistor
b. vacuum tube
c. integrated circuit
d. computer

16.
a. Programs
b. Radios
c. People
d. Books

17.
a. software
b. wetwear
c. hardware
d. memory

18.
a. one type
b. two types
c. three types
d. four types

19.
a. hardware
b. integrated circuits
c. paper
d. instructions

20.
a. lost
b. messed up
c. changed
d. altered

21.
a. books
b. articles
c. stories
d. programs

22.
a. computers
b. plumbers
c. companies
d. people

Key Word Definitions

Directions
The column on the left contains the key words from this chapter. The column on the right contains a scrambled list of phrases that describe what these words mean. Match the correct meaning with each word. Write your answer on a separate sheet of paper.

Key Word	*Description*
1. Computer	a) Replaced the transistor
2. Computer program	b) Electrical Morse code
3. Transistor	c) CPU, monitor, keyboard, disk drive, or joystick
4. Integrated circuit	d) Program that acts like a disease
5. CPU	e) Software
6. Data	f) Random access memory
7. Input device	g) Smallest piece of information
8. Output device	h) Replaced the vacuum tube
9. Monitor	i) Facts such as numbers and symbols
10. Computer hardware	j) The programs
11. Computer software	k) Read only memory
12. Floppy disk	l) Keyboard, mouse, or joystick
13. Operating system	m) Eight bits together
14. ROM	n) A device that processes information
15. RAM	o) Instructions that tell a computer how to run its systems
16. Binary code	p) Television screen
17. Computer bit	q) The computer's brain
18. Computer byte	r) Recording material that programs are on
19. Computer virus	s) Printer or monitor

Programming a Computer to Control a Machine

Did You Know?

The United States landed the first people on the moon using less computer power than is found in today's automobiles. On the Apollo moon mission, computers controlled many systems on the mother ship and moon lander, such as navigation, cabin atmosphere, and rocket-firing sequences. Today computers control the fuel system, engine, and other parts of your car as well as many other technology systems.

Chances are that you have already used a computer as a game machine, word processor, and teaching machine at home or in school. Have you ever programmed a computer to control a motor-powered machine?

Problems To Solve

In this activity you will build a motorized machine using the LEGO DACTA® Control Lab. The motors and sensors that are a part of your machine communicate with your computer through the LEGO DACTA Serial Interface.

You will learn how computers control motors and sensors.

Working on the LEGO DACTA Program Setup Page, you will identify and test the input and output devices of your machine. You will write a computer program on the LEGO DACTA Program Procedures Page and document your activity on the program's Project Pages.

Resources You Will Need

- **Macintosh LC (68030) or newer computer with System 7 or higher, available modem or printer port, and HD 3.5" disk drive**
  OR
- **486/66 MHZ PC with 8 MB of RAM and Windows 95 operating system. The Control Lab software can be run in a DOS format on older 386 PCs that have an EGA or better monitor and 4 MB of RAM. The PC must have available serial port and 3.5" or 5.25" HD disk drive.**
- **LEGO DACTA® Control Lab Starter Pack, either MS-DOS/Windows 95 or Macintosh version**

A.

A Procedure To Follow

1. Read the Quick Start Guide that came with your LEGO DACTA Control Lab. You might want to complete the fan control experiment before trying to build and program a machine on your own.
2. Each construction brochure in the 9701-1 to 9701-7 series provides a step-by-step procedure for a different computer-controlled project.
3. Pick one machine and follow the directions for assembly.
4. Test motors and sensors following the Quick Start Guide procedure.
5. Use the LEGO Dacta manuals to learn Logo commands for controlling the motors and sensors on your machine. Logo is a programming language that lets you "talk" to the computer using phrases that the machine (through the software) can understand.
6. Plan out the sequence of commands that will tell the computer how to control your machine.
7. Test each command one at a time in the Command Center.
8. Program the computer to control your machine. Make certain that you properly save your project before exiting the program.
9. Demonstrate the operation of your machine to the class.

What Did You Learn?

1. In computer control systems you often find that the machine under control is equipped with optic sensors, touch sensors, and motors. How do these subsystems enter into the control of the machine?
2. In the Logo language, when would you use *talkto, tto, on, rd, off, wait, waituntil, onfor,* and *repeat* in your program?

B.

C.

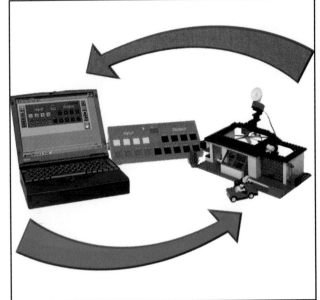

Energy

— **Key Words** —

In this chapter you will learn and study the meanings of the following important technology terms.

ENERGY
POWER
ELECTRICITY
PETROLEUM
NUCLEAR REACTION
RADIOACTIVE
CALORIE
COAL RESERVE
SOLAR HEATING
SOLAR CELL
WIND FARM
HYDROELECTRIC
GEOTHERMAL
ENERGY CONSERVATION
RECYCLE
GREENHOUSE EFFECT

Do you have a little brother or sister or know someone who does? When you see the youngster running around, do you think, "This kid sure has a lot of *energy*"?

What is this stuff called energy? Is it strength? Is it speed? Is it motion? Is it action? What do you think energy is?

Because we use the word *energy* so often, we sometimes confuse it with other words. One of your teachers might seem *energetic* as he or she quickly moves around the classroom, but energy is not the same as speed. An energetic football player might be big and strong, but energy is not the same as strength.

Can you point to something and say "*That* is energy"? Even this seemingly simple task can be difficult. Nature doesn't usually give us energy in a form we can use directly. It's as if the energy were hiding—waiting for someone to find it. Coal, for example, is a source of energy, but it's just a black rock. It doesn't do anything by itself. You can't see or use the energy. Or can you?

Technology makes it possible to find the energy hidden in nature. Technology also makes it possible for us to put that energy to work. Technologists are constantly looking for new and better ways to make use of nature's energy. For these reasons, learning about energy is an important part of your technology education.

The Difference Between Energy and Power

How would you describe the difference between energy and power?

Many people use the words *energy* and *power* to describe the same things. Energy and power are related to each other, just like you are related to a cousin. But, the two words have different meanings, just as you and your cousin are different people. However, if you happen to look like your cousin, people might get the two of you mixed up. That happens with energy and power because they certainly do resemble each other.

Energy

What is Energy?

Energy is the *ability* to do work. Simply having ability doesn't mean, however, that work will get done. A talented young man or woman certainly has ability, but he or she might take a midday nap and not do any work.

All the energy in nature can be grouped into six types. These types are:

- Mechanical energy, as in a kicked soccer ball
- Heat energy, as in a hot air balloon
- Electrical energy, as in an electric curling brush
- Chemical energy, as in a wristwatch battery
- Nuclear energy, as in the power source for a US Navy ship
- Light energy, as in the solar cells of a calculator

As you read at the beginning of the chapter, coal is a source of energy. To accomplish some work with that energy, however, technology must do something with that coal. Let's say, for example, that we want to use the coal to produce electricity, which we can use in many ways. To do this, we would need to change the form of the energy many times. The path from black rock to electricity would take many turns.

1. Burning coal develops heat.
2. Heat changes water to steam.
3. Steam operates a generator.
4. The generator produces electricity.

Power

What is Power?

Power measures the work done when energy is used. In some early factories, a waterwheel turned a shaft that was located along

energy
the capacity or ability to do work. There are six types of energy: mechanical, heat, electrical, chemical, nuclear, and light.

power
the measurement of the work done by using energy; energy under control

---

Tech Talk

How much power can your body develop? James Watt conducted tests to see how much power a horse could develop. He found that a horse pulling a rope over a pulley could lift 550 pounds 1 foot in 1 second. That's the technical definition of a *horsepower*. If you're in very good physical condition, you might be able to develop about 0.2 horsepower for several minutes.

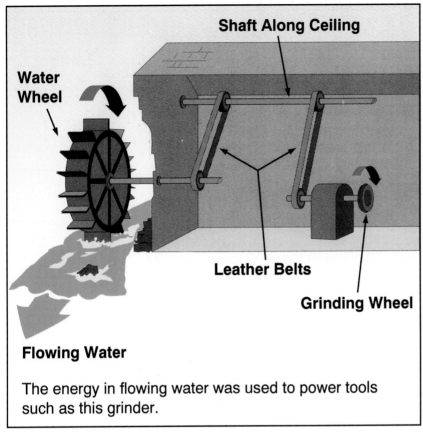

Shaft Along Ceiling

Water Wheel

Leather Belts

Grinding Wheel

Flowing Water

The energy in flowing water was used to power tools such as this grinder.

■ **Figure 4-1** ■ An average water wheel in colonial America produced about four horsepower and operated all the machines in a factory.

the ceiling. Circular leather belts connected the turning shaft to tools at floor level. Figure 4.1 shows an example of a belt-operated grinder. A worker used the grinder to smooth or reshape metal. The *energy* of the flowing water provided *power* to operate the grinder. We sometimes say that the grinder was powered by the flowing water.

Power is also a way of rating how quickly an engine can perform work. You can probably cut grass more quickly with a powerful riding mower than with a less powerful push-type lawn mower. In both cases, the work is the same: cutting the same area of grass. However, the more powerful engine in the riding mower allows you to do it more quickly — and more comfortably.

One common unit for measuring power is the *horsepower*. A 3-horsepower lawn mower engine produces about as much power as three horses. The horsepower is one way we remember animals and their contribution to technology. See Figure 4.2 on the next page. You'll learn more about power in Chapter 13.

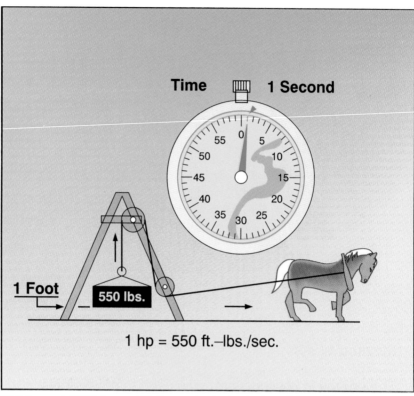

Figure 4-2 James Watt conducted a test and found that the average horse could lift 550 pounds, one foot, in one second.

— Learning Time —

Recalling the Facts

1. Does nature usually give us energy in a form we can use directly? Explain with an example.
2. Name the six types of energy.
3. Are energy and power the same? Explain.
4. Can a 5-horsepower lawn mower do more work than a 3-horsepower lawn mower? Explain.

Thinking for Yourself

1. What do you think energy is?
2. Does it take much more power to lift an elevator holding four people than an elevator holding two people? Explain.
3. Give two examples of natural energy that we *can't* use directly, and explain.
4. Give an example of natural energy that we *can* use directly, and explain.

Applying What You've Learned

Make a device to create fire from wood-on-wood friction heat.

The History of Energy

"You have to play the game according to the rules." Coaches and referees say that again and again. That statement also applies to the use of energy over the ages. People have always used whatever energy was available to them.

The rules: You can only use the type of energy available during your own time in history.

To play the game: To accomplish work, you must use energy. Under the rules 500 years ago, people used animal power, wood fires, and candles. Under today's rules, we also use energy from gasoline and electricity. The rules will certainly be different 500 years in the future. Can you imagine what kind of energy we might be using in the twenty-fifth century?

Muscles

What was the first source of energy?

The first source of energy used by people was their own muscles. They carried food on their backs and moved logs with their arms. Dogs were the first animals trained for help around the home. Dogs hunted food with people and transported items on simple wooden frames they dragged along the ground.

Fire

How did people use the heat energy produced by burning wood and coal?

The next source of energy people used was fire. The first fires were started by lightning, but people learned to make their own fires about 500,000 years ago. Those first flames were probably created from a spark caused by striking rocks together. See Figure 4.3 on the next page.

With heat energy from burning wood, people kept warm, cooked food, hardened pottery for bowls, and made tools from metals such as bronze. Fire also provided light to brighten dark nights. Wood was the most important fuel until the 1700s. People used so much wood that it became scarce. Coal soon took its place.

Wind and Water

What have been the most common uses of wind and water energy?

Wind and water have been used as energy sources for many years. The ancient Egyptians invented sails about 5,000 years ago and used wind to propel their ships. Waterwheels came a little later,

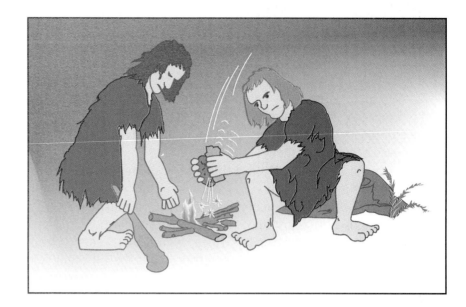

Figure 4-3 Prehistoric people discovered that sparks from hitting stones together could easily start a fire.

about 2,500 years ago. Those early waterwheels used the energy in moving water to grind corn and wheat for flour. Waterwheels soon powered factories like the New England textile mill. The mills made cotton thread and cloth during the 1700s.

The Sun

Why is the sun the most important source of energy?

Practically all of the energy on earth comes from the sun. The sun's rays are needed for plants to grow. People and other animals depend on plants for food. The sun evaporates water, which later falls as rain. The rain fills rivers and lakes, which we use for water transportation. Because of cloud cover, the sun heats the earth unevenly. The uneven heating causes winds, which once powered ancient oceangoing boats.

People have used the sun for warmth since the beginning of time. Many of today's new houses are built to use the sun's warming rays for wintertime heat. Sunlight enters windows to help heat the inside.

Electricity

What is static electricity?

Electricity is a form of energy that comes from invisible electronic particles called electrons. As long ago as 600 B.C., Greek scientists noticed that bits of straw were attracted to certain stones that had been rubbed with cloth. The straw was attracted by *static electricity*. The word static means "nonmoving". Static electricity "sits" on the outside of an object. It does not move down a wire like the electricity used to light your desk lamp. You sometimes experience static electricity after you rub your shoes on a rug. The rubbing causes static electricity to build up on the outside of your body. When you touch another person or a doorknob, you might feel a

---

Tech Talk

Rubbing a piece of *amber* with a cloth makes static electricity. Amber is tree resin that hardened over the ages. The Latin word for amber is *electrum,* which is the origin of our words *electron* and *electricity.*

---

electricity
a form of energy that comes from the movement of invisible particles called electrons through an electrical conductor. Electric current is used as a source of power.

■ Figure 4-4 ■ Using a comb can create static electricity that "charges up" your hair and makes it stand on end.

small shock and see a spark. See Figure 4.4.

The first person to seriously investigate electricity was Benjamin Franklin. He conducted a famous, but quite dangerous, kite experiment in 1752 to find out if lightning was electricity or fire. As a thunderstorm approached his hometown of Philadelphia, Franklin and his son sent a kite into the air. The kite had a pointed wire on top that was connected to the kite string. A key dangled from the kite string near Franklin's hand. As the storm cloud gathered, a light rain moistened the kite string. Lightning crackled in the distance. It sent a small amount of electricity through the air, to the pointed wire, and down the string. Franklin drew a spark from the key to his hand and proved that lightning was electricity. As the rain increased, Franklin wound the string and brought the kite back to the ground.

It's a fascinating experiment, *but don't you ever try it*! It's a killer. The next two persons who tried it were electrocuted by the lightning. Franklin was just plain lucky that he wasn't hurt.

Electricity held the promise of being a very useful energy source. Important contributions were made by people all over the world. Here are some of them:

- 1794 Alessandro Volta, Italian, made the first battery. The *volt* is named after him.
- 1823 André Ampère, French, was the first to apply advanced mathematics to electrical theory. The *amp* is named after him.
- 1827 Georg Simon Ohm, German, developed the theory of electron flow. The *ohm* is named after him.
- 1831 Michael Faraday, English, made the first transformer. The *farad* is named after him.
- 1831 Joseph Henry, American, made the first electric motor. The *henry* is named after him. See Figure 4.5.

■ Figure 4-5 ■ Joseph Henry came from a poor family, worked his way through school, and made the first electric motor while he was a high school teacher.

Tech Talk

All electrical measurement units came from the names of people. Those named are good examples. The *volt* is electrical pressure. The *amp* (or *ampere*) is electrical flow rate. The *ohm* is electrical resistance. The *farad* is electrical charge. The *henry* is a magnetic effect caused by electricity.

— Learning Time —

Recalling the Facts

1. From what source do we receive practically all of the energy on earth?
2. How did people 500,000 years ago probably start their fires?
3. What was the first form of power for early New England textile mills?
4. How do modern houses take advantage of the sun for heat?
5. You can build up static electricity by rubbing your shoes on a rug. Where is that electricity stored?
6. Would it be a good idea to duplicate Franklin's kite experiment? Explain.

Thinking for Yourself

1. Name some types of energy we might be using in 100 years that are not important sources now.
2. Name some types of energy we might *not* be using in 100 years that are important sources now.
3. Besides sailing ships, how else did wind power provide power in the past?
4. Franklin's kite experiment would probably not have worked with a dry kite string. Explain.
5. List some electrical measurements named after people not mentioned in this section.

Applying What You've Learned

1. Use a protractor and a simple frame to measure the sun's angular height. Do it once a week for several weeks. Draw some conclusions for solar homes in your region.
2. Blow up frankfurter-shaped balloons. Stick them to a wall after rubbing them with different materials. See which stick better by adding weights such as paper clips.
3. Make a simple electrical circuit with a flashlight battery and a lamp. Use cotton string in the circuit. Slowly dampen the string to determine the minimum wetness needed to conduct electricity.

Petroleum

Where were the first oil wells drilled in the United States?

People traveling through Pennsylvania in the 1600s saw native Americans scooping oil from surface pools. The native Americans used the liquid for fuel and medicine. We now call that liquid crude oil or **petroleum**. Petroleum is any liquid that comes from the ground and can burn. The word comes from *petra*, meaning "rock," and *oleum*, meaning "oil." Its ability to burn has been known for centuries.

petroleum
a natural liquid that comes from the ground and can burn; a fossil fuel known as "crude oil"

The oil industry started in 1859 in the small town of Titusville, Pennsylvania. That's where the first oil well was drilled. Many others were soon drilled nearby, and America became the world's first oil-producing country. See Figure 4.6. In 1859, 2,000 barrels of oil came from the ground. People soon found so many uses for the liquid fuel that 64 million barrels came from the ground in 1900. One barrel is equal to a little more than a half-filled bathtub, 42 gallons.

Nuclear Energy

What were some of the first experiments leading to the discovery and use of nuclear energy?

Our newest and most powerful form of energy is nuclear energy. It comes from the *nucleus*, or center, of an atom. To create nuclear energy, atomic particles must hit each other and split apart, or react. We call this a **nuclear reaction.**

Around 1900, scientists were investigating new materials that seemed to contain a large amount of energy. Polish-born Marie Sklodowska Curie was one of those scientists. See Figure 4.7 on

nuclear reaction
the process by which atomic particles hit each other and split apart, or react, thus creating nuclear energy

Sklodowska Curie
(sklə-'dȯv-skə) (kyu̇-'rē)

■ **Figure 4-6** ■ Edwin Drake's 1859 oil well was the first one in America to obtain oil from under the ground.

Benjamin Franklin and Electricity

How would you describe Benjamin Franklin? Would you use words like *old*, *overweight*, *short*, or *bald*? That might do for Franklin the politician, the diplomat our country sent to France during the Revolutionary War. He was sent to France because he, more than any other American, commanded respect in Europe.

Franklin earned that respect because of his brilliant experiments with electricity. In logical, step-by-step fashion, which was unusual for that time in history, he was able to figure out how electricity was stored in special glass jars known as Leyden jars. He also showed that pointed lightning rods worked better than blunt ones. And he proved that electricity was either positive or negative. Franklin was the first to describe electricity using words like *plus*, *minus*, *conductor*, *charge*, *battery*, and *electric shock*.

Franklin was famous in Europe not because he was a successful printer and not because he was one of the best writers in America. It was because he was an excellent scientist. The man who earned that reputation was not the dumpy, stringy-haired figure from paintings done almost 100 years after his death. Franklin the scientist was a vigorous athlete just entering his forties.

In 1968 the International Swimmers Hall of Fame in Fort Lauderdale, Florida added Franklin's name to their honor roll as "America's first famous swimmer and coach." Clearly, there was much more to Benjamin Franklin than some people might think.

---

radioactive
describes certain materials that give off invisible, high-energy rays or radiation

page 101. She was the first person to be awarded two Nobel prizes. Both were for her work with these **radioactive** materials. She used the word *radioactive* to describe certain materials, like uranium, that give off invisible rays, such as X rays. Madame Curie won the Nobel Prize in physics in 1903 and the Nobel Prize in chemistry in 1911.

The first practical nuclear reaction was in 1942 at the University of Chicago. It was in a makeshift laboratory in a room under the seats of the football stadium. The reaction took place inside a large pile of special bricks. The scientists called it an *atomic pile*. See Figure 4.8. There was no noise, no fire, no motion. Only measuring instruments indicated what was happening inside that pile of bricks.

The power of the atom is the mightiest force in nature. That's why technologists, politicians, and ordinary people are concerned that it's used properly. An atomic explosion is an example of an *uncontrolled* nuclear reaction. See Figure 4.9. However, *controlled*

Figure 4-8 An atomic pile, an early name for nuclear reactors, was made of many graphite bricks. Pencil lead is also made of graphite.

Figure 4-7 Marie Sklodowska Curie shared a 1903 Nobel Prize with her husband, Pierre. Their daughter, Irene, shared a 1935 Nobel Prize with her husband, Frederic.

nuclear reactions safely provide power for ships and make electricity. The sun's energy comes from nuclear reactions taking place inside the sun.

The first controlled use of nuclear energy was the nuclear submarine *Nautilus*. It was launched by the U.S. Navy in 1954. The vessel was named after Captain Nemo's fictitious submarine in Jules Verne's 1870 book *Twenty Thousand Leagues Under the Sea*.

The first nuclear power plant produced electricity in 1956 at Calder Hill, in England. The first American plant started the following year in Shippingport, Pennsylvania, near Pittsburgh. Nuclear reactors that produce electricity are carefully controlled by highly trained professionals. There are about 435 nuclear power reactors in the world.

—— A Closer Look at ——
Today's Energy Sources

All of today's energy sources are often grouped into what three categories?

You use energy every day in many different ways. You use electricity for lights, gasoline to travel by car, and food to fuel your body. Let's take a closer look at today's sources of energy. Because

Figure 4-9 Atomic explosions were as powerful as 40 million pounds of TNT (tri-nitro-toluene), one of the strongest explosives in the world.

of our concern with the amount of energy available for our increasing needs, these sources are often divided into three groups: renewable, nonrenewable, and unlimited.

Renewable Sources of Energy

What is the main advantage of renewable energy?

Renewable energy sources are those that come from plants and animals. They can be replaced, or renewed, when we need more. For example, we can replant trees to replace those used for fuel in wood-burning stoves. Renewable energy sources include food, animals, wood, and alcohol.

Food. Your body requires food for the energy you use in walking, blinking your eyes, thinking, and all your other activities. See Figure 4.10. Food also keeps you warm. It's burned inside your body to produce heat. You may have heard the phrases "burning up energy" or "burning up calories." There is no flame. Your body uses chemicals instead of a fire to consume food and keep your temperature at its normal 98.6° F.

Putting Knowledge To Work

Benjamin and His Franklin Stove

Early American colonists used inefficient fireplaces to ward off the bitter cold of New England winters. Much of the heat from their burning wood went up the chimney. At today's energy costs, the average house used about $3,000 to $4,000 worth of fuel each winter.

Enter Benjamin Franklin. He designed and built a metal stove that kept houses warmer while using only one-fourth as much wood. Franklin was a remarkable early American personality. He was intelligent, fun to talk with, and a wonderful inventor. He also invented bifocal glasses and lightning rods, and he suggested adding lime to the soil to improve crops.

Franklin knew that he could get more heat from a metal stove by increasing the time the smoke and heat stayed inside. He placed a large piece of metal inside the stove, almost to the top. The smoke from the fire had to go up and over the piece of metal. The smoke left the stove at the bottom and exited the house through a much smaller chimney.

Others began to copy and sell inferior versions of his stove. The inside metal piece was left out by some because it was difficult to make. Nonetheless, all box-shaped metal stoves came to be known as Franklin stoves. The Franklin stove of today does not even resemble what Franklin invented, and no originals are known to exist.

Figure 4-10 Everything you do requires energy but physical activity uses the most.

Food energy is measured in **calories**. The word comes from the Latin word *calor*, which means "heat." Some foods contain more energy than others; more food energy means more calories. A bran muffin and a small glass of milk, for example, contain about 300 calories. A large banana has about 170 calories. If you are a normally active person, your body requires between 2,000 and 3,000 calories of energy every day.

Animals. Muscles were our first energy source. Since horses, oxen, and mules have stronger muscles than people, they were often used to pull plows and transport people. Animals were an important source of energy until the development of engines. They are not used much in the United States today, but animals are important sources of energy in other parts of the world. See Figure 4.11 on the next page.

Wood. Most homes are heated with electricity, fuel oil, or natural gas. The rising cost of these fuels has made wood a popular fuel alternative. Some homes have metal stoves to burn wood for heat.

calories
the measure of food energy in a person's body. Also, the energy required to raise the temperature of one gram of water one degree Celsius.

103

Figure 4-11 Elephants can lift 600 pounds with their trunks and are used in some undeveloped countries.

A wood-burning stove does not heat the inside of a house evenly, but it does provide a warm location. When you come inside on a cold day, the stove can quickly warm you up. However, not many people have this kind of stove, and very little of our country's energy comes from wood.

Alcohol. Alcohol is a liquid fuel made from agricultural crops such as corn and sugarcane. It can be used in special car and truck engines. Many automobiles in Brazil use alcohol. Their production of fuel-grade alcohol is the largest renewable energy operation in the world. See Figure 4.12. Besides using a renewable fuel, alcohol-fueled cars produce less air pollution than those that use gasoline.

State and federal governments in the United States are experimenting with cars that can operate with alcohol. Alcohol is also added to gasoline to extend supplies as well as help reduce air pollution. A mixture of 10 percent alcohol and 90 percent gasoline is sold at some service stations and can be operated in all modern gasoline engines.

Figure 4-12 Straight alcohol can't be used in American automobiles because it has only half the energy of gasoline. It must be mixed with gasoline.

— Learning Time —

Recalling the Facts

1. What was the world's first oil-producing country?
2. Who was the first person to be awarded two Nobel prizes? In what two years were they awarded?
3. How did scientists in 1942 know that a nuclear reaction was taking place inside their atomic pile?
4. Why are animals no longer an important source of energy in the United States?
5. Wood is being used more and more as an alternative fuel for heating a house. Why?
6. What country has the largest renewable energy operation in the world?

Thinking for Yourself

1. Is coal also known as petroleum? Explain.
2. Is a nuclear reaction a *mechanical* activity or a *chemical* activity? Explain.
3. Suppose you're marooned on a small island that has 30 banana trees. You estimate that there are 100 large bananas in each tree. They are your only source of food. For how many days can the bananas provide your normal number of daily calories?
4. List some reasons why wood is not used much as a heating fuel.
5. Alcohol is a more labor-intensive fuel than gasoline. What does that mean?

Applying What You've Learned

1. Show that oil and water don't mix. See if diesel fuel mixes with oil. Let the mixtures stand overnight to verify results.
2. Make a list of all U.S. Nobel Prize winners since 1945.
3. Investigate early nuclear scientists, such as Marie Sklodowska Curie, her husband Pierre, and her daughter Irène Joliot-Curie; Albert Einstein; Enrico Fermi; and J. Robert Oppenheimer.
4. Determine the moisture content of a piece of fuel wood. Construction-grade lumber has a maximum moisture content of 19 percent, and fuel wood is typically 25 to 33 percent. Moisture content = weight of water in dry wood ÷ weight of dry wood × 100 percent. Weigh a piece of wood and then heat in a microwave or other oven to drive off all moisture. Weigh again and calculate the moisture content.

Nonrenewable Sources of Energy

Why must we be concerned about conserving energy?

Some things can't be replaced once they're gone. If a dentist pulls out one of your permanent teeth, another won't grow back to replace it. Natural teeth are nonrenewable. Like pulled teeth, nonrenewable energy sources are those that cannot be replaced. Coal, oil, and natural gas are three of them. Coal is a solid fuel, oil is a liquid fuel, and natural gas is a gaseous fuel. Uranium is also an example of a nonrenewable energy source.

Coal, oil, and natural gas are known as *fossil fuels*. A fossil is what remains from a plant or animal that lived many years ago. Coal, oil, and natural gas are formed from once-living plants and animals.

Coal. When plants died as far back as 400 million years ago, their remains formed thick layers. Wind and water moved soil over the layers of dead plants. Pressure and heat over millions of years slowly changed the plant material into coal. See Figure 4.13.

Those layers are now *coal seams*. A seam is a strip of coal that can be close to the surface or deep underground. Many seams are only 2 or 3 feet thick. See Figure 4.14. Regions of the United States that contain large amounts of coal are called *coal fields*. One large coal field stretches from western Pennsylvania into eastern Kentucky and western West Virginia. Another covers much of Illinois,

HOW FOSSIL FUELS WERE FORMED

  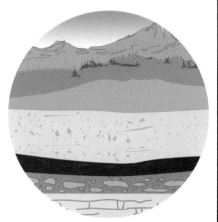

The development of fossil fuels began when plants and animals died and their remains formed thick layers on the bottom of swamps.

Eventually, the remains were covered by thick layers of earth.

Pressure and heat over millions of years slowly changed the remains into fossil fuels.

 **Figure 4-13** Prehistoric plants created most of our coal and prehistoric animals created most of our oil.

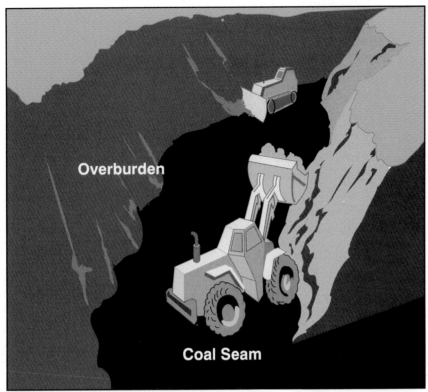

Figure 4-14 Almost two-thirds of our coal comes from surface mining a coal seam after the overburden (dirt) is removed.

and still another is in the Wyoming, Montana, North Dakota region. There are many other smaller coal fields. Most of the coal is used by power companies to generate electricity.

The United States has a large amount of coal still underground. This is called a **coal reserve**. At the present rate of use, experts estimate that worldwide coal reserves will last about 300–500 more years. However, much of that coal is located in areas that are difficult to get to. Some of our coal reserves are far from roads or in mountainous regions. Others are located on public land, where it is sometimes illegal to dig coal because it could affect wildlife or damage the environment.

Oil. You can thank the dinosaurs that lived about 100 million years ago for the gasoline used in today's cars. When they died, their remains combined with those from plants to form crude oil just like plants alone formed coal. The fuels used by cars, trucks, locomotives, airplanes, and ships come from oil. More of our energy comes from oil than from any other single source. Oil is also turned into useful products, such as plastics, paint, and asphalt for roads.

Large oil fields are located in northern Alaska, the Texas-Louisiana region, and Pennsylvania. See Figure 4.15 on the next

reserve
the amount of coal which has not been mined and is still underground

107

Tech Talk

Are you ever confused by the word *gas?* We use it to mean gasoline as well as natural gas. The British avoid the problem by using the word *petrol* to mean gasoline.

page. Offshore drilling platforms are located in the Gulf of Mexico and off the coast of California. Underground supplies of oil are much more limited than those of coal. Experts estimate that the world's oil reserves may run out in as little as fifty years. This is why oil companies are constantly searching for new supplies.

Natural Gas. Natural gas is a flammable gas that usually forms in underground caverns above a pool of crude oil. The slow transformation of plant and animal matter into oil also produces natural gas. As a result, it is often found near oil deposits.

The United States produces more natural gas than any other country. This large supply has encouraged people to use it. Many industrial processes use natural gas, but it is mostly used as a fuel

Figure 4-15 The 800-mile Trans-Alaska pipeline snakes across 20 large rivers and three mountain ranges between Prudhoe Bay and Valdez.

Offshore Oil Rigs

Did you ever think that your school might be located on land that was once underwater? The earth looked quite a bit different millions of years ago. Plants and animals lived where oceans are now located. The remains of those life forms created oil deposits that are now underwater. Ancient sea life also formed underwater oil deposits. Getting that oil requires an offshore oil rig. A *rig* consists of an offshore platform and all the machinery necessary to drill an oil well.

Oil platforms are built at a seaside city. The platforms can float and are towed into position several miles offshore. Legs are jacked down until they hit the sea bottom. The platform is then jacked higher until it's well above the highest waves. These *self-elevating platforms* are used in water up to 300 feet deep. Most offshore platforms are of this type. Other platforms float in the water and are anchored to the seafloor with many cables.

The largest self-elevating oil platform went to the North Sea oil fields off the coast of Scotland in 1982. Built in England, the main steel supporting tower was taller than the Eiffel Tower in Paris. Similar platforms are used in the Gulf of Mexico.

Working on an offshore oil rig is both exciting and just a bit dangerous. High waves and winds are a constant threat to the outdoor workers. A fire can cause serious problems. A worker on a drilling crew is commonly called a *roughneck*. Perhaps you can figure out why.

Because offshore oil drilling is expensive, each rig drills more than one hole into the ocean floor. The rigs have the ability to drill at an angle. As many as forty-two wells can be drilled from one fixed platform.

for home heating and cooking. Over half the homes in America are heated with natural gas. What kind of energy is used to heat your home?

Uranium. Like fossil fuels, uranium is also a nonrenewable source of energy. Unlike fossil fuels, however, uranium does not come from plants and animals. It's a radioactive, rocklike mineral that is dug from the surface of the ground. See Figure 4.16 on the next page. Uranium is used for fuel in nuclear power plants that produce electricity. It is also used to power some U.S. Navy ships. Most of the U.S. supply of uranium comes from New Mexico and Wyoming.

Uranium develops a large amount of heat during a controlled nuclear reaction. The heat changes water into steam, and the steam operates generators that produce electricity.

The amount of energy in uranium is amazingly high. One pound of the material can produce as much electricity as 3 million

Figure 4-16 Large trucks carry surface-mined uranium ore to a plant that separates uranium oxide from the rock.

deuterium
(dσo-tîr'ē-əm)

pounds of coal. The 110 nuclear reactors in the United States produce almost 20 percent of our total amount of electricity.

One major problem with nuclear energy is that the used uranium must be kept out of the environment at all costs. Enough radioactivity remains in used-up fuel that it continues to be dangerous for thousands of years. Current plans call for reducing the used fuel to solid blocks resembling glass. The blocks will be buried at sites away from large population centers.

Uranium isn't the only fuel that can be used in nuclear reactors. Scientists are experimenting with another type of fuel called *deuterium*. It's a liquid made from ordinary water, and the largest power plant would use less than 1 quart per day.

— Learning Time —

1. Name a solid fossil fuel, a liquid fossil fuel, and a gaseous fossil fuel.
2. How is most of the coal in the United States used?
3. What is a serious problem with coal reserves?
4. How long ago did the dinosaurs live that formed our crude oil?
5. Describe the steps in producing electricity from nuclear energy.
6. What percentage of U.S. electricity is produced by nuclear reactors?

Thinking for Yourself

1. Surface mining of coal is sometimes called *strip mining*. Why?
2. Dinosaur remains helped create our oil supply. Could any of those dinosaurs have been killed by prehistoric people?
3. Cars, trucks, and airplanes use different fuels. How do we get them from crude oil?
4. How is home heating with natural gas different from home heating with electricity?
5. An eighteen-wheeler semitrailer can carry about 50,000 pounds. How many truck loads of coal would equal the energy in 1 pound of uranium?

Applying What You've Learned

1. Obtain some coal and investigate its physical properties. Will it float? What is its density? What is its general hardness? Does it break apart easily?
2. Some types of coal have more energy than others. Think up a device that could burn coal to measure its energy content.
3. Construct a model of an offshore oil-drilling rig. Put it on display with a description of how it operates.

Unlimited Sources Of Energy

What are four sources of energy that we will never run out of?

Although we are using up our nonrenewable sources of energy, we have several sources that will never run out. They are unlimited. We will never use up all our solar energy, wind, flowing water, or geothermal energy. They are more plentiful than fossil fuels, but are also quite difficult to use.

Solar Energy. Solar energy comes from the sun. Unlike coal or oil or wood or many other sources of energy, solar energy is available

Tech Talk

The prefix *sol* often tells you that a word has something to do with the sun. Some examples are *solar energy* (energy from the sun), the *solar system* (planets with the sun at the center), and *solarium* (a glassed-in porch that captures the sun's rays). The *solar plexus* is a group of nerves near your stomach. The nerves fan out and resemble the sun's rays. In Roman mythology, *Sol* was the sun god.

all over the world. We use the sun's rays for light, electricity, and heat. Some homes are heated with **solar heating systems**.

An *active solar heating system* uses mechanical devices to take sun-heated water into a house. The most important parts of the system are large flat panels, called *solar collectors*, on the roof of a house. See Figure 4.17. Water flows through tubes in the solar collectors and is warmed by the sun. The water continues into the house and is used to heat the interior. Some homes in your community may have black solar collectors on the roof. See if you can find them.

A *passive solar heating system* uses only windows and walls to take advantage of the sun's warming rays. It is called passive because there are no mechanical devices to collect the solar heat. There are no pumps or flowing water. A passive solar home usually has large windows on the south side. Sunlight streams in and warms the interior. Some of that warmth is absorbed by the walls. The walls radiate heat back into the room at night.

Sunlight is a cost-free method to heat a house. Most new homes use at least a small amount of passive solar heating. The biggest problem with solar heating is that there are fewer hours of sunlight in the winter than in the summer. Of course, there's none at night and only a little sunlight on cloudy days. Extremely large solar collector panels or large window areas are required to obtain even a small amount of the heat required for a house during an entire day. Except in southern states like Florida, homes cannot usual-

■ **Figure 4-17** ■ Roof-mounted solar collectors use the sun to heat water which, in turn, heats the inside of the house.

ly be comfortably heated by solar energy alone. Most homes also use electric heating or have a furnace.

Solar energy can also produce electricity. This happens when sunlight strikes wafer-thin **solar cells** and releases electrons locked up in the material. The effect is similar to rubbing a piece of amber with a cloth. Solar cells are made from specially treated sand and other materials. Orbiting satellites obtain their electricity from solar cells. Solar cells are not often used on earth because their electricity costs about 100 times more than electricity from a power plant. The cells are also known as *photovoltaic cells* or *photocells. Photo* means "light," and *voltaic* means "producing voltage."

Every day the sun delivers 20,000 times as much energy as we use all over the world. However, it's been difficult to use solar energy. How can we inexpensively harness it to operate cars and trucks, or to light homes at night, or to completely heat a home? Perhaps you will be the person who shows the way.

solar cells
devices that convert sunlight into electrical energy. Solar cells are most commonly used in space vehicles, but commercial, industrial and residential use is increasing.

photovoltaic
(fō'tō-vōl-tā'ĭk)

Putting Knowledge To Work

Blowing in the Wind

"Peter, Paul, and Mary" was a popular 1960s singing group. They once recorded a song titled, "The Answer Is Blowing in the Wind." Some modern clean-energy advocates have used that title to describe their feelings about generating electricity.

The United States has about 16,000 electricity-generating wind turbines, the most of any country in the world. The entire continent of Europe, for example, has only about 2,700 wind turbines.

Here's how wind turbines work. Electrical alternators (ac generators) sit on top of tall towers. The wind spins the three-bladed rotors, 58 to 62 feet in diameter, at 120 revolutions per minute. The rotors are directly connected to alternators and controlled by advanced computers.

Only certain regions of the country provide enough regular wind to make wind turbines practical. California is one state that has several such locations. Its major wind farms are in three locations. They are the Tehachapi Pass (about 5,000 high-capacity wind turbines), the Altamont Pass (about 6,000 wind turbines), and the San Gorgonio Pass (about 3,500 wind turbines). Each high-capacity alternator in the Tehachapi Pass produces enough electricity for 100 people.

Figure 4-18 The Tehachapi Pass, which is about 80 miles north of Los Angeles, is the world's largest producer of wind-generated electricity. It uses over 5,000 wind turbines like these in the photograph.

Wind. The motion of air across the earth has filled the sails of ships and turned windmills for centuries. More recently, wind has been used to turn propellers connected to generators that produce electricity. Unfortunately, the wind is not a dependable source of energy. Some days it's stronger than others. Some days it doesn't blow at all.

It is not usually practical for one house to have its own windmill. The windmill and other necessary equipment (like large batteries for storing electricity) are quite expensive. The propeller makes noise when it turns. Windmills also interfere with television reception, and some people find them unpleasant to look at.

Experiments have been conducted with **wind farms** in California and other places. See Figure 4.18. Wind farms are made up of a

wind farm
a large number of windmills located in an area that has a fairly constant wind speed

114

great many windmills. They are located on land once thought to be useless and in areas known to have a fairly constant wind speed. Wind energy provides about 1 percent of California's electricity.

The world's largest propeller-driven wind generator is in Hawaii. Its blade is longer than a football field, and it provides power to 1,200 houses.

So far, these experiments have not shown windmills to be practical for generating large amounts of electricity. Windmills are expensive, and wind remains an undependable source of energy. However, these are not strong enough reasons to stop trying. Perhaps windmill builders have yet to find the one design that will solve the problems. Do you have any ideas?

Flowing Water. We use flowing water to generate electricity in **hydroelectric** dams. *Hydro* means "water." A controlled amount of water flows through pipes in the dam and into a *turbine*. A turbine is a modern version of the waterwheel. A spinning turbine, connected to a generator, creates electricity. See Figure 4.19. The Grand Coulee Dam near Spokane, Washington, is the largest hydroelectric dam in the United States. It is as tall as a forty-six-story building. About 8 percent of our electricity comes from hydroelectric dams.

Another example of energy from flowing water is tidal energy. The rising and falling of the ocean tides can produce electricity with generators. During high tide, water collects behind a dam. During low tide, the water flows through turbines connected to generators

hydroelectric
describes the electricity generated by turbines propelled by flowing water

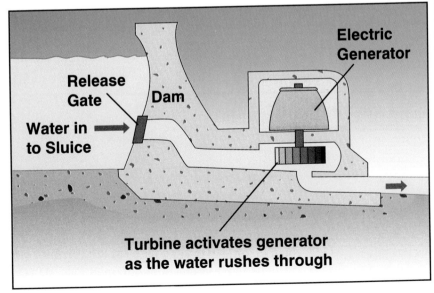

■ **Figure 4-19** ■ Water flowing through a pipe in a dam, spins a turbine connected to a generator that makes electricity.

Release Gate

Dam

Electric Generator

Water in to Sluice

Turbine activates generator as the water rushes through

Figure 4-20 One common type of geothermal energy is underground steam used for heat or to make electricity with turbo-generators.

and produces electricity. The first tidal power plant was operated on the Rance River in northwestern France in 1966. It produces about one-tenth as much electricity as a large coal-fired power plant.

Geothermal Energy. You may already know that a volcano flows molten rock from its center. This is a form of **geothermal** energy— heat from the ground. Hot water or steam is created when underground water comes in contact with hot materials. The steam is used to produce electricity. There are some geothermal electric power plants along the west coast of America. See Figure 4.20.

Iceland has many hot springs. Practically all the homes in the capital city of Reykjavik are heated with this geothermal energy.

geothermal
describes thermal, or heat, energy in the ground that is caused by the radioactive decay of certain elements

Reykjavik
(rā′kyə-vēk′)

— Learning Time —

Recalling the Facts

1. Where are a house's solar collectors located?
2. What tells you that a house uses passive solar heating?

Thinking for Yourself

1. Some solar houses have no windows on one side. Explain.
2. Name some everyday uses for solar cells.
3. Suppose that your home obtained

(Continued on next page)

3. How do orbiting satellites obtain their electricity?
4. Why isn't the wind a dependable source of energy?
5. How do dams produce electricity?
6. What kind of unusual energy gave Hot Springs National Park its name?

electricity only from the wind. What would you do if the wind stopped blowing for several hours?
4. Dams that hold back large quantities of water can provide a more dependable source of electricity than those that are only very tall. Explain.

Applying What You've Learned

1. Build a model car to run on solar cells.
2. Build the waterwheel/dam shown under *dam* in the *World Book Encyclopedia*.
3. Measure your body's instantaneous power by running up a flight of stairs and timing yourself. Power = [Your weight (pounds) × vertical height of stairs (feet)] ÷ [Time (seconds) × 550 foot-pounds per second per horsepower.]

Energy Conservation

What can we do to maintain a supply of energy for our future needs?

Our earth has a limited supply of energy sources. However, we use more and more energy as the world's population grows. Also, each person uses more now than in the past, partly because of all the electrical conveniences.

We have to be sure that there is enough energy to provide for human needs. One way to do that is through **energy conservation**. Energy conservation is the management and efficient use of energy sources. See Figure 4.21 on the next page.

Until other sources of energy are fully developed, we must conserve our present sources to make them last longer. Here are some ways you can help conserve energy:

• Set your home thermostat at or below 65° F in the winter. You will use less energy because the heating system will operate for less time.

energy conservation
the management and efficient use of all energy sources

■ **Figure 4-21** ■ More aluminum is recycled than any other metal (left). Turning a thermostat down in the winter is a good way to save both energy and money.

recycle
to reuse all or part of substances, such as metal, glass, paper, plastics, etc.

- If your home is air-conditioned, set the thermostat at or above 78° F in the summer. You will use less energy because the air conditioner will operate for less time.
- Use less hot water by spending less time in the shower.
- Turn off all unnecessary lights.
- Walk or ride a bike.
- Whenever you can, use buses, trains, or subways instead of automobiles.
- Use renewable energy sources whenever possible.

Energy conservation should be a personal goal for every citizen of the United States. We could greatly extend our supplies of energy if everyone helped just a little. Suppose that each person saved just 1 gallon of gasoline a month by taking public transportation, car pooling, or driving a smaller car. One gallon isn't very much, but it would result in a nationwide saving of about 250 million gallons of gasoline each month!

Another way to conserve energy is to **recycle** the materials and products we use. To recycle means to use again. Metal, glass, paper, and some plastics can be recycled. Reusing aluminum cans is one of the biggest energy savers. Recycled cans require only 80 percent of the energy required to produce cans from raw materials. All glass can be recycled. Paper is usually recycled into bags, paper towels, and packaging materials. Plastics that can be recycled have a special triangular symbol on them.

Recycling reduces our use of natural resources. It also reduces the amount of solid waste sent to landfills. Most cities and towns have a recycling center or a civic organization that arranges routine collections. Check to see how recycling is handled in your community.

Environmental Effects

What have been the positive and negative effects of our energy consumption?

You may have heard that Americans have one of the highest *standards of living* in the world. Standard of living refers to how comfortably people live. We have good transportation systems, countless electrical devices, excellent housing, and an abundance of food. Our comfortable way of life has been made possible by advancements in technology. The driving force of technology has been our use of energy. Unfortunately, our energy consumption has created the serious problem of pollution.

Greenhouse Effect

What is the greenhouse effect?

Over 90 percent of our energy comes from burning fossil fuels. All that burning creates a great deal of air pollution. It also contributes to the **greenhouse effect**—the heating of the earth's atmosphere. If we produce too much heat and it can't escape, the temperature of the earth will increase.

On the positive side, warmer temperatures might mean longer growing seasons for crops. More food could be produced for the world's increasing population. On the negative side, the polar caps could melt slightly. That would raise the ocean level and cause flooding of some ocean-side cities. A 1-foot rise in the Atlantic Ocean, for example, would flood Florida 200 to 1,000 feet from the coast. Louisiana would be flooded several miles inland. These are very serious possibilities.

To reduce the greenhouse effect, delegates from 159 countries met in Japan in December 1997. The group included mostly industrialized countries from North America, South America, Europe, and Asia. The United States agreed to further discussions to reduce emissions to about 7 percent below what they were in 1990. That goal is to be met by the year 2012. The reductions were aimed at carbon dioxide, nitrous oxides, and other gases.

Air Pollution

What are some different causes of air pollution?

You may already know that burning produces a great many pollutants. Burning fills the air with haze and sometimes make your eyes burn. It can also be a serious health threat. Each fuel produces many pollutants but there is usually one that is particularly serious.

All coal has a small amount of sulfur. When coal burns, it creates sulfur dioxide. The sulfur dioxide combines with water vapor

greenhouse effect
the gradual, steady increase in the temperature of the earth's atmosphere caused by the rise in the amount of industrial gases in the earth's atmosphere

Tech Talk

The expression *greenhouse effect* was first used in 1822 by the French mathematician Jean Fourier. He used it to compare the earth's atmosphere to the glass of a greenhouse.

and oxygen in the air to form a weak sulfuric acid. This acid falls to the earth as a liquid called *acid rain*, which accumulates in lakes and on the ground. See Figure 4.22. It can kill fish, crops, and trees. Acid rain also damages monuments and statues. There is no total solution to the problem. However, acid rain can be reduced if power companies and industries use a lower-sulfur coal. They can also install special equipment to remove up to 90 percent of the sulfur dioxide from their smoke.

Gasoline forms carbon monoxide when it burns. Carbon monoxide is an odorless, colorless, and poisonous gas. When released outside and breathed into your lungs, it reduces the ability of your blood to carry oxygen. If a person already has a lung problem, too much carbon monoxide can make the problem even more serious. More people die from carbon monoxide poisoning each year than from any other deadly poison.

The warm temperature from a wood-burning stove provides a cozy living room atmosphere. However, the burning wood provides

Figure 4-22 Acid rain is created when certain pollutants in the air mix with rain which falls to earth.

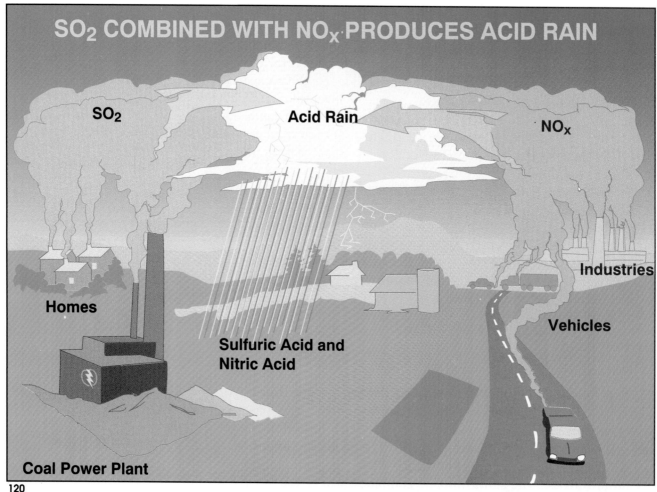

a quite different type of outdoor atmosphere. Wood smoke contains large particles, or particulates, that can make breathing difficult. Some cities have banned wood burning when air pollution reached dangerous levels. Many new stoves are designed to recycle the smoke and greatly reduce particulates.

Nuclear Waste

Why is nuclear waste dangerous?

Nuclear pollution may be the most threatening pollution of all. Nuclear, or radioactive, waste is a solid left over after nuclear fuel is used up. The waste remains dangerous for many years. Proper disposal of this material is one of our most serious problems. The waste is placed in special concrete containers that are buried where they can cause no harm. The area is constantly checked for radioactive leakage. Some people think that nuclear waste should be sent into space with rockets. What do you think?

What You Can Do

How can you help conserve our energy supply?

We can't stop using energy just because it causes pollution. It's also true that we can't ignore pollution just because we want energy. We have to find a proper balance between the two. We can help control pollution in four ways:

1. We can pass laws and obtain newspaper publicity. This will make everyone aware of the problems and what is being done by our political leaders.
2. We can control how much pollution enters the environment from each energy source. We can never remove all the pollution, but we can always reduce the amount.
3. Little pollution is created when we use solar, wind, water, or geothermal energy. We need to develop more and better ways to use these unlimited energy sources.
4. We can reduce our energy use.

What can you do about pollution? You might think that since you can't vote and don't have much money, your opinion doesn't count. That's not true. Whether or not you know it, you can greatly influence the adults around you. They will listen to you and even follow your advice if they think that you're sincere and that you know what you're talking about.

Keep informed: Know what's happening in your community, state, country, and world. Read newspapers and weekly magazines. Talk to your parents, relatives, teachers, and other adults, such as Scout leaders, civic group leaders, and religious leaders. Have discussions with your friends. Adults have always paid attention to well-informed young men and women.

— Learning Time —

Recalling the Facts

1. How do you save energy by setting your home thermostat at 65° F or lower in the winter?
2. How much energy is saved by recycling aluminum cans?
3. If ocean levels rose 1 foot, how would the state of Louisiana be affected?
4. Why does low-sulfur coal reduce acid rain?
5. What deadly poison kills the most people?
6. What type of pollution may be the most dangerous?

Thinking for Yourself

1. We use more and more energy every day. Can conservation really help?
2. What are some ways you can save energy that are not included in this section?
3. How much metal, glass, paper, and plastic is recycled in your community?
4. Why do we need to be concerned about the greenhouse effect?
5. Why is it a bad idea to dispose of nuclear waste by dumping the concrete containers in the ocean?

Applying What You've Learned

1. Conduct a type of acid rain test by planting bean seeds in four paper cups of dirt. Water all the seeds until each bean plant is a half inch high. It will take seven to ten days. Continue watering one with a mixture of one ounce of vinegar and five ounces of water. Water another with a one-to-ten solution and another with a one-to-twenty solution. Continue watering the fourth with ordinary water. Measure the plants each day and draw some conclusions.
2. See how difficult it can be to clean up an oil spill. Place a small amount of cooking oil in a large bowl of water. Try and soak up the oil with paper towels, tissue, or a sponge. Break up the oil spill with a few drops of detergent. Determine the best way to deal with your oil spill.

—— An Energy Career ——

What is an example of a possible career in energy?

You couldn't find two closer friends than Alex Gapinski and Michael Moniz. They grew up in the same neighborhood, played soccer together, belonged to the same Scout Troop, and were in many of the same classes in middle school and high school. Michael's father owned a construction business and built houses for a living. He hired Alex and Michael during the summer after their junior year in high school. The two spent most of their time at a passive solar house Mr. Moniz was building. Neither one had any construction experience, so they did basic labor work like carrying lumber, shoveling dirt, and nailing up wall paneling. See Figure 4.23. They both enjoyed the outdoor work, but Alex seemed to enjoy it the most.

During the following year in school, Alex read all he could about passive solar houses and wrote to several companies. He learned about proper selection of a building site, the latest construction techniques, new insulating materials, and many modern approaches to passive solar house construction.

After graduation from high school, Michael's father again hired the two boys. Mr. Moniz said that it was "just until they decided what to do with their lives." Unlike the previous summer, Alex had some ideas he wanted to try out. Without first getting permission, he built a wall with a much larger window than shown in the plans. He didn't take the time to check and see that the windows had already been purchased. Mr. Moniz was upset when he found out. He and Alex came close to having an argument.

The next day, Alex had a long talk with Mr. Moniz. It was only then that Mr. Moniz found out that Alex had a sincere interest in the construction business and was quite knowledgeable about passive solar houses.

A few days later, Mr. Moniz found Alex installing a door and asked to speak with him. Mr. Moniz had done some checking at the town library. He told Alex that *Peterson's Two-Year Colleges* listed 230 community college programs in construction technologies. One community college noted for its emphasis on solar house construction was not far away. Mr. Moniz offered to help Alex attend the college. Alex was amazed. It was the last thing he expected to hear.

■ **Figure 4-23** ■ Passive solar houses have many windows on their south sides to collect sunlight and transfer it inside as heat.

That was five years ago. Alex and Michael are still close friends, and they still see each other almost every day. Michael is the accountant for his father's solar homes company. He makes sure that all the bills are paid. He went to the community college with Alex and majored in business. Alex is the solar house expert. He designs the houses and makes sure that each one is properly constructed. His time is split between office work and the construction site.

The two friends are talking about starting their own business building solar houses. Mr. Moniz has offered to help them. He feels they have a good chance at succeeding.

There are many different organizations that can help you obtain information about careers in energy. Here are just a few of them.

- Wood: Wood Heating Alliance
  1101 Connecticut Avenue NW
  Washington DC 20036
- Alcohol: Alternative Sources of Energy
  107 South Central Avenue
  Milaca MN 56353
- Coal: National Coal Association
  1130 17th Street NW
  Washington DC 20036
- Oil: National Petroleum Refiners Association
  1899 L Street NW
  Washington DC 20036
- Natural Gas: American Gas Association
  1515 Wilson Boulevard
  Arlington VA 22209
- Nuclear: American Nuclear Energy Council
  410 First Street SE
  Washington DC 20003
- Solar: American Solar Energy Society
  2400 Central Avenue
  Boulder CO 80301
- Wind: American Wind Energy Association
  1730 North Lynn Street
  Arlington VA 22209
- Water: Edison Electric Institute
  1111 19th Street NW
  Washington DC 20036
- Geothermal: Geothermal Resources Council
  P.O. Box 1350
  Davis CA 95617
- Environment: Environmental Policy Institute
  218 D Street SE
  Washington DC 20003

- Conservation: Conservation and Renewable Energy Inquiry and
  Referral Service
  U.S. Department of Energy
  P.O. Box 8900
  Silver Spring MD 20907
- Recycling: Aluminum Recycling Association
  1000 16th Street NW
  Washington DC 20036

  Plastics Recycling Foundation
  1275 K Street NW
  Washington DC 20005

Summary

Directions

The following sentences contain blanks. For each numbered blank, pick the answer that makes the most sense in the entire passage. Write your answer on a separate sheet of paper.

Learning about energy is an important part of your technology education. ____1____ is the ability to do work. People sometimes confuse energy with power, which is the measurement of energy. There is a difference between energy and ____2____ .

The first source of energy used by people was their own muscles. The sun, wind, water, and fire were sources of energy from ____3____ .

Other forms of energy put to use by people were electricity, petroleum, and nuclear energy. A pioneer in electricity, Ben Franklin conducted his famous kite experiment in 1752. The oil industry began in the mid-19th century in Pennsylvania. The first controlled use of nuclear energy occurred almost 100 years later when the US Navy launched a/an ____4____ submarine.

Our sources of energy are often divided into three groups: renewable, nonrenewable, and unlimited. Renewable energy sources are those that come from plants and animals. These energy sources can be ____5____ or renewed when we need more.

Nonrenewable sources of energy cannot be replaced. Four important ____6____ sources are coal, oil, natural gas, and uranium. All of these sources except ____7____ are called fossil fuels

1.
a. Power
b. Strength
c. Energy
d. Force

2.
a. speed
b. strength
c. power
d. motion

3.
a. civilization
b. nature
c. history
d. space

4.
a. underwater
b. energized
c. nuclear
d. electric

5.
a. replaced
b. manufactured
c. purchased
d. eliminated

6.
a. renewable
b. synthetic
c. unlimited
d. nonrenewable

7.
a. coal
b. oil
c. natural gas
d. uranium

(Continued on next page)

Summary

(Continued)

because they developed from once living plants and animals.

The unlimited sources of energy will never run out. These sources, such as solar energy, wind, flowing water, and geothermal energy, are more plentiful than the ___8___ fuels, but they are difficult to use.

Solar energy, which comes from the sun, is used to heat our homes. It also produces electricity when sunlight strikes wafer thin ___9___ cells. Wind farms in California produce about 1% of that state's electricity. Flowing water is used to generate electricity in hydroelectric dams. ___10___ from the ground, known as geothermal energy, produces steam which can, in turn, produce electricity.

To make sure that there is enough energy for human needs, we must practice energy conservation. Energy ___11___ is the management and efficient use of energy sources. One way of conserving energy is to recycle the materials and products we use. Recycling reduces our use of natural ___12___ .

Americans have one of the highest standards of living in the world. This has been made possible by ___13___ in technology and our use of energy. Unfortunately, our energy consumption has created serious pollution problems.

Over 90% of our energy comes from the ___14___ of fossil fuels. This burning contributes to the greenhouse effect, which is the heating of the Earth's atmosphere.

Burning fuels also produce many pollutants that can be serious threats to our health and our environment. One such ___15___ is acid rain, which results from the burning of high sulfur coal. Another pollutant is carbon monoxide which is a/an ___16___ gas that results from burning gasoline. But perhaps the most threatening pollution of all is nuclear pollution. ___17___ wastes must be placed in special concrete containers and buried where they can cause no harm.

8.
   a. automobile
   b. fossil
   c. commercial
   d. expensive

9.
   a. solar
   b. animal
   c. body
   d. prison

10.
   a. Roots
   b. Ores
   c. Plants
   d. Heat

11.
   a. exploration
   b. conservation
   c. use
   d. education

12.
   a. creatures
   b. plants
   c. resources
   d. trees

13.
   a. setbacks
   b. meetings
   c. theories
   d. advancements

14.
   a. melting
   b. burning
   c. freezing
   d. compressing

15.
   a. fuel
   b. pollutant
   c. system
   d. mineral

16.
   a. poisonous
   b. plentiful
   c. powerful
   d. pink

17.
   a. Household
   b. Recyclable
   c. Electronic
   d. Nuclear

18.
   a. eliminate
   b. reduce
   c. increase
   d. ignore

(Continued on next page)

Summary

We can all help control pollution by making everyone aware of the problems. We can ___18___ the amount of pollution that enters our environment, and we can develop more and better ways to use energy. We can also reduce our energy use.

Key Word Definitions

Directions

The column on the left contains the key words from this chapter. The column on the right contains a scrambled list of phrases that describe what these words mean. Match the correct meaning with each word. Write your answer on a separate sheet of paper.

Key Word		*Description*	
1.	Energy	a)	Many windmills in one location
2.	Power	b)	Amount of unmined coal
3.	Electricity	c)	Efficient use of energy
4.	Petroleum	d)	Measure of food energy
5.	Nuclear reaction	e)	Ability to do work
6.	Radioactive	f)	Uses sun-heated water
7.	Calories	g)	Heating of the earth's atmosphere
8.	Coal reserve	h)	Splitting apart atomic particles
9.	Solar heating	i)	Electricity from flowing water
10.	Solar cells	j)	To use again
11.	Wind farm	k)	Liquid from the ground that burns
12.	Hydroelectric	l)	Energy from invisible electronic particles
13.	Geothermal	m)	Heat from the ground
14.	Energy conservation	n)	Produces electricity directly from sunlight
15.	Recycle	o)	Measures the work done by using energy
16.	Greenhouse effect	p)	Invisible rays from certain materials

Building a Solar Heating System

Resources You Will Need

- Two 8 x 12-inch sheets of Styrofoam plastic
- One 6 x 10-inch sheet of corrugated cardboard
- One 6 x 10-inch sheet of black (or very dark) paper
- Clear plastic wrap
- 5 feet of small-diameter flexible clear plastic tubing
- Small plastic bottle and bowl
- Thin wire and wire cutters
- Drill
- Razor knife
- Rubber bands
- Tape
- Red food dye
- Thermometer
- Clothespin
- Wristwatch

Did You Know?

Solar collectors trap heat from the wintertime sun and transfer the heat into a building. Most solar collectors are black because dark colors absorb the sun's rays. Light colors reflect the sun's rays.

The simplest kind of solar collector has a black heating plate and a glass or plastic cover. Sunlight strikes the collector's plate, and the plate becomes hot. Heat is trapped inside by the glass or plastic cover. The trapped heat is transferred to water flowing through tubing. The warmed water is then sent inside the building to a location where the heat can be used.

Problems To Solve

You will make a small solar collector from Styrofoam plastic, plastic tubing, and other miscellaneous materials. The collector will be the most important part of your solar heating system. You will complete the system and test it to see how well it heats water.

A Procedure To Follow

Look over the simple plans shown in Figures A and B. Become familiar with the general procedure you will follow.

1. First make a flat plate solar collector. Use the razor knife to cut the Styrofoam pieces to the sizes shown in Figure A. Save the

A.

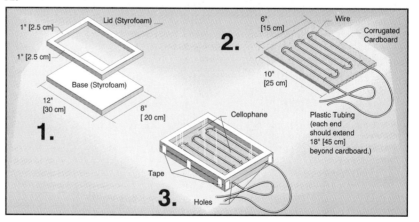

center piece from the lid.

2. Make a heating plate by first placing the black paper on top of the corrugated cardboard. Arrange the plastic tubing into S-shaped curves on the black paper. Make as many S curves as will fit. Leave about 18 inches of extra tubing at each end. Fasten the tubing to the black paper and corrugated cardboard with short pieces of wire. Poke a piece of wire through the cardboard, and twist its ends together on the underside.

3. Use tape to attach the clear plastic wrap over the hole in the Styrofoam lid.

4. Place the heating plate onto the Styrofoam base. Notch holes in the lid for the tubing, and place the lid over the base. Tape the assembly together, and your solar collector is finished.

5. Now complete the solar heating system. Drill a hole in the bottom of a plastic bottle. The hole should be just large enough so that the plastic tubing will fit through.

Force one of the free ends of the tubing into the hole. If necessary, seal the connection with waterproof glue.

6. Make a bottle stand with the piece of Styrofoam from the lid center.

7. Move the solar collector to a place where the sun's rays will strike the heating plate. Raising one side to a 45° angle will improve the heating effect. Put the free end of the tubing in the bowl, and clamp with the clothespin. Fill the bottle with water colored with red food dye. The dye will help you see the water. The bottle must be at least 6 inches above the collector so that the water can easily flow through the heating plate.

Your system is complete. Now let's test it.

8. Open the clothespin to allow the water to flow through the tubing in the heating plate. Measure its temperature as it flows into the bowl. Close the clothespin.

9. Allow the water to remain in the collector for 2 to 5 minutes. Open the clothespin and measure the water's temperature by flowing water onto the thermometer.

10. Compare your results with the results of others in your class.

B.

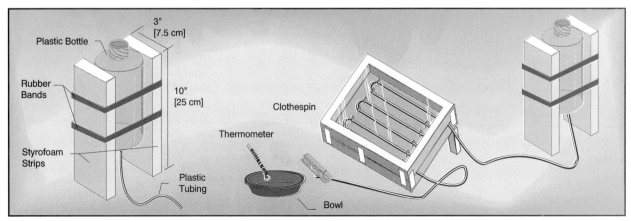

Plastic Bottle
Rubber Bands
Styrofoam Strips
Plastic Tubing
3" [7.5 cm]
10" [25 cm]
Thermometer
Clothespin
Bowl

Hello!
Computers, telephones, and satellites make it possible for us to communicate almost instantly with people around the world. How might this technology affect the way we think about people in other countries? How might it change the way we learn, do business, or spend leisure time?

Communication

Chapter 5

Understanding Communication Technology

The word *communication* comes from the Latin word *communis*, which means "to share." In a way, communication also means to share — it means sharing messages. To have communication, you need four elements:

1. A sender
2. A message
3. A receiver (who understands the message)
4. Feedback

If you cut yourself and scream out in pain, you are sending a message. Two of the four needed elements exist. You are the sender, and your scream is the message. However, if no one hears your call for help, no communication is taking place. Do you see why?

If you are heard by someone who can't understand what you are trying to say, communication still isn't taking place. Your message must be sent, and your message must be received, understood, and answered for the communication process to be completed.

Try defining communication right now. Don't be afraid to put things in your own words.

As long as you mentioned a sender, a message, a receiver, and feedback, you probably came up with a very good definition. Communication is sending, receiving and responding to messages.

— Chapter — Outline

Look over the headings below. Think about what you will learn as you read this chapter.

I. Introducing Communication Technology

II. Types of Communication Systems

III. Communication: People and Machines

IV. Communication and the Spread of Technology

V. The Impact of Communication Technology

— Key Words —

In this chapter you will learn and study the meanings of the following important technology terms.

COMMUNICATION TECHNOLOGY
COMMUNICATION SYSTEM
GRAPHIC COMMUNICATION
WAVE COMMUNICATION
SOUND CARRIER WAVE
TELECOMMUNICATIONS

communication technology
the use of knowledge, skill, tools, machines, and materials to enhance communication through the development and use of communication devices

Introducing Communication Technology

What is the difference between communication and communication technology?

Imagine yourself standing in the hallway of your school talking to your friends. Are you using communication? Yes, you are communicating. Are you using communication technology? No. When you are talking face-to-face, you are using communication but not communication technology.

If, on the other hand, you communicate using a public address system, a microphone, or a written note, then you are using communication technology. Whenever your ability to communicate is enhanced by technology such as pictures, printed messages, or machines, you are using a specific type of technology. This type of technology is called **communication technology**. See Figure 5.1.

Your telephone is probably your favorite *communication device*. Communication technology includes all the knowledge, skills, and tools that led to the development of all of the communication devices ever invented—including the telephone.

Communication Systems

What are some examples of communication systems?

A **communication system** involves a plan, or route, to take your message to its receiver. As you learned in Chapter 2, systems

Figure 5-1 When these teenagers talk face-to-face they are using communication. When they use the telephone they are using communication technology.

134

can be charted on a universal systems diagram. This diagram breaks all systems into input, process, output, and feedback.

Newspapers, magazines, books, and photographs are four different communication systems. Without the writers, photographers, publishers, printers, and delivery people, these systems wouldn't work.

Your radio and television are the output devices of their communication systems. Without the machines and people at the radio and television stations, these devices wouldn't work. Computers and recording devices (tape recorders and video recorders) are also communication systems that often contain the input, process, and output devices all in one unit.

In any communication system, your message is the input, how you move your message is the process, and the reception of your message at the other end is the output. The quality of the output (received message) must be checked for clarity, and this checking process is the feedback. See Figure 5.2 on the next page.

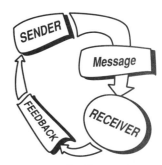

communication system
a plan or route that takes a message from the sender to the receiver. Includes the steps of input, process, output, and feedback.

How You Use Communication Technology

What is an example of input, process, output, and feedback in a communication system?

You use communication technology all the time. When you read a book or magazine, you are at the output end of the communication system, receiving the message. When you use a telephone to talk to a friend, or use a computer for word processing, or shoot a movie with your video recorder, you are controlling both the input and output parts of these communication systems. What parts of a system are you involved with when you play a video game, watch television, or build a telegraph system?

When you play a video game, you are sending a signal that affects the outcome of your game. See Figure 5.3 on page 137. You are therefore sending and receiving information (input and output). The way you move the game controllers will be influenced by what you see on the screen. Each move you make is a reaction to what you see on the screen and therefore a good example of the feedback part of the system.

When you watch television, you are receiving the message that is being sent from the studio. The popularity of the show and the sales success of its advertisers serve as feedback for the producers of the program.

When you build a telegraph system, you are assembling the process part of the system. When your system is built, it will be ready to carry messages.

135

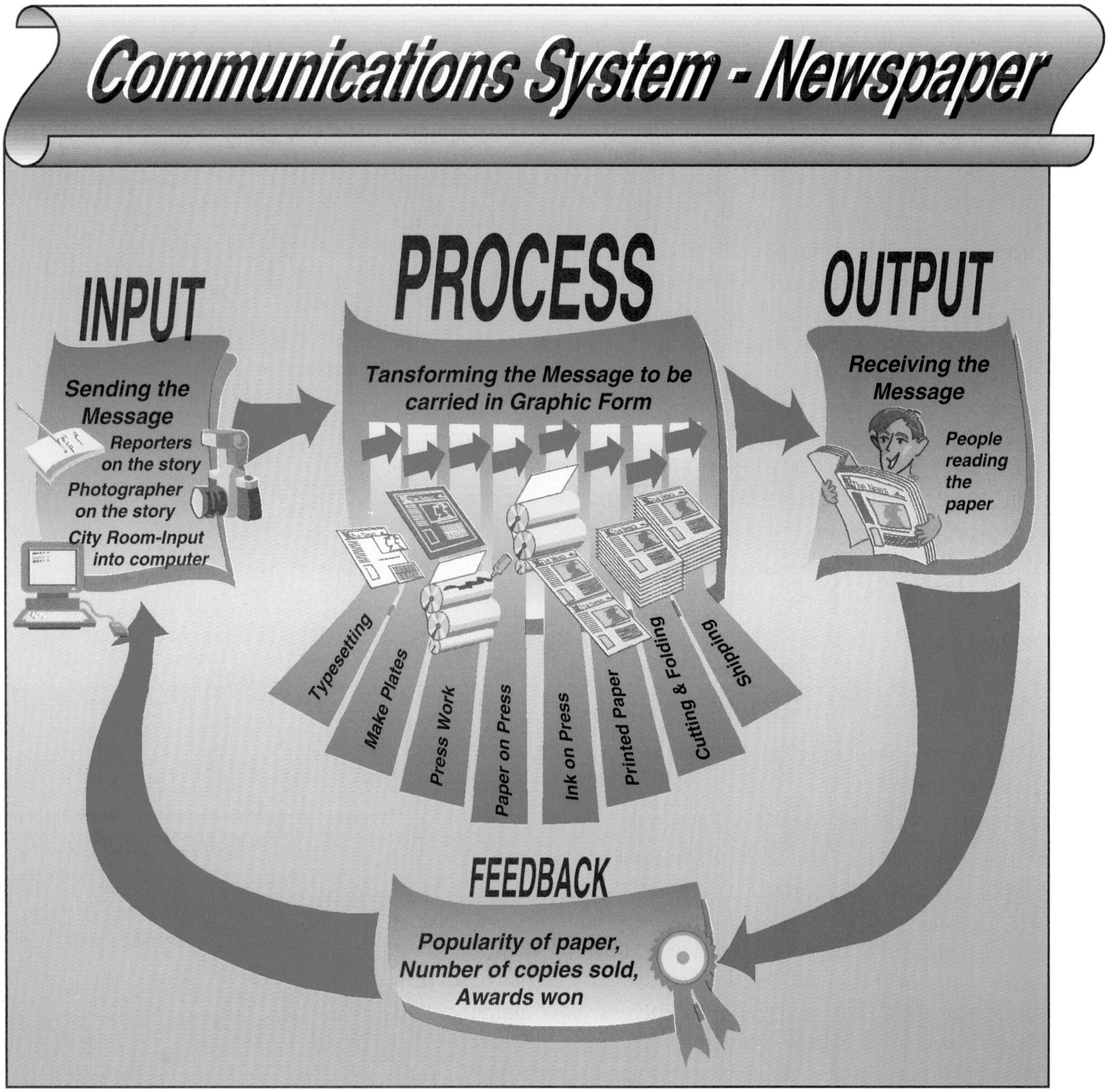

Communications System - Newspaper

INPUT

Sending the Message

Reporters on the story

Photographer on the story

City Room-Input into computer

PROCESS

Tansforming the Message to be carried in Graphic Form

- Typesetting
- Make Plates
- Press Work
- Paper on Press
- Ink on Press
- Printed Paper
- Cutting & Folding
- Shipping

OUTPUT

Receiving the Message

People reading the paper

FEEDBACK

Popularity of paper, Number of copies sold, Awards won

■ **Figure 5-2** ■ In this diagram you can see the different steps that must be completed to produce a newspaper.

— Types of — Communication Systems

▼ *What is one way of grouping communication systems?*

Communication systems can be grouped by the way they carry messages. Let's look at the different ways you can move a message from one location to another.

Biological Communication

What type of communication occurs without technology?

Biological communication isn't part of technology. You will study it in depth in science. It is important for you to realize that communication exists outside of technology and that most living things communicate biologically.

Biological communication includes all forms of communication that use natural body parts, such as the brain, voice box, ears, arms, and hands, to transmit and receive messages. Examples of this form of communication include speaking (language), smiling, frowning, and hand signals. See Figure 5.4 on the next page.

Graphic Communication

What things are used to make graphic communication possible?

Graphic communication includes all forms of communication that send and receive messages visually through the use of stationary pictures and symbols. We find such symbols and pictures in books, magazines, newspapers, photographs, messages printed on clothing, billboards, paintings, drawings, and even graffiti. See Figure 5.5 on page 139. In these cases, people send and receive information

Figure 5-3 People who play video games are actually communicating with computers.

graphic communication
the methods of sending and receiving messages using stationary, visual images and printed words or symbols. Includes books, magazines, newspapers, photographs, etc.

Putting Knowledge To Work

Low-Tech Telecommunication

When you drive on a highway, you often see very large billboards with advertisements on them. Some of these billboards are so large that they can be seen very far away. A sign or flag that is large enough to be seen over long distances is an example of low-tech telecommunication through a graphic communication system.

Long-distance communication between ships takes place when flags are raised high above the sails. A flag message known to every sailor is the skull and crossbones of the pirate ship. In a pirate movie or television show, seeing that flag on a ship means that a lot of action is about to begin.

Many other messages were sent this way

before the ship-to-ship wireless telegraph was invented. During periods of radio silence, military craft often sent and received messages through the use of flags by day and flashing lights by night.

Today private boat owners still use flags for telecommunication. Other examples of telecommunication systems that are not part of electronic communication include smoke signals, flare guns, and very large letters spelling out messages like the Hollywood sign. Remember that telecommunication is long-distance communication, not necessarily high-technology communication.

■ **Figure 5-4** ■ You often communicate your feelings through facial expressions. Can you identify the happy, sad, and surprised people in this illustration?

through reading, writing, drawing, and photography. You'll learn more about graphic communication in Chapter 7.

When you go to the movies, the large posters in the lobby are considered part of graphic communication. Is the motion picture part of graphic communication? The movie is not considered part of graphic communication because the picture is in a state of motion. Read the definition of graphic communication again and you will see that it consists only of *stationary* pictures and symbols. The motion picture is part of wave communication.

— Learning Time —

Recalling the Facts

1. What four elements must exist for communication to take place? Why is each one necessary?
2. What is biological communication?
3. What is graphic communication?

Thinking for Yourself

1. What is the difference between communication and communication technology?
2. When you write a note, are you using graphic communication?

Applying What You've Learned

1. Prepare a universal systems diagram for the communication system of your choice. Use magazine pictures, drawings, xerox prints from books, and computer graphics to show what the input, process, and output of your system looks like. Figure 5.2 will get you started by showing you the input, process, and output of a communication system.

Wave Communication

What is wave communication?

Wave communication refers to all forms of communication that have their messages carried by energy sources that move through air, water, outer space, or solids in a wave form. The signals that carry our messages are made up of sound, electrical, or light waves.

In studying wave communication, you will find some of the newest and oldest technological inventions. Smoke signals were used by our early ancestors. People sent puffs of smoke into the air to create a coded signal. The laser is a way of sending messages with light.

Figure 5-5 This magazine display shows many magazines that are of interest to different people. Which magazines do you find interesting?

sound carrier wave
communication signals carried by air

Tech Talk

Have you noticed that several communication technology terms, such as *telephone*, *television*, *telegraph*, and *telecommunication*, begin with the prefix *tele*? Can you guess the meaning of this prefix? If you guessed *distance*, you were right.

telecommunication
the methods of communicating over long distances; includes telephone, television, radio, etc.

Our early ancestors used hollow logs as drums to send messages through the air. Their banging on these logs caused the air to vibrate.

The vibrating air sent the drum's signal to people far away. You can't see the air vibrate, but it does. A stone dropped into a pool creates a very similar wave in the water. See Figure 5.6.

Since the air carried the drum's signal from one place to another, we refer to it as a **sound carrier wave**. All musical instruments depend on sound carrier waves for us to hear their music. All *face-to-face communication* depends on our atmosphere as a sound carrier wave.

Our telephone, radio, and television depend on electrical carrier waves to bring messages to us. Here the signal is being carried by electricity in the form of electrical signals. How this is done will be covered in detail in Chapter 9.

It is important to realize that messages leave your telephone or radio speaker as sound carrier waves carried by our atmosphere. Your television signal finishes its journey when it leaves the T.V.'s speakers as sound waves and the T.V.'s picture tube as light waves.

Telecommunication

What are some devices commonly used in telecommunication?

Long-distance communication is called **telecommunication**. Today most telecommunication systems use electronic devices. Did you ever use a telecommunication machine? You most certainly did if you ever used a phone, watched television, or listened to a radio. They are all examples of telecommunication.

Satellites are also telecommunication devices. A satellite placed 22,300 miles above the earth can travel at the same speed as

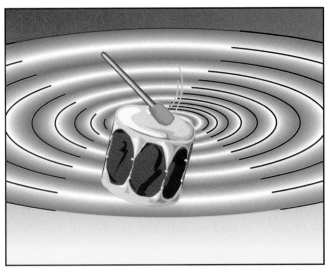

Figure 5-6 The wave created by a stone hitting a pool of water is visible. The wave created by hitting a drum is invisible.

the earth spins. This is called a *geosynchronous orbit*, which means that the satellite always stays above the same ground location. Its lack of movement in relationship to the ground could almost give the impression that it was attached to the end of a very long pole. When a satellite is in a geosynchronous orbit, what movement does it appear to have in relationship to objects that are on the ground?

Satellites can also spy or map what is going on around the world. The United States had at least six spy satellites looking down at the Mid-East during the 1991 Desert Storm conflict. Some of these spy satellites were taking photos using the same design as the Hubble telescope.

In science you learned that the Hubble telescope is a very large observatory floating in space and aimed at the stars. The pictures taken by spy satellites that use the Hubble design are very detailed. Imagine having a very powerful telescope, 188 miles up in space, aimed down at the earth. With these telescopes, it is possible to see an object as small as a grapefruit sitting on the desert sand. See Figure 5.7.

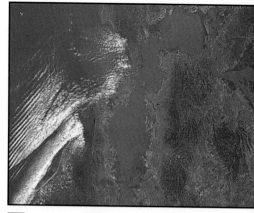

Figure 5-7 This photo, taken by an astronaut during a space shuttle flight in 1991, shows the San Francisco Bay area of California.

geosynchronous
(jē′ō-sĭn′krə-nəs)

—— Communication: ——
People and Machines

Can people communicate with machines?

At one time all communication was face-to-face without the aid of technology. The development of communication technology has done more than give us new ways to talk to our friends. Have you used communication technology to talk with a nonhuman lately?

People-to-People Communication

What progress have people made in their ability to communicate with each other?

People-to-people communication wasn't always as it is today. At one time people had to communicate with each other within the limits of their own physical makeup (biological communication).

We can understand how our early ancestors communicated by looking at the way a very young child communicates. Do you have a baby brother or sister or know someone who does? Have you noticed how the baby communicates its needs to its mother? It points, cries, grunts, stamps its feet, grabs what it wants, or speaks just enough "baby talk" to be understood.

Our earliest ancestors communicated in this very same way. They communicated this way before they had an alphabet, printing

If I Could Talk to Animals

In the movie *Dr. Doolittle*, Professor Doolittle could talk to all kinds of animals. Do you think that we will ever be able to talk to animals?

Do monkeys, lions, dogs, cats, and other animals speak to each other? They do make sounds that communicate certain messages to their relatives and neighbors. These messages help them to live together and survive in the wild.

Do you find it hard to believe that animals can communicate? Did you ever approach a strange dog and have it tell you to stay away from its home? That growl was certainly a form of animal speech that could be understood by other dogs as well as people.

Today communication technology is being used by scientists to talk with some animals. Scientists have taught chimpanzees to use sign language and computers to talk to their trainers.

Computers have also been designed that let porpoises communicate with their trainers. By recording animal sounds along with the animal's behavior, scientists hope to one day use the animal's own language for communication. A computer with a synthesizer would duplicate these sounds to indicate a particular behavior.

press, or even cave paintings. Just like other species in the animal kingdom, our ancestors communicated using simple speech and nonverbal movements.

We differ from other animals, however, because we have learned to create new and more powerful methods of communication. How does a baby improve its communication skills? It learns new skills from other people. We are the only animal that can surpass its biological capacities. Through a learning process, our earliest ancestors developed new methods of communication. They then passed this knowledge on to their children and other members of their clan, or family.

In time, people gained the knowledge and skill needed to build complex communication devices. They created all kinds of graphic communication systems to carry their messages through the printed word. Finally, they developed electrical forms of communication.

People-to-Machine Communication

How can a person communicate with a machine?

Up until the development of electronic communication, people were always talking to other people. The machines that they had built could only carry the message, and this message could only be understood by other people. Today, however, we have people communicating with the machines that they have created. Examples of people communicating with machines include a computer programmer typing a program into a computer, a person using a joystick to control a video game, and someone setting an electric timer to turn on a lamp.

Machine-to-People Communication

How does a machine send a message to a person?

Today machines also send messages to people. Examples of machines delivering their own messages include a whistling teapot that tells you the water is boiling, an alarm system that tells you someone has entered a protected store or home, and a smoke detector that senses a fire and warns you of the danger. See Figure 5.8. Can you think of other examples of machines communicating with people?

Machine-to-Machine Communication

What is one example of machine-to-machine communication?

In the movie *The Terminator*, a machine from the future is sent back in time to help destroy humanity. In this movie, the machines are out to create a society that only includes machines. These machines communicate with each other in a people-to-people manner. In real life, we don't, at this time, have that type of machine-to-machine communication. However, machines communicating with other machines is quite common today.

Your computer, for example, gives instructions to your printer, telling it to print your report. In the automated factory, computers attached to sensors control the flow of raw materials and the operation of the machines. The assembly and finishing processes that need to be performed on the product are also controlled by machines communicating with other machines. Finally, the packaging of the finished product may be handled by machines under the direction of—you guessed it—still other machines.

■ Figure 5-8 ■ Here we see a computer billboard warning motorists that a traffic accident has occurred. This is an example of machine to people communication.

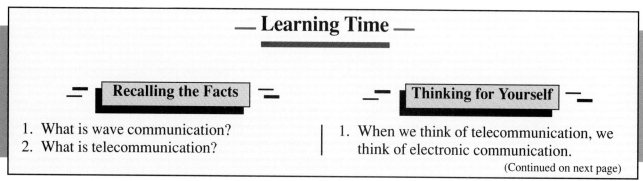

— **Learning Time** —

Recalling the Facts	**Thinking for Yourself**
1. What is wave communication? 2. What is telecommunication?	1. When we think of telecommunication, we think of electronic communication.

(Continued on next page)

— Learning Time —

Continued

3. Explain people-to-people, people-to-machine, machine-to-people, and machine-to-machine communication. Give an example of each.
4. What kind of carrier wave makes it possible for us to hear a musical instrument?
5. Describe the system that technologists are using to communicate with animals.
6. Name three types of signals that carry wave communication messages.

Explain and give examples of how telecommunication can also be low-tech wave or graphic communication.

2. If you look at movie film, you see lots of individual photographs. Each picture is slightly different from the one next to it. When shown as a motion picture, these pictures are part of wave communication rather than graphic communication. Why is a motion picture considered wave communication?

Applying What You've Learned

1. Construct a small musical instrument.
2. Use your artistic talents, computer graphics program, and magazine pictures to create a large poster. The theme of the poster should be, Look Who Is Talking to People and Machines.
3. Construct a model of a machine that can send messages to people. You could construct a warning system to alert a family that their basement is flooding; or a system to tell you that someone has entered your room or opened your drawer or closet; or a device that tells you that the toast is done or the water is boiling; or any other machine-to-people communication device you like. Prepare a chart of your device and be prepared to demonstrate or explain it to the class.

—— Communication and the —— Spread of Technology

What has been the key to the advancement of technology?

How did you learn to talk, walk, play a video game, read, or do any of the thousands of other things that you do? You learned from others, just as your parents and your grandparents did before you.

Your great-grandfather might have learned how to use a new communication tool called a fountain pen. Your grandfather might have learned how to use a ballpoint pen or typewriter. You are learning how to use the computer, laser disk, cellular phone, and fax machine. See Figure 5.9. What new things do you think your children will learn to use as tools of communication as they grow up?

All knowledge was at one time passed on from one person to another by word of mouth. As new things were learned, there was

■ **Figure 5-9 a–d** ■ What communication devices were your parents, grandparents, and great-grandparents using when they were teenagers?

more to teach to the next generation. Our ancestors only shared their knowledge with their own people. People who lived in other areas had to form and pass along their own knowledge. The key to the advancement of our technology was the development of communication. It allowed us to learn from the achievements of people who lived far away from us, or even to learn from people who lived and died long before we were born.

Before the invention of printing, only churches, princes, kings, and the very wealthy owned books. Printing made it possible for more people to own books. As more people learned to read, they became more concerned with the problems of their times.

Do you remember our definition of technology? Technology is the knowledge and skill of how to use raw materials, tools, and energy to create the products and services that we want. Can you see why the growth of the printing industry was so important to the spread of technology? With the development of printing and the increased numbers of people who could read, more and more people could learn about raw materials, tools, and the discoveries of others. They learned to build where others had stopped.

——— The Impact of ———
Communication Technology

= *What parts of our society have been affected by advances in communication technology?*

When people talk about our shrinking world, they don't mean that our planet is really getting smaller. Our technology has made it possible for us to communicate almost instantly with any person located at any point on our planet. Our political institutions, cultures, and environment have all been affected by advances in communication systems.

Political Impact

= *How has communication technology affected our political knowledge?*

Today you receive information on what is happening all over the world at the same time the events are taking place. Princess Diana's death in 1997 and the public mourning of the world unfolded instantly on your television screen.

You were able to see the tearing down of the Berlin Wall as it was happening in Germany. See Figure 5.10. When world leaders speak, you hear their statements instantly through satellite communication. Mass communication systems report events and help form world opinion.

Your knowledge of world events will help you decide who can best lead our country. You may not be old enough to vote, but you probably have much more political knowledge than your grandparents had when they were of voting age.

Are you familiar with the book *1984* by George Orwell? The people in this book were watched, and even controlled, by machines. The expression "Big brother is watching" came from this book.

Our political leaders have tried to make sure that communication technology can never be used by people to watch or control our lives. But banking, stock market transactions, and credit card purchases are all monitored by communication systems. It is even possible to create a full profile of a person by accessing the National Computer System. This particular impact of communication technology may be good—as long as it is controlled by people who believe in freedom of expression and democracy.

Social and Cultural Impact

What good might come from communication technology's effect on our understanding of different cultures?

Our communication systems have made us more aware of the different cultures that share our planet. Through motion pictures and television, you have been invited into the homes of families of different cultures, religions, and nationalities. Hopefully, your awareness of the customs and traditions of others will lead to less prejudice and a greater acceptance of others. Information itself is not good or evil. It is what people do with this information that can help or hurt us.

Environmental Impact

How has communication technology affected our environment?

You have already learned that all systems are broken down into input, process, output, and feedback. Unfortunately, many of our processes not only produce what we want, but also produce what we *don't* want. They form products that can harm our environment.

Most communication systems use paper as if it grew on trees. The problem is that paper doesn't grow on trees—it is made out of trees. See Figure 5.11. At one time trees were cut down to feed the ever-hungry paper mills without any concern about future needs. Today paper manufacturers have replanting programs to guarantee trees for the future. Papermakers also must take great care not to contaminate nearby rivers and streams with the waste products of the manufacturing process.

■ **Figure 5-11** ■ This picture shows trees being harvested for paper production. Paper doesn't grow on trees, it is made from trees.

The computer, copying machine, and fax machine have increased our demand for paper. Do you use a computer for your school reports? How many paper printouts go into the garbage before you get your final report printed?

Many chemicals, metals, and plastics are used to manufacture communication systems. Still more chemicals and paper are used in some of these machines to make the books, magazines, and newspapers that you read. Our environment is harmed when these materials are not disposed of properly. What's more, workers can be injured when they work with some of these materials.

The power lines and microwave transmitters that carry our communication signals visually affect the look of our communities and can physically affect our bodies.

When we think about environmental impact, we usually think about what is wrong with technology. In developing new systems, cost and profit are usually most important to the manufacturers and distributors of the new system. Sometimes people aren't aware of the dangers of the new system until after the product is being used. Government agencies and consumer protection groups try to protect us from these dangers.

— Learning Time —

Recalling the Facts

1. Before the development of printing, who owned books?
2. How has communication technology made our world smaller?
3. What are some of the negative and positive impacts of communication technology?

Thinking for Yourself

1. World events appear on your television screen as they happen. How does this instant communication affect our world?
2. If you could talk to an animal, which animal would you want to communicate with, and why?

Applying What You've Learned

1. Design and construct a special container that could be used for the collection of recycled paper products.
2. Design an environmental impact study to determine what your community is doing about recycling paper. Interview school, town, and civic leaders, and then set up a schoolwide recycling program with their assistance.

Summary

Directions

The following sentences contain blanks. For each numbered blank, pick the answer that makes the most sense in the entire passage. Write your answer on a separate sheet of paper.

Communication is the sharing of information. To have ___1___ you need a sender, ___2___, receiver, and feedback. Whenever your ability to communicate is enhanced by technology such as pictures, printed messages, or machines, you are using a specific type of technology called communication technology.

In any communication ___3___, your message is the input, how you move your message is the process, and the reception of your message at the ___4___ is the output. People who use a communication system provide ___5___ to the system. They rate quality of service on surveys or by ___6___ telephone companies, television stations, and newspapers when they aren't satisfied.

Communication systems can be grouped by the way that they carry your ___7___. Biological communication occurs without technology. You communicate by speaking, smiling, frowning, and using hand signals. In graphic communication you ___8___ messages visually through the use of drawings, paintings, photographs, printed words and pictures, books, magazines, newspapers, and even graffiti.

Wave communication includes all forms of communication where messages are carried by energy that ___9___ through air, water, outer space, or solids as sound, electrical, or light waves. Messages leave your ___10___, telephone, and radio as sound carrier waves carried by our atmosphere. Messages are sent to your telephone, radio and television on electrical carrier waves.

Long distance communication is called telecommunication. With the assistance of satellite telecommunication it is possible to instantly whisper to a person in ___11___.

People have always communicated with each other. Thanks to communication

1.
a. technology
b. communication
c. reaction
d. government

2.
a. message (information)
b. textbook (printed material)
c. telephone (phone book)
d. computer (disk)

3.
a. project
b. textbook
c. system
d. company

4.
a. beginning
b. microphone
c. keyboard
d. receiver

5.
a. feedback
b. money
c. machinery
d. time

6.
a. purchasing
b. praising
c. recommending
d. switching

7.
a. books
b. belongings
c. messages
d. problems

8.
a. send and receive
b. handle and transmit
c. see and listen to
d. forgive and forget

9.
a. flies
b. crawls
c. hops
d. travels

10.
a. ears
b. fingers
c. house
d. mouth

11.
a. your classroom
b. another country
c. your house
d. your car

(Continued on next page)

Summary

Continued

technology, people today are also able to communicate with machines, machines with people, and machines with other machines. One example of machine-to-people communication is ___12___ . Your computer giving instructions to its printer is an example of ___13___ communication. Can you think of other examples?

Communication technology has made the entire world your ___14___ . Your television brought you the war with Iraq as it was taking place. You watched the Soviet Union turn away from ___15___ . You have seen how people of different cultures, religions, and nationalities live. Communication technology has made you a citizen of the ___16___ . Do you ever wonder what communication devices your children will use when they are teenagers?

12.
   a. a joystick controlling a game
   b. a whistling teapot
   c. programming a computer
   d. the automated factory

13.
   a. machine-to-people
   b. people-to-people
   c. machine-to-machine
   d. wave

14.
   a. enemy
   b. next-door neighbor
   c. friend
   d. rival

15.
   a. democracy
   b. socialism
   c. friendship
   d. communism

16.
   a. city
   b. state
   c. nation
   d. world

Key Word Definitions

Directions

The column on the left contains the key words from this chapter. The column on the right contains a scrambled list of phrases that describe what these words mean. Match the correct meaning with each word. Write your answer on a separate sheet of paper.

Key Word		*Description*	
1.	Communication technology	a)	Visual messages through pictures and printed words
2.	Communication system	b)	Messages carried through air vibration
3.	Graphic communication	c)	Input, process, and output of messages
4.	Wave communication	d)	Long-distance communication
5.	Sound carrier wave	e)	Enhanced communication
6.	Telecommunication	f)	Messages carried by energy through air, water, or solids

PROBLEM-SOLVING ACTIVITY

Developing an Advertisement for a Communication Invention of the Past

Resources You Will Need

- Video camera
- Tape recorder
- Markers, pens, pencils
- Poster board
- Scissors
- Glue
- Computer and printer
- Graphics software
- Electrostatic copier
- Reference books with appropriate illustrations of inventions

Did You Know?

Our total communication system is made up of many independent graphic and wave communication systems. They have become an important part of our lives. Do you take books, newspapers, magazines, and photographs for granted? Do you take your telephone, radio, television, and video recorder for granted?

Graphic communication brought books and literacy to the masses. Wave communication made it possible for people to be aware of the world around them even if they couldn't read. It is unlikely that the telephone, radio, and television, which are all wave communication systems, would have ever been developed if it weren't for graphic communication.

Problems To Solve

What makes an invention important? If it meets people's needs and wants, will the invention automatically become important? Printing from movable clay type was invented by Pi Sheng 400 years before Gutenberg was born. The Koreans were casting metal characters at least 50 years before Gutenberg developed his mold. Still, Gutenberg received the credit for the greatest graphic communication invention of recorded history. Why? For any invention to become important, people must become aware of its existence. People must also be convinced that they need the

A.

new invention in their lives.

If you wanted to inform everyone about a new invention, what mass communication system would you use? If your answer is television, then you are ready to start production on your first television commercial. If radio, newspapers, or magazines seem the best way to go, then be prepared to design a commercial for the graphic or wave communication system of your choice.

With your teacher's help, pick a communication invention. Working in groups of no more than four people, develop a commercial for this invention. Your advertisement should be a 1-minute commercial for wave communication or a full-page ad for graphic communication.

The invention that you advertise could be the first printing press, radio, or television. Your commercial can still be prepared to run on television, radio, or a newspaper. You will use a modern communication system to advertise an invention of the past.

What Did You Learn?

1. What makes a new technological development financially successful?
2. What is the input, process, and output phases of the communication system that carried your commercial?
3. If you were to make a new commercial, what would you do differently to improve your final product?

A Procedure To Follow

1. Determine the invention that your group is going to advertise.
2. Determine the communication system that you will use.
3. Brainstorm the theme of your commercial. Use commercials that you have seen as role models. Be creative and have fun.
4. Prepare props and/or script if you are going to do a wave communication commercial. Prepare illustrations, text, and headings if you are going to do a graphic communication commercial.
5. Your commercial should state who invented the device, when the invention was invented, what it does, and why it is important to the development of technology.
6. Produce your commercial using the appropriate equipment. This means tape-recording your radio commercial, videotaping your television commercial, or pasting up your artwork and text for your newspaper ad. See Figures A and B.
7. Share your commercial or ad with the class.
8. Make copies of the videotape or audiotape for each member of your team, so that they all have a copy to take home.

B.

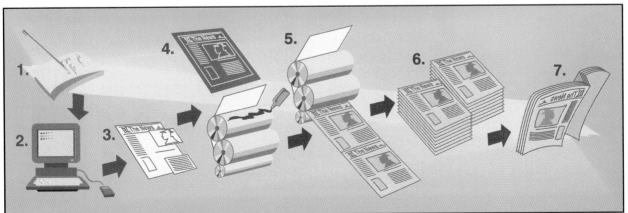

Electrostatic CAD Plotting
Blue Print Service Co.
149 Second St.
San Francisco, CA 94105
Continental U.S.
800-221-PLOT

California ONLY
800-521-PLOT

Chapter 6

Drafting

You share so much with your friends, including your food, clothing, homework, and feelings. You talk about what you do, how everyone looks, and what everyone wears. Can you remember ever saying "I wish I had a picture to show you what I'm talking about"? You say this when an object or event that lives on in your memory just can't be described fully with words alone.

When you only use words to describe an object, it can be difficult getting other people to understand what the object actually looks like. Sometimes a picture says it better—which is why we have the expression, "A picture is worth a thousand words."

Drafting refers to drawing techniques used by architects, engineers, technicians, tradespeople, and many others to describe the size, shape, and structure of objects. All areas of technology rely on drafting plans to construct the machines, buildings, highways, products, and systems of our technological world. These drawings can be created by freehand sketching, technical drawing, and computer-aided design and drafting.

— Chapter — Outline

I. Freehand Sketching
II. Technical Drawing
III. Computer-Aided Design and Drafting (CADD)
IV. Technical Drawing Applications
V. Drafting Careers

— Key Words —

In this chapter you will learn and study the meanings of the following important technology terms.

DRAFTING
TECHNICAL DRAWING
DRAFTER
MECHANICAL DRAWING
PICTORIAL DRAWING
ISOMETRIC DRAWING
OBLIQUE DRAWING
PERSPECTIVE DRAWING
ORTHOGRAPHIC PROJECTION
MULTIVIEW DRAWING
SCALE
SECTIONAL DRAWING
COMPUTER-AIDED DESIGN (CAD)
COMPUTER-AIDED DRAFTING (CAD)
WORKING DRAWING
TOLERANCE
SCHEMATIC DIAGRAM

Freehand Sketching

How are freehand sketches involved with technology?

Do you or your friends like to do freehand drawings of cartoon characters or cars? When these drawings accurately express the size and shape of objects, your drawings are called *freehand sketches*. Freehand sketching can be the first stage in the development of a working plan that will be used to construct a building, bridge, home appliance, machine part, or something that you will build. Designers often make many freehand sketches of possible designs to get their ideas down on paper. Then they use the sketches to evaluate their designs and pick the best one.

Freehand sketches are drawn using only pencil and paper. To make your first sketching experience easier, try drawing on graph paper. This type of paper has ruled lines printed on it that can guide you in drawing straight and angled lines. See Figure 6.1.

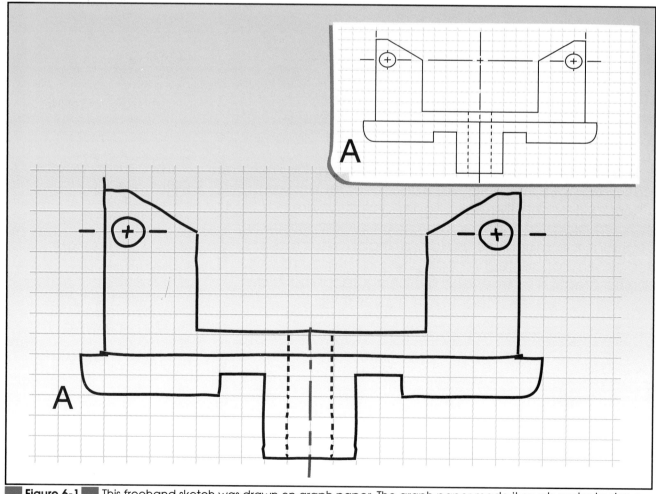

Figure 6-1 This freehand sketch was drawn on graph paper. The graph paper made it much easier to draw straight, angled, or curved lines.

In the technology laboratory, you will often use freehand sketching to show the size and shape of a project that you plan to construct. Your drawing helps you size and form the material that your project is made of. Drawing is one of the simplest ways of communicating ideas to other people.

— Learning Time —

— Recalling the Facts —

1. Name several occupations where drafting is used.
2. What are some reasons for making freehand sketches?

— Thinking for Yourself —

1. What do we mean by the expression, "A picture is worth a thousand words"?
2. Is drawing a form of communication? Why or why not?

— Applying What You've Learned —

1. Freehand sketch the object shown in Figure 6.2 on the next page onto an enlarged grid. Start your drawing at point A.
2. Freehand sketch the symbol shown in Figure 6.3 on the next page onto an enlarged grid.
3. Design and sketch your own symbol.

—— Technical Drawing ——

How does a technical drawing differ from a freehand sketch?

Can you sketch an object accurately enough to build it? Many simple, one-person construction projects can be done effectively from simple freehand sketches. As the project becomes more complex, however, it becomes necessary to have more accurate plans. If the project consists of a number of parts that will be made by different people, the plans must be extremely accurate.

Technical drawing is graphic drawing that uses mechanical or electronic tools to accurately size and shape the objects that are drawn. In freehand sketching, you use only pencil and paper to do what will now be done with all kinds of drawing tools.

When you produce a technical drawing, you use pencils, pens, paper, drawing board, rulers, triangles, compasses, and other tools

technical drawing
the drawing techniques that accurately represent the size, shape, and structure of objects. *Drafting* and *mechanical drawing* are other terms with the same meaning.

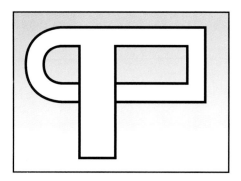

Figure 6-3 On a separate sheet of graph paper freehand sketch this symbol.

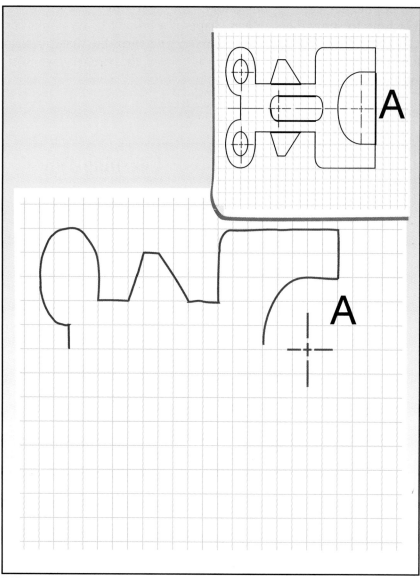

Figure 6-2 On a separate sheet of graph paper freehand sketch this object.

Tech Talk

Draft is an important word in technology. It means a drawing, sketch, or design. A draft is also an early copy of anything written. Your English teacher probably has you do first *drafts* of important papers in English class. *Draft* has many other meanings, including an air current, selection of people (military draft), and the transfer of money (bank draft).

drafter
the person who prepares technical drawings

mechanical drawing
the drawing techniques that accurately represent the size, shape, and structure of objects. *Drafting* and *technical drawing* are other terms used to describe the same techniques.

and machines to produce a very accurate drawing. See Figure 6.4. All these tools give the **drafter**—the person doing the drawing—the ability to draw perfectly straight, round or curved lines. Because drawing tools are used, **mechanical drawing** is another popular name for technical drawing. You now know that freehand sketching, technical drawing, mechanical drawing, and drafting are all used to describe the same activity. When this activity is done with the aid of a computer, it is called computer-aided design or computer-aided drafting, usually abbreviated CAD.

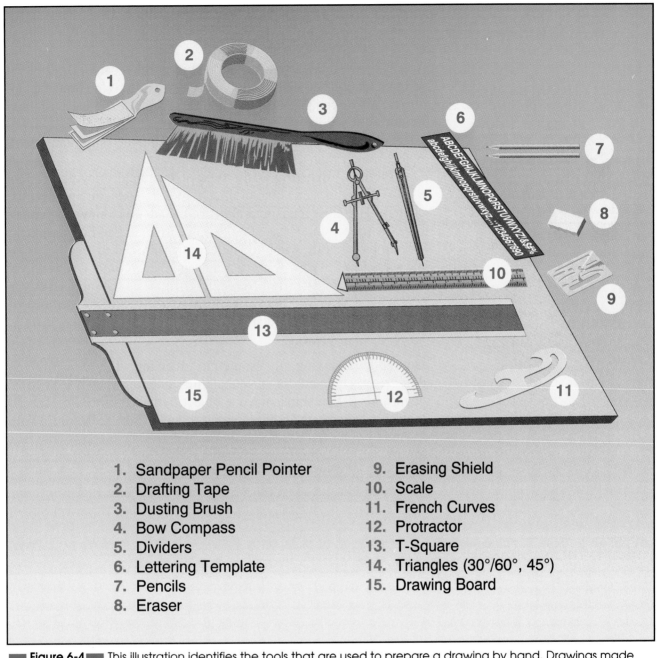

1. Sandpaper Pencil Pointer
2. Drafting Tape
3. Dusting Brush
4. Bow Compass
5. Dividers
6. Lettering Template
7. Pencils
8. Eraser
9. Erasing Shield
10. Scale
11. French Curves
12. Protractor
13. T-Square
14. Triangles (30°/60°, 45°)
15. Drawing Board

■ **Figure 6-4** ■ This illustration identifies the tools that are used to prepare a drawing by hand. Drawings made with these tools can be referred to as mechanical drawings, draftings, or technical drawings.

Drafters produce technical drawings for architects, engineers, manufacturers, construction companies, and map makers. Each area of technology has its own drawing rules. Drafters become specialists in architectural, structural, map, engineering, manufacturing, electric, or electronic drawing. You will explore the different drawing techniques used by drafters working in these fields later on in the chapter.

Now let's take a closer look at the different types of technical drawings.

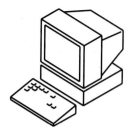

pictorial drawing
a drawing that shows a three-dimensional object so accurately that it resembles a photograph

isometric drawing
a pictorial drawing that shows three, equally distorted sides of the object

Pictorial Drawing

What is a pictorial drawing?

When you take a picture of an object with a camera, you often see partial views of a number of the object's sides. Why draw the object when you can just take its picture? You must remember that these drawings are probably being made for objects that don't exist yet. Also, these drawings will sometimes describe the shape and size of objects better than a photograph.

Pictorial drawing is the type of technical drawing drafters use to create this same kind of three-dimensional picture without the camera. The most realistic 3-D pictorials are now drawn on a computer using computer-based modeling software. See Figure 6.5.

In pictorial drawing, the object is positioned to create exactly the same level of *distortion* (wrong shape) each time. Did you ever see your image distorted by a mirror in an amusement park? In mechanical drawing, there are three types of pictorial drawings, and they range from no distortion of the front view to distortion of all views and surfaces.

Isometric Drawing. In **isometric drawing**, the object being drawn is rotated and drawn so that three sides are shown with equal distortion. If you think of your object as being in a glass box, it looks like you rotated it around on its edge and then tilted it forward 30°. See Figure 6.6 on page 162. The edges of the box that now face toward you all form the same angle. Your object is drawn as if it were cut out of this tilted, angled glass box. See Figure 6.7.

A drafter draws an object as an isometric drawing if this type of drawing will give the best picture of what the object is going to

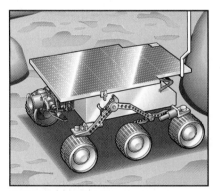 

Figure 6-5 The 3-D pictorial on the left looks almost as lifelike as the photograph on the right. However, the drawing contains size-description information that can be used to draw other views of the Mars rover called *Sojourner*.

■ **Figure 6-8** ■ Lettering that has been shadowed is an oblique drawing.

■ **Figure 6-7** ■ In this isometric drawing of a toy locomotive all round parts are distorted. Round objects appear oval in shape.

look like. Certain shapes are better represented by one of the other two pictorial drawing systems.

Oblique Drawing. In **oblique drawing**, you draw the object so that you see a perfect, nondistorted front view. If your object has circular parts on one surface, it is easier to draw the object as an oblique pictorial drawing than as an isometric pictorial drawing.

Have you ever drawn your name in balloon lettering and then added a top and a side to each letter? The word *technology* in Figure 6.8 is actually an oblique drawing. Figure 6.9 shows an object drawn as an oblique pictorial drawing. Notice that the holes drawn in the front view of the object are not distorted. Most objects present a clearer picture when they are drawn by the other pictorial methods. Therefore, oblique drawings are the least drawn pictorials.

Perspective Drawing. In **perspective drawing**, you draw the object as it would appear in real life. This means that the parts of the

oblique
(ō-blēk′)

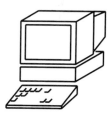

oblique drawing
a pictorial drawing that shows one side of the object without any distortion

perspective drawing
a pictorial drawing that makes the object look realistic. The parts of the object that are farther away look smaller.

■ **Figure 6-9** ■ In this oblique drawing of the toy locomotive, parts that are viewed straight on aren't distorted. Notice how the front of the locomotive is a perfect circle. Compare this with the oval shapes found in the isometric drawing in Fig. 6.7.

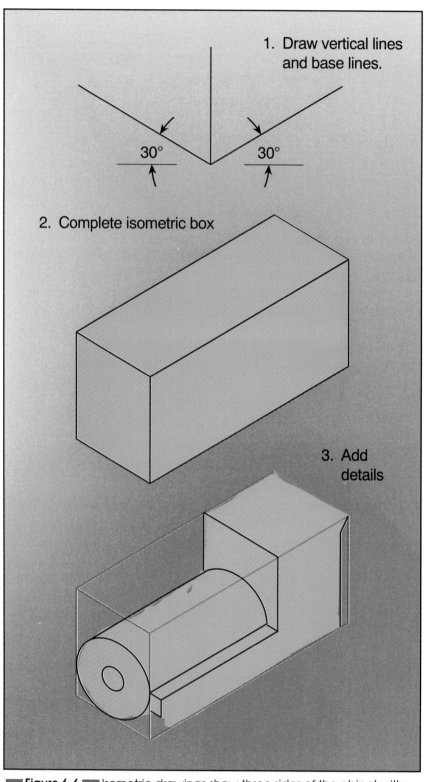

1. Draw vertical lines and base lines.

30°   30°

2. Complete isometric box

3. Add details

■ **Figure 6-6** ■ Isometric drawings show three sides of the object with each side equally distorted.

object that are farther from your eyes will appear smaller. If the object is very long, it will appear to vanish at the end of your drawing.

This is the hardest of the pictorial drawings to do, yet with certain objects, the easiest to understand when viewing. Figure 6.10 shows railroad tracks that have been drawn in isometric, oblique,

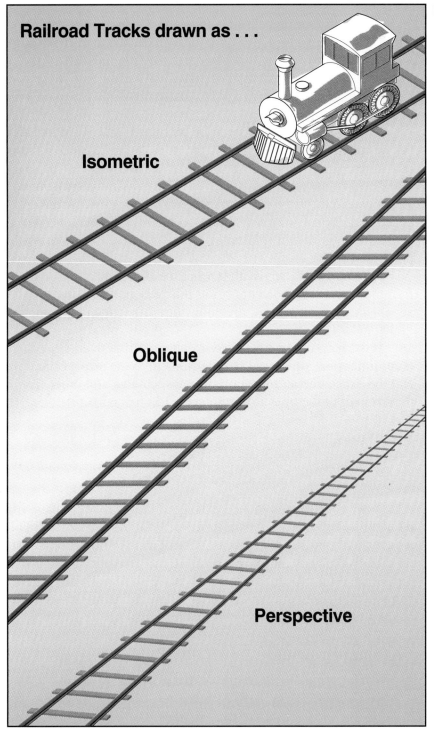

Railroad Tracks drawn as . . .

Isometric

Oblique

Perspective

Figure 6-10 The pictorial that you draw should be determined by the shape of your object. Railroad tracks are best represented as a perspective drawing.

and perspective pictorials. Can you see how your positioning of the object before it is drawn affects your final drawing? Can you see why isometric, oblique, or perspective drawings are better with some objects than they are with other objects?

Multiview Drawing

☰ *How do multiview drawings differ from pictorial drawings?*

Think of your object as being inside a square box that is made out of glass. If you took a picture through each side of the box, you would have six pictures that show each individual side of the object. You can also draw the six individual views that show every side of the object. See Figure 6.11. These drawings are often called **orthographic projection.**

Although six drawings can be produced, most objects can be fully explained by drawing the front, top, and right side of the object. These three views are drawn on the same piece of paper and together are called a **multiview drawing**. This arrangement of views is called *third-angle projection*. When necessary, the drafter creates additional views of a part so that every detail needed for construction is shown. Multiview drawings are easier to draw than pictorials.

It is very hard to visualize what the object actually looks like from its multiview projections. For this reason, when drafters show an object as a multiview drawing, they often include a pictorial drawing.

It is easier to place *dimensions* on a multiview drawing than it is to dimension a pictorial. Dimensions give the size of the object and the exact size and location of holes, cutouts and other features that are part of the object. See Figure 6.12 on page 166.

Line Work

☰ *Why must drafters have very precise rules about the types of lines they can use in their drawings?*

If all your friends looked exactly alike, you would have trouble identifying your friends in a crowd. If all the lines of a drawing were exactly the same thickness, your object would become lost in its dimension lines. Drafting therefore has rules on how dark and how long each type of line should be. Figure 6.13 on page 167 shows a multiview drawing with the different line thicknesses printed out in red. It also shows how each technical line is drawn.

Sectional Drawings

☰ *What is the purpose of sectional drawings?*

The drafter might need to show what the inside of an object looks like. How could the drafter do this? If you wanted to see what

orthographic projection
a set of drawings that shows a straight-on view of the six individual sides of an object

multiview drawing
a drawing that includes the top, front and right side views of an object. The three views are drawn on the same piece of paper.

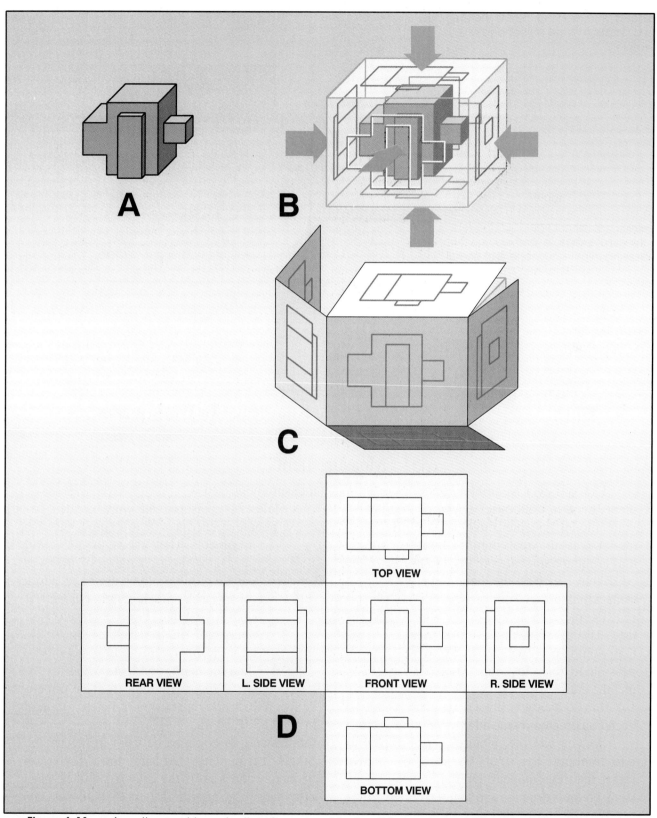

A

B

C

TOP VIEW

REAR VIEW

L. SIDE VIEW

FRONT VIEW

R. SIDE VIEW

D

BOTTOM VIEW

Figure 6-11 An orthographic projection shows an object as if it was photographed through each side of a glass box.

scale
objects or drawings made where the size is smaller or larger than the real thing. For example, one inch on the drawing might equal one foot of real distance.

Putting Knowledge To Work

Measurement

Did you ever assemble a scale model airplane, boat, or dollhouse? If you photographed your model, people might think that your model was the real thing. Objects that are made to **scale** are constructed so that 1 inch on the model represents a real distance on the object.

People who make movies often use scale model sets to film scenes that would be very expensive or impossible to do on location. For example, the entire airport set in the movie *Die Hard II* was actually a scale model that had been constructed inside a large building.

As you can see, pictures can fool people into thinking that small objects are actually larger than they appear. Drafters place dimensions on their drawings so that the construction people will know what size the object is supposed to be. Drafters need to use a system of

measurement.

Rules of measurement were created because people needed a way to express the size of objects. They used a person's foot as a measure of length. To measure smaller things, they used the width of a thumb or the spread of a hand. If these standards of measurement were still in use today, you would want to buy from a salesperson who had big hands and feet. Why?

Today, drafters indicate dimensions on their drawings using the metric system or the U.S. Customary Measurement system. If the drawing is going to be used in the United States, it is dimensioned using the U.S. system of inches and feet. If the drawing will be used elsewhere, the metric system or both systems are used.

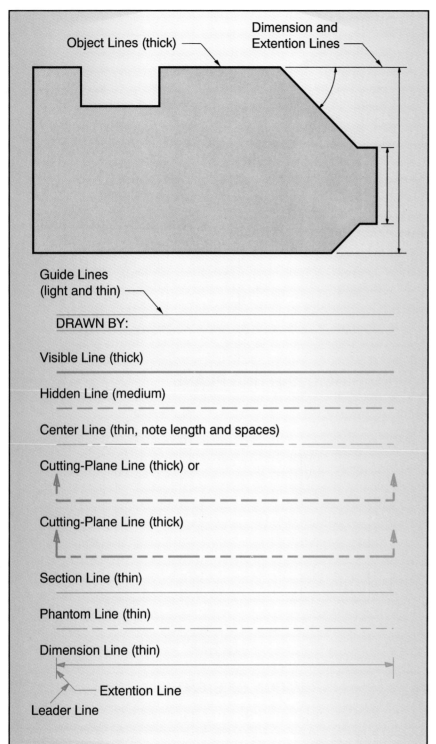

Object Lines (thick)

Dimension and
Extention Lines

Guide Lines
(light and thin)

DRAWN BY:

Visible Line (thick)

Hidden Line (medium)

Center Line (thin, note length and spaces)

Cutting-Plane Line (thick) or

Cutting-Plane Line (thick)

Section Line (thin)

Phantom Line (thin)

Dimension Line (thin)

Extention Line

Leader Line

Figure 6-13 The lines of your drawing must vary in thickness or it will be difficult to tell the object from the dimension lines.

sectional drawing
a drawing that shows the interior of an object

was inside a candy bar, what would you do? You would cut open the candy bar to see what was inside. In the same way, the drafter can show the inside of objects by cutting them open and drawing a view of their interior. Drawings that show the inside of an object are called **sectional drawings.** See Figure 6.14 on the next page.

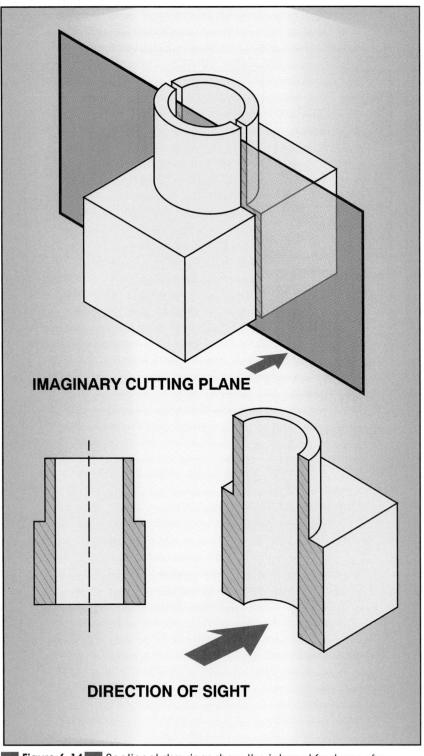

IMAGINARY CUTTING PLANE

DIRECTION OF SIGHT

Figure 6-14 Sectional drawings show the internal features of an object.

— Learning Time —

Recalling the Facts

1. What is the difference between technical drawing and drafting? Between technical drawing and mechanical drawing? Between technical drawing and freehand sketching?
2. Name and describe three types of pictorial drawings.
3. Describe the difference between a multiview drawing and a pictorial drawing.

Thinking for Yourself

1. Describe object shapes that can best be represented by isometric pictorials, oblique pictorials, and perspective pictorials.
2. Would a sectional view of a candy bar always tell you what was inside the candy? Why or why not?
3. In what way would a pictorial drawing of an object differ from a photograph of the same object?

Applying What You've Learned

1. Complete the isometric sketch using the information supplied in the multiview drawing in Figure 6.15 on the next page.
2. Complete the two missing multiview projections using the information supplied by the isometric drawing in Figure 6.16 on page 171.
3. Sketch your name as an oblique drawing in balloon letters.
4. Match the multiview drawings to their pictorials in Figure 6.17 on page 172.

— Computer-Aided Design and Drafting (CADD)

How does CADD differ from traditional design and drafting?

In **computer-aided design (CAD),** designers use the computer to test and develop their designs. In **computer-aided drafting (CAD),** the drafter uses a computer instead of a drawing board, paper, T-square, and triangles to make the drawing. It is confusing that CAD can stand for two different computer systems.

Computer-aided design and drafting (CADD) is a combination of the two terms you just learned. In this process, the person operating the computer is designing and drawing the object at the same time. The computer serves as a very powerful tool because it lets you change and even test your design while you are still drawing it. For example, an automotive design can be tested for wind resistance right on the computer. Even an expensive computer, however,

computer-aided design
a computer system that is used, instead of a drawing board and drawing tools to design an object

computer-aided drafting
the process of using a computer instead of regular drafting tools (T-square, paper, triangles, etc.) to make the drawings used by engineers, architects, and craftspeople

169

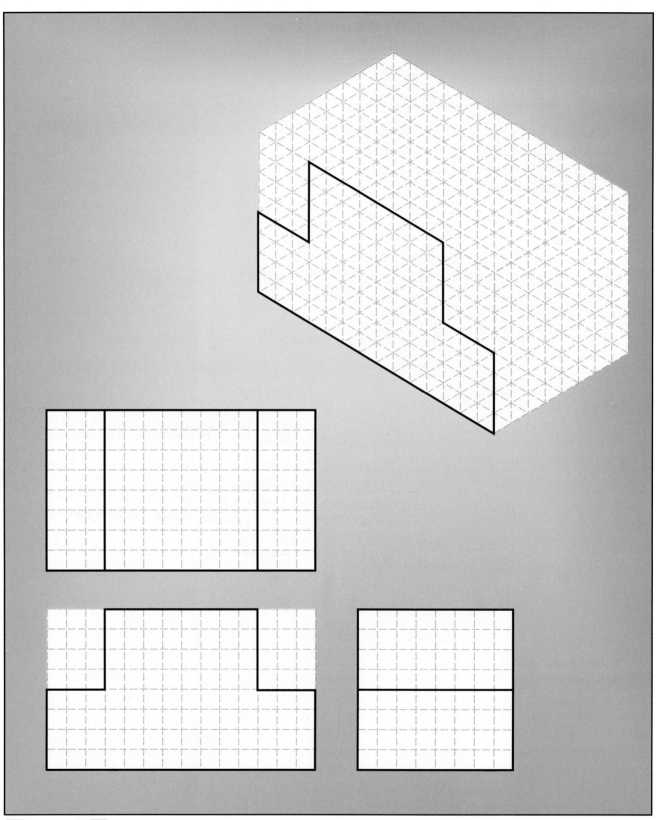

Figure 6-15 On a separate piece of graph paper sketch the isometric pictorial for the object that is shown as a multiview drawing.

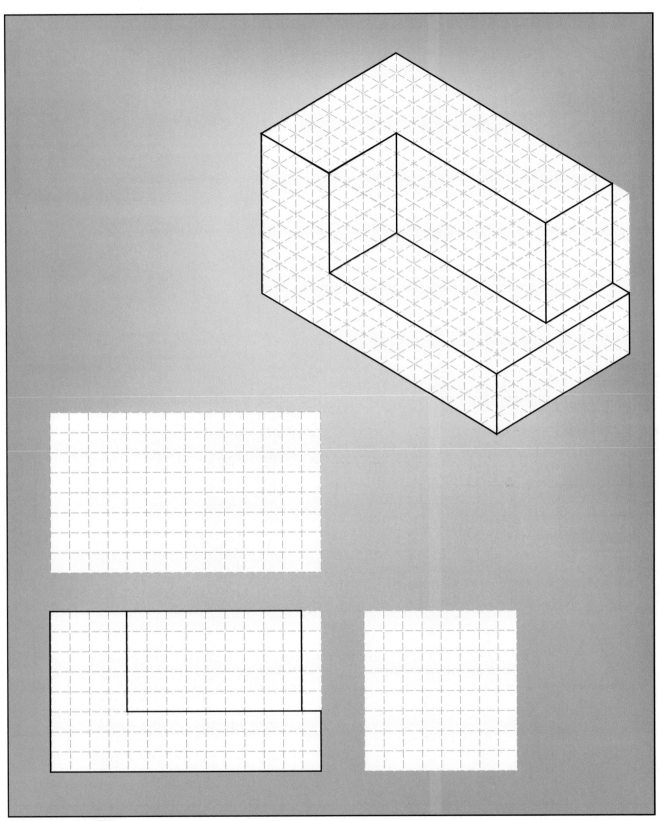

Figure 6-16 On a separate piece of graph paper complete the two missing multiview projections using the information supplied by the isometric drawing.

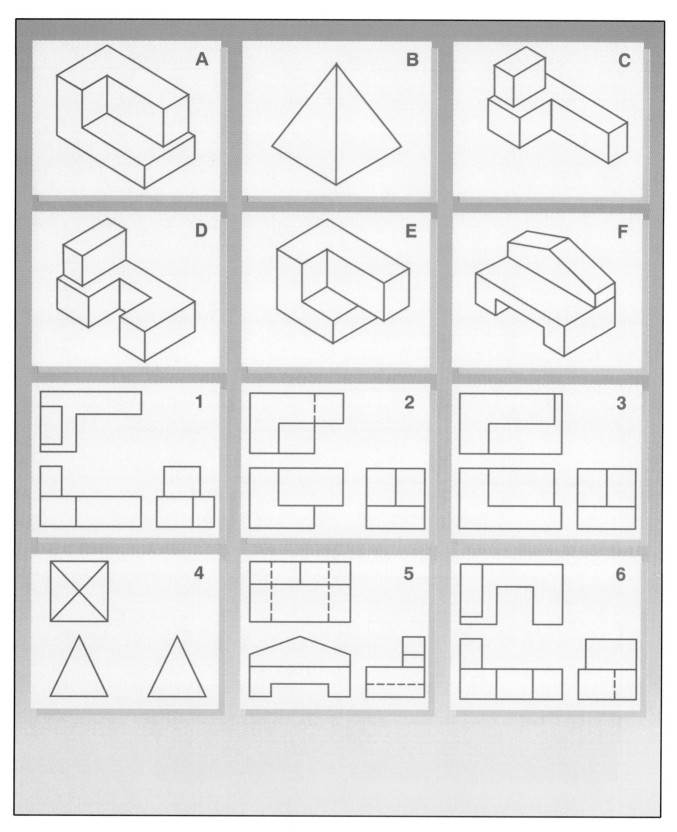

Figure 6-17 Match the multiview drawings to their isometric pictorials. Is it easier to understand the shape of the object when you view a pictorial or a multiview drawing?

can't turn poor designs or drawings into good ones, just like expensive sports equipment won't help a poor player become a superstar.

In computer-aided design and drafting, the drafter selects the type of lines to be drawn by choosing from an assortment of drawing tools. The drafter feeds the location of the line into the computer by indicating its starting and ending point. This is done by clicking a button on an electronic mouse or by drawing on a special electronic drawing tablet. See Figure 6.18.

The computer program contains many predrawn symbols that the drafter might want to include in the drawing. The drafter picks what is needed and then *cuts and pastes* it into the drawing using the mouse or drawing tablet. This is called *cut and paste* even though it is done without ever touching glue or scissors.

Computer-aided drafters increase or decrease the size of their drawings at will. For fine detail work, they magnify areas of their drawing, add the details, and then reduce the section back down to size.

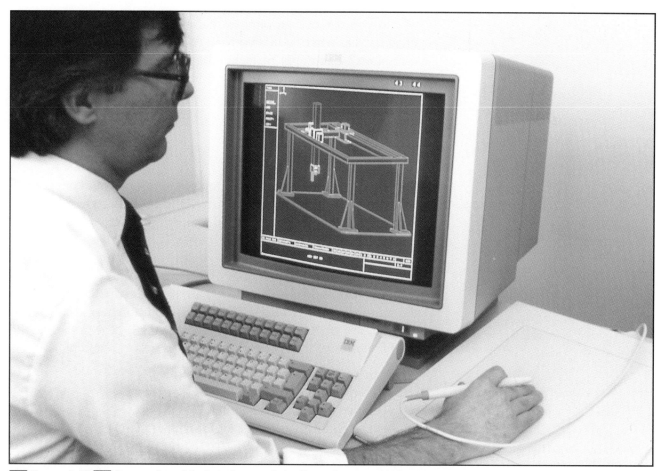

■ **Figure 6-18** ■ Computer aided design and drafting allows a drafter to use the power and speed of the computer to prepare technical drawings.

When the drawing is finished, the computer will add the dimensions, draw the pictorial, rotate the object so that you can see what the other side looks like, and even send the drawing to the printer. The completed plans are now ready for the construction workers or craftspeople to use to make bridges, buildings, or consumer products.

Why can't you bypass the construction crew and have the computer direct the construction itself? Today *computer-aided manufacturing (CAM)* is being used to take the plans directly from a CAD system and make the finished product. This means that some jobs once done by highly skilled workers no longer exist. The telephone is a perfect example of a product that is being created on the computer (CADD) and then produced without much human contact.

Technical Drawing Applications

Why do different groups have their own unique languages?

Do you use certain phrases that are only understood by other teenagers? Do you find that you can communicate faster using these expressions? Each area of technology has also developed its own language to speed up and improve communication. This is reflected in the language of drafting too.

Industrial Drawings

What are working drawings?

Without technical drawing, modern industry would never have developed. The machines of industry, and all the products that are made by these machines, started out on the drawing boards of drafters.

Working drawings are a set of drawings that gives all the information necessary to build an object without any further written or verbal instructions. This information would include complete shape and size description, as well as additional instructions in notes that are written right next to the drawing. These notes and the bill of material include the following kinds of information:

working drawings
scale drawings of an object that contain all of the information needed by the engineers, architects, and builders to construct the object

- What material the object is made of
- How the object will be finished
- How many objects are to be made
- How much **tolerance** (room for error) there is

tolerance
the allowable difference in size (smaller or larger) that a part can have from the design size and still be useable

The complete working drawing package consists of multiview drawings, assembly drawings and pictorial drawings. See Figure 6.19. Did your house, furniture, streets, automobile, television, radio, calculator, bike, and Nintendo game need a set of drawings for their construction? All objects are constructed only after they go through a design and drafting stage.

Electrical Drafting

Drafters need to provide a set of drawings that contain all the information necessary to build an appliance. This information includes complete shape and size description of the case, electrical circuits, electric components, electronic components, and the brackets that hold everything together.

Just as you have your own way of expressing things, drafters working in the electric/electronic field phrase things in their own way. They call their drawings *diagrams* because they show electrical paths and circuits. Drawings that show how the electric and

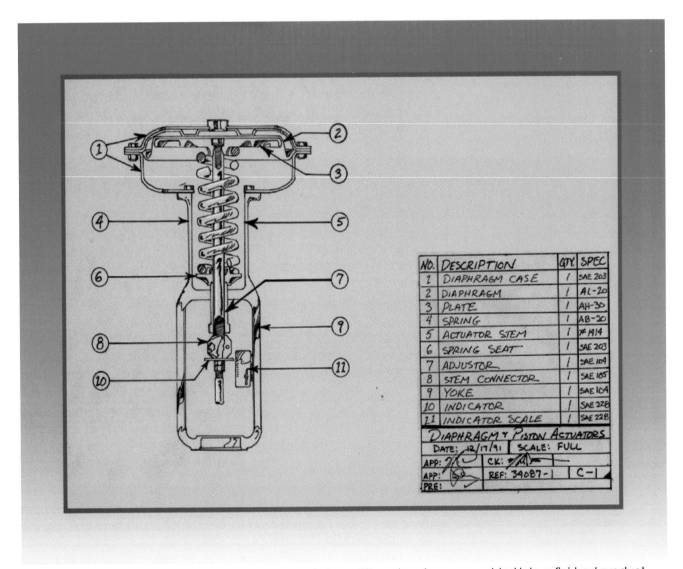

NO.	DESCRIPTION	QTY.	SPEC
1	DIAPHRAGM CASE	1	SAE 203
2	DIAPHRAGM	1	AL-20
3	PLATE	1	AH-30
4	SPRING	1	AB-20
5	ACTUATOR STEM	1	# M14
6	SPRING SEAT	1	SAE 203
7	ADJUSTOR	1	SAE 104
8	STEM CONNECTOR	1	SAE 105
9	YOKE	1	SAE 104
10	INDICATOR	1	SAE 228
11	INDICATOR SCALE	1	SAE 228

DIAPHRAGM Y PISTON ACTUATORS
DATE: 12/17/91   SCALE: FULL
APP: CK:
APP: REF: 34087-1   C-1
PRE:

■ **Figure 6-19** ■ Drawings are used to show people how different parts are assembled into a finished product.

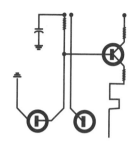

schematic diagrams
drawings prepared for the electric/electronic industry that show the circuits and components of electrical appliances

electronic components are connected together are called **schematic diagrams.** See Figure 6.20.

Symbols have been developed to represent all the electrical parts, such as the battery or plug, switch, light bulb, and socket. Symbols have also been developed to represent all the electronic parts, such as tubes, transistors, and integrated circuits.

Architectural, Structural, and Civil Engineering Drawings

What kind of information is included in a set of working drawings for a construction project?

Architects and engineers use design and construction knowledge that has developed over centuries. The first designers probably outlined plans for their prehistoric structures on the ground so that helpers could work together.

Today a set of working drawings is prepared for the construction of a home, bridge, dam, skyscraper, or any other construction project. The information on the drawings includes complete shape and size description of the foundation of the inside and outside of the structure, of the electrical systems, and of the plumbing. Pictorials are included that show how the structure should look three dimensionally. See Figure 6.21 on page 178. Depending on the type of structure, the working drawings might also include driveways, parking lots, landscaping, ground elevations, and connecting roads.

Construction Codes and Zoning. The drawing and construction processes must also take into consideration various codes and zoning restrictions. Federal, state, and local codes determine legal methods of construction, the kinds of materials that must be used, and even the size of a structure that people can build on their land. Townships usually have zoning laws that divide their area into industrial, light industrial, commercial, and residential zones. Without a zone change, you can't put up stores in a residential area, even though the land belongs to you.

To protect us from poor or life-threatening construction techniques, building codes are checked before construction permits are issued. These codes are enforced by planning boards and building inspectors, who make certain that the architect and builder use accepted planning and building technology.

Designing a Home. Designing a home involves more than just deciding what it will look like. Certain areas are more susceptible to earthquakes, termites, floods, and ground slides. When building in these areas, it is necessary to plan and build a structure that can survive the elements.

Designing a home should be done from the inside out. First you determine the activities that will go on in the house. Then you

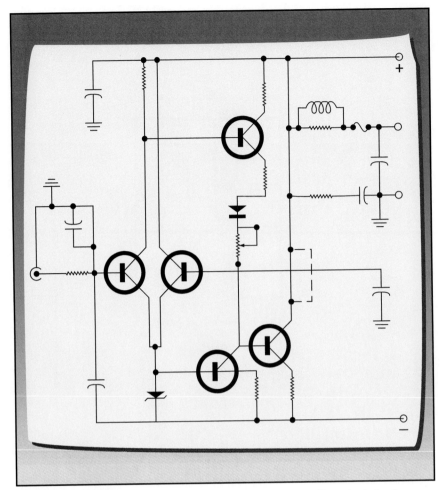

Figure 6-20 A schematic diagram uses lines and symbols to show electrical circuits.

decide on the types and sizes of rooms that will be needed. Must every room be enormous? Remember that a larger house costs more—and larger doesn't necessarily make a house better.

The architect's plan is always to design a house big enough to meet the client's needs and small enough to fit the client's pocketbook. The house must be divided up into activity areas, which include the living, sleeping, and service areas of the house. In very large houses, you will find more rooms in all of these areas. In a small house, you will find fewer rooms and also rooms that are being used for more than one activity.

Drawing Reproduction Systems

What are some of the different groups of people needing copies of working drawings?

The construction plans that are used today are copied by a number of different systems. Why would most companies insist on

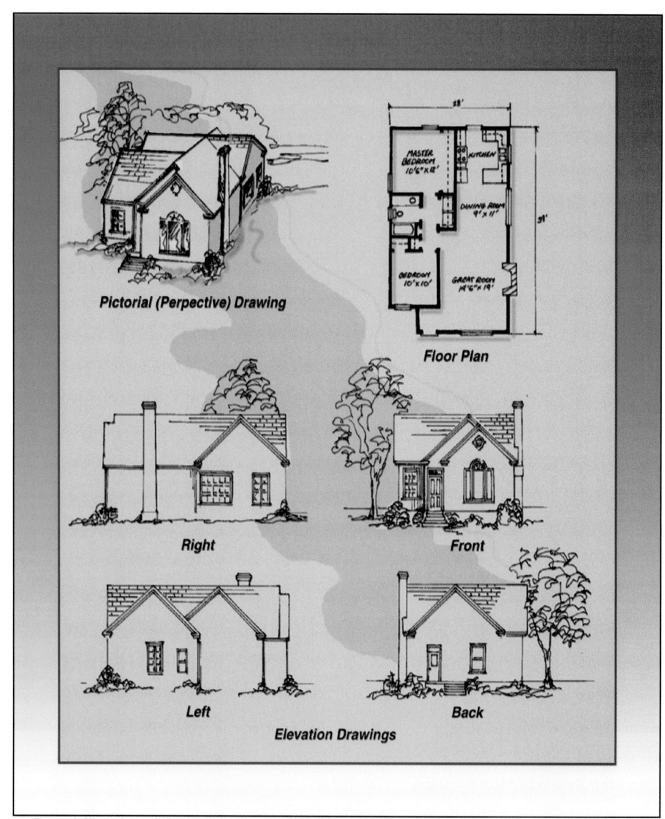

Floor Plan labels:
- MASTER BEDROOM 10'6" X 12'
- KITCHEN
- DINING ROOM 9' X 11'
- BEDROOM 10' X 10'
- GREAT ROOM 14'6" X 19'

Pictorial (Perpective) Drawing

Floor Plan

Right

Front

Left

Back

Elevation Drawings

Figure 6-21 An architectural drawing of a house gives the size and shape description of the structure.

using copies instead of the original drawings? In almost all situations, many copies of the same set of drawings are needed. Some sets are given to people to check for design problems. Other sets are filed with the correct government officials so that building permits can be obtained. And many sets will be around for use at the construction site. When necessary, copies of older drawings can be scanned into a computer system. Once in the computer, these drawings can be altered to meet new specifications. See Figure 6.22. Computer drawing systems can be hooked up to large printers and plotters that can quickly draw copies of the plans from the computer's memory.

Large printers and plotters all "draw" the CAD drawing onto paper using different printing processes. The different types of plotters and printers that are described here are all shown in Figure 6.23 on page 181.

Pen plotters are usually flatbed or microgrip in design. These plotters use different color ink pens to physically draw the lines onto the paper. *Flatbed pen plotters* hold the paper flat while the print head moves across as well as up and down the sheet of paper. *Microgrip pen plotters* roll the paper in and out of the machine while the print head is moving back and forth across the paper. This combination of movement allows the pens to draw angled, curved, and straight lines.

Most large plotters and printers have drums, where the paper moves slowly through the machine during the printing process. The image is created by printing a series of dots so close together that the lines look solid. On an *ink-jet printer*, these dots are created by very small ink guns that squirt the image onto the paper. *Laser* and *electrostatic printers* form their images by fusing a toner to the paper. A full description of how electrostatic and ink-jet printers work can be found in Chapter 7.

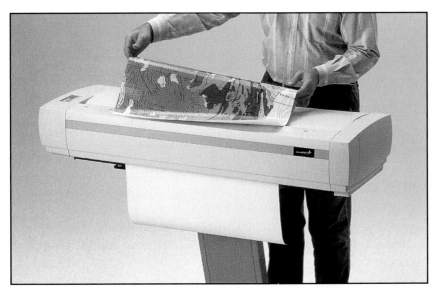

■ **Figure 6-22** ■ Scanners can produce clean images from old drawings. With proper software, these images can then be altered to make a new design.

Paper in Space

A visible enemy is attacking our military, our government, and our technological society. On land, sea, air, and space, this enemy has been left behind to allow our technology the freedom to move.

People in the military call this enemy mechanical drawings, and its existence is a weight that everyone must carry. Every machine part, structure, consumer product, and transportation system had to have drawings made for every part in its systems.

The weight of these drawings, in many cases, exceeds the weight of the system itself. The engineering and maintenance drawings that were drawn up to produce a nuclear submarine take up more space than the submarine. The manuals for one 9,600-ton Ticonderoga cruiser weigh 26 tons.

The more complex our technology becomes, the more drawings our drafters have to draw. The space shuttle is 122 feet long. Its booster rockets are the size of giant railroad cars. It took 20,000 engineers twenty years and over 1 billion hours to design and build the shuttle's systems. For each engineer that worked on the space shuttle, five technicians worked at his or her side, creating many of the needed drawings.

Could a future space system generate enough drawings that its stack would reach into orbit? The federal government has found a solution to its paper problem. Its weapon is called CALS, or computer-aided acquisitions and logistics support. In a nutshell, this system will require that all new records, drawings, and service manuals be issued in a computer format.

By the turn of the century, the Pentagon expects to have most of its paper files converted to a space-saving computer format. The systems that must be developed for the government are expected to be marketed to business and industry.

Does this mean goodbye to the paper shredder? Only time will tell.

Drafting Careers

What are some different career possibilities in drafting?

Have you ever thought about a career in drafting? If so, your level of education will determine your entry-level skills and therefore what entry-level job you will be qualified for. People enter this field from vocational schools, technical institutes, community and junior colleges, and four- and five-year colleges.

As you gain experience, you can be promoted from one position to another. Many companies will pay your college costs if you decide to further your schooling.

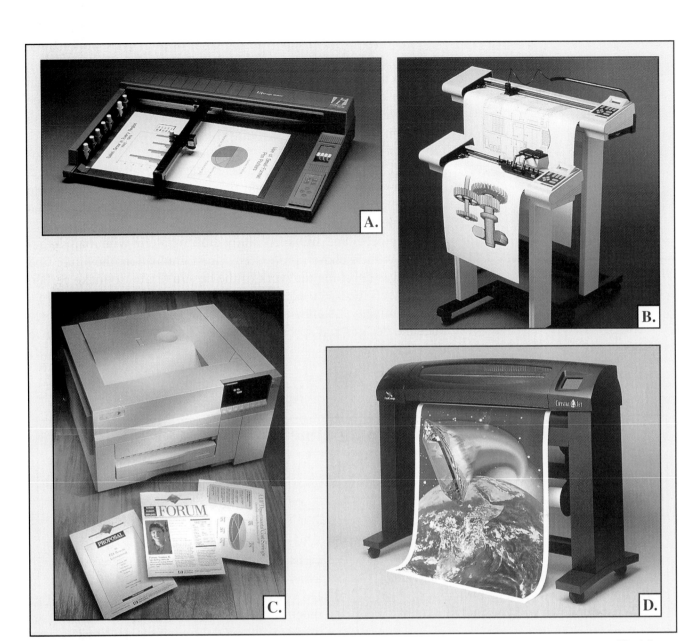

■ **Figure 6-23** ■ These are some common printers and plotters. (A) The flatbed printer holds the paper still while the pens travel across, up, and down during printing. (B) The microgrip pen plotter rolls the paper in and out of the machine to draw angled pen lines. (C) The highest-quality printing is created by laser or electrostatic printers. (D) The ink-jet printer squirts ink to form the picture.

Tracers are the newest drafters in the company. Tracers copy drawings onto special papers and film. Their copies are then used to produce duplicates of the drawings. This is usually your first job straight out of a vocational school or technical institute.

Junior detailer is a promotion up from being a tracer. Junior detailers produce drawings from sketches, existing drawings, and written or verbal instruction. They are given simple drawings to work on, which gives them the opportunity to improve their skills while producing drawings that are needed by the company.

Detailer-drafters report to the chief drafter. They produce the same kind of drawings as the junior detailer, but from much more complex sketches. People who have graduated from a vocational program or technical institute will be promoted to this level once they have proved that they can do very accurate drawings with a minimum of supervision.

Design drafters are usually graduates of two-year college programs. They usually also have five years of drafting experience. Because of their experience and field knowledge, they are capable of designing new parts and products and training and checking the work of less skilled drafters.

Engineers and architects study drafting along with many other courses in their chosen profession. Upon graduation, they may find themselves checking the work of the design drafter to make certain that it meets the specifications of a senior engineer or architect. In a small company, or after a few years of experience, they will produce the designs and specifications that drafters will use to produce their drawings.

— Learning Time —

Recalling the Facts

1. Describe the difference between computer-aided design, computer-aided drafting, computer-aided design and drafting, and computer-aided manufacturing.
2. What drawings will you find in a set of working drawings?
3. Name and describe three methods that are used to reproduce a drawing.

Thinking for Yourself

1. What would probably happen if all construction and zoning codes were done away with?
2. Do you feel it is easier to produce a mechanical drawing on a computer or using mechanical drawing tools? Why?

Applying What You've Learned

1. All kinds of graphic illustrations are used in advertising. Technical drawings are usually included in instruction booklets for products that must be assembled. Look through magazines, newspapers, and instructional booklets for examples of technical drawings. Prepare a poster that shows four labeled examples.
2. Using a computer architectural drawing program or conventional drawing tools, design a floor plan for a single-family home.

(Continued on next page)

— Learning Time —

3. Many popular movies have scenes that show a person doing computer-aided design, computer-aided drafting, or computer-aided manufacturing. Bring in such a videotape advanced to the specific scene for class viewing.

Summary

Directions

The following sentences contain blanks. For each numbered blank, pick the answer that makes the most sense in the entire passage. Write your answer on a separate sheet of paper.

Drawings are often used instead of words to describe the shape and size of objects. These ___1___ are often used by scientists, architects, engineers, designers, and art illustrators to show what objects look like.

When a freehand sketch is made to describe the ___2___ and size of the object, the drafter works without mechanical tools. It is very ___3___ to draw an accurate picture without the use of rulers, compasses, and other ___4___ drawing tools.

In technical drawing, the drafter draws many kinds of drawings. In orthographic projection, the drafter draws the front, top, and right side views of the ___5___ . Extra ___6___ are drawn if they are needed. When the internal structure must be seen, it is drawn as if part of the object were cut away.

It is often difficult to picture the object from these multiview drawings. Therefore, the drafter often draws a pictorial drawing of the object.

In this drawing the ___7___ shows a number of sides of the object at once. These ___8___ represent what the object will look like after it is ___9___ .

1.
   a. tools
   b. drawings
   c. objects
   d. photographs

2.
   a. material
   b. shape
   c. pictures
   d. type

3.
   a. difficult
   b. pleasurable
   c. strange
   d. significant

4.
   a. computer
   b. large
   c. mechanical
   d. artistic

5.
   a. picture
   b. object
   c. section
   d. orthographic

6.
   a. pictorials
   b. views
   c. schematics
   d. floor plans

7.
   a. engineer
   b. image assembler
   c. illustrator
   d. drafter

8.
   a. pictorial drawings
   b. schematic drawings
   c. photographs
   d. purchases

9.
   a. made
   b. drawn
   c. designed
   d. sold

(Continued on next page)

Summary

Drafters represent the object by drawing different kinds of lines. Each ___10___ thickness and length is determined by the rules of drafting. These rules also govern how ___11___ are placed onto the drawing. Dimensions give the size and location of the object's features.

Illustrators create artwork that is so perfect that it looks like it was originally ___12___ rather than having been drawn by hand. Many company logos (company emblems) that have been produced by the ___13___ appear on the products that you ___14___ to wear or own.

Today many ___15___ are being produced by people who work directly on a computer. In computer graphics the ___16___ draws directly on the screen using special computer ___17___ .

In computer-aided design (CAD), the designer creates and tests his or her designs directly on the computer. The drafter working on a computer-aided drafting system uses a mouse and a drawing tablet instead of a drawing board and T-square. The drafter uses a ___18___ to remove parts of his or her drawing that need to be changed.

Today, computer-aided design is being combined with ___19___ . The final drawings can be manufactured by machines that can work directly from the computer's plan without the need of ___20___ working the machines.

Drawings that are made by hand or machine are often duplicated so that ___21___ will exist for the people that need them. There are many different ___22___ systems that can be used to produce these copies.

Drafters produce different types of drawings for people designing and building electrical circuits, machine parts, and construction projects. Almost every object that is made must first be drawn, so drafters and illustrators will always be needed.

10.
   a. part's
   b. object's
   c. line's
   d. item's

11.
   a. parts
   b. circles
   c. squares
   d. dimensions

12.
   a. copied
   b. photographed
   c. traced
   d. computerized

13.
   a. drafter
   b. engineer
   c. computer program
   d. illustrator

14.
   a. hate
   b. have
   c. like
   d. plan

15.
   a. photographs
   b. games
   c. reports
   d. drawings

16.
   a. photographer
   b. planner
   c. drafter
   d. engineer

17.
   a. rulers
   b. compasses
   c. T-squares
   d. tools

18.
   a. pencil eraser
   b. cleaning solution
   c. mouse
   d. new program

19.
   a. computer games
   b. computer graphics
   c. computer drafting
   d. illustrating

20.
   a. robots
   b. computers
   c. artists
   d. people

21.
   a. copies
   b. photographs
   c. books
   d. tools

22.
   a. typing
   b. drawing
   c. writing
   d. duplicating

Key Word Definitions

Directions

*The column on the left contains the key words from this chapter. The column on the right contains a scrambled list of phrases that describe what these words mean. Match the correct meaning with each word. Write your answer on a separate sheet of paper. (**Note: Three of the key terms have the same meaning. Indicate the same letter for these three terms.**)*

Key Word	Description
1. Scale	a) Technical drawing technique that gives size, shape, and structure of objects
2. Computer-aided drafting	b) Shows the inside of an object
3. Drafter	c) Computer used instead of drafting tools
4. Tolerance	d) Drawings showing all six sides of an object
5. Pictorial drawing	e) All drawings needed to build an object
6. Isometric drawing	f) A pictorial with one view that isn't distorted
7. Oblique drawing	g) Room for error
8. Perspective drawing	h) A pictorial that shows three equally distorted sides of the object
9. Orthographic projection	i) Designing objects using a computer
10. Multiview drawing	j) A drawing that shows a three-dimensional view of the object
11. Drafting	k) The person who produces draftings
12. Sectional Drawing	l) Drawings of electric and electronic circuits
13. Computer-aided design	m) Three views on one sheet of paper
14. Technical drawing	n) Drawn or made less than full-size
15. Working drawing	o) A pictorial that looks most life-like
16. Mechanical drawing	
17. Schematic diagram	

The Back-Massaging Vehicle

Did You Know?

The sign on the board reads, "Plan ahead." The comic drawing of this sign shows the last few letters of the sign squeezed in. When preparing a drawing that will be used to construct the parts of a machine, there is no room for poor planning. Drawings with insufficient or incorrect information will lead to failure. To build the machine, you need drawings that list the correct size and shape of every part.

Drawing or reading these plans isn't always easy. People must learn to see three-dimensional objects when they are represented as flat lines on paper.

Problems To Solve

When you are learning how to prepare a mechanical drawing, you must learn how to break down an object into a front, top, and right side view. You must also learn how to determine the appearance of an object with only the information supplied by its front, top, and right side view.

The van shown in the multi-view drawing in Figures A–C has hexagonal wheels. This and other shaped vehicles are sold in novelty stores as back massagers. When you roll them on a person's back they give a massage.

Your problem is to design your own back-massaging car or truck. Your vehicle can have small or large wheels. The only restriction on your design is that the wheels can't be round.

Resources You Will Need

- Drawing board
- Drawing tools
- Paper
- Pencils
- Computer
- Computer printer
- Graphics software
- Toy Factory program
- Modeling clay
- 2 x 4 lumber
- Wood
- Dowels
- ¾-inch plywood
- Thumbtacks
- Sandpaper
- Drill set
- Drill press
- Wood glue
- Scroll saw
- Woodworking tools
- Wood vises

A.

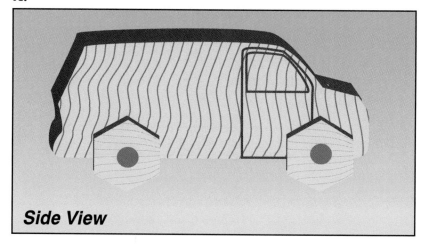

Side View

Prepare a drawing that will give you the size and shape description you will need to build your vehicle. The design and drawing phase of this project can done as a sketch, as a mechanical drawing, or by computer. The computer program *Toy Factory* or a graphics program can turn this phase of this activity into a computer-aided design and drafting (CADD) project. Construct the vehicle that you have designed.

A Procedure To Follow

1. Prepare sketches of different car or truck shapes, or design model shapes using modeling clay.
2. Prepare a full-size front, top, and right side view of your final design. These views should give you a complete description of your vehicle's shape.
3. Add the needed dimensions to give size description.
4. Prepare a bill of materials that lists all the parts that you will need to construct your vehicles.
5. Prepare a pictorial drawing of your vehicle.
6. Fabricate the body of your car out of 2 x 4 lumber.
7. Make your wheels out of plywood.
8. Cut the axles for your wheels from the dowels.
9. Locate and drill the body location for the axles.
10. Locate and drill the wheel axle joints.
11. Sand all parts.
12. Slip the axles through the body, and slip the wheels onto the end of the axles.
13. The holes in the body must be large enough for the axles to freely turn.
14. Glue the axle ends into the wheels.
15. Be certain that no glue gets into the body of your vehicle, or your wheels won't be able to turn.

What Did You Learn?

1. Could you build your vehicle using only the size and shape description provided by your drawing?
2. How could you have improved your multiview drawing?
3. If another person was to build your vehicle from your plans, would they end up with exactly the same finished product? Why?

B.

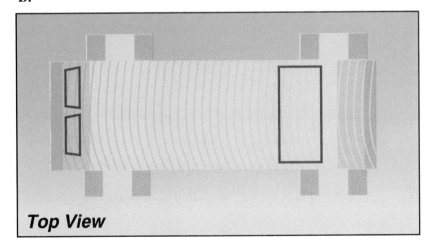

Top View

C.

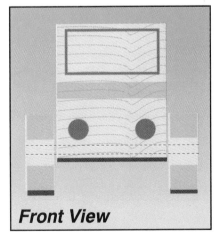

Front View

187

Chapter 7

Printing

If everything printed was to suddenly vanish, how would your world change? Without printing there would be no books, magazines, or newspapers to read. A supermarket would contain thousands of items, all packaged in boxes and cans that had no labels. All sales slips, stock certificates, and paper money would disappear. The designs on our clothing and the signs along our roads would suddenly be gone.

The electronic printed circuit boards in your telephone, television, radio, and car would no longer exist. Since these boards make these devices work, all of these modern appliances would stop working. Life as we know it would cease to exist, and civilization would be thrown back into the dark ages.

Graphic reproduction includes all the operations and procedures that must be performed to change original artwork, photographs, and text into printed material. People often refer to graphic reproduction as **printing** which is the eighth largest industry in the United States. All the tasks performed by a number of related industries that produce the materials used in graphic communication make up the printing industry. In this chapter you will begin to learn about this essential part of communication technology.

— Chapter — Outline

Look over the headings below. Think about what you will learn as you read the chapter.

I. Typesetting
II. Reproducing Illustrations
III. Printing Processes
IV. Desktop Publishing
V. Careers in the Printing Industry

— Key Words —

In this chapter you will learn and study the meanings of the following important technology terms.

GRAPHIC REPRODUCTION
PRINTING
HALFTONE
FOUR-COLOR PROCESS PRINTING
LETTERPRESS PRINTING
FLEXOGRAPHY
GRAVURE PRINTING
LITHOGRAPHY
SERIGRAPHY
XEROGRAPHY
ELECTROSTATIC PRINTING
INK-JET PRINTING
DESKTOP PUBLISHING

Typesetting

What is typesetting?

The original item to be printed is called the *copy*. From this copy the type will be set exactly the way it is going to look when printed. *Text* (words) must be set in type before the printing plates can be made and the printing presses can run.

Hot-Type Composition

What is hot-type composition?

Typesetting is divided into *hot-type composition* and *cold-type composition*. Hot-type composition is any typesetting that uses letters cast out of metal. See Figure 7.1. Very little commercial typesetting is now done with hot-type.

When you set type in your technology laboratory, you are taking previously cast *foundry type* out of a *California job case* and setting the metal letters into a *composing stick*. See Figures 7.2 and

Figure 7-1 This typecasting machine casts the type in hot metal as it is needed.

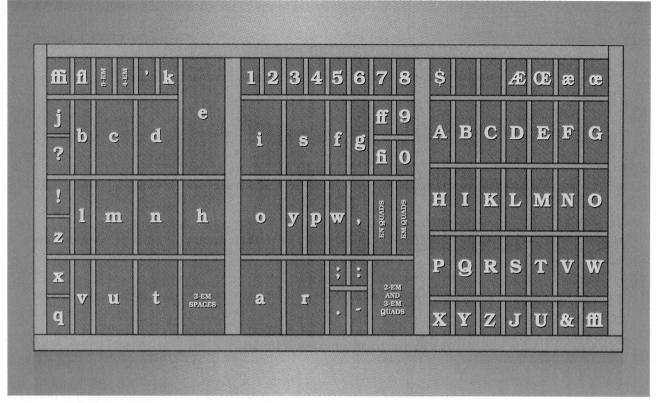

Figure 7-2 The California job case is used to store foundry type when it is not in use.

7.3. Even though these letters aren't hot at the time you use them, they are considered part of hot-type composition because they were once cast in hot metal.

Cold-Type Composition

What is cold-type composition?

Cold-type composition includes all methods of setting type other than hot-metal composition. Computer typesetting, typing on a typewriter, and photographic typesetting are all examples of cold-type composition.

When the operator of a phototypesetting machine presses down on a letter, a negative of that letter is moved into position and is then photographically printed to a sheet of photographic film or paper. See Figure 7.4 on the next page. These individual letter pictures are lined up so that a photographic image of the entire text is produced.

Computers store the images of the letters as part of the typesetting program. The computer sends printing directions to the printer.

In a laser printer a laser beam exposes the film or photographic paper that is mounted on the machine's rotating drum. The laser light forms individual letters as it moves. The computer turns the beam on and off so that only the places that contain the image will be exposed to the laser's light.

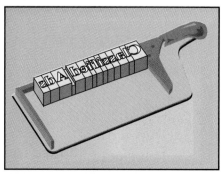

Figure 7-3 Foundry type is set by hand into a composing stick.

191

Figure 7-4 On a phototypesetting machine the images of the letters are created photographically.

— Learning Time —

Recalling the Facts

1. What effect has the development of the computer had on the growth of printing? Why?
2. What is the purpose of the laser beam in a laser printer?
3. What is the difference between cold-type and hot-type composition?

Thinking for Yourself

1. If all printed material and processes disappeared, how would our world be changed?
2. Why is hot-type composition being re-placed by cold-type composition?
3. If printing disappeared, your telephone, tele-vision, radio and car wouldn't work. Why?

Applying What You've Learned

1. Set your name, design or slogan using cold-type composition.
2. Set your name using foundry type. Print this line of type at the proof press. Lock up this line of type so that it can be made into a flexographic printing plate (rubber stamp).

Reproducing Illustrations

What kind of printing plates are made to reproduce photographs and illustrations?

Printing usually involves printing photographs and illustrations as well as text. Regardless of which printing process is used, a printing plate must be prepared from the original photograph or illustration.

When the original picture is made up entirely of solid lines and shapes, a *line printing plate* is made. A line picture has no shading. The entire picture is made up of solid lines and areas. See Figure 7.5.

Figure 7-5 A line photograph has no shading. The entire picture is made up of solids or lines that are all of equal density.

halftone
a picture produced by using a series of dots. The picture is rephotographed through a special screen. A continuous tone picture is produced when the halftone is printed.

If the original is a photograph, then the *continuous tone* of the picture must be reproduced. When you look at a black-and-white photograph, for example, you see black areas, white areas, and shaded gray areas. See Figure 7.6. The printer reproduces these tones by printing the picture as a series of dots called a **halftone**. A halftone is a picture made up completely of dots. The spacing and size of the dots will determine if you see black, white, or gray.

To turn the photograph into a halftone, a printer makes a *halftone negative*. To produce a halftone negative, the printer rephotographs the photograph through a *halftone contact screen*. See Figure 7.7. Photographs also can be turned into halftones through the use of computer scanners.

When you look at a picture printed in a newspaper, does it remind you of the way the world looks when you look out through a window screen? Newspapers use fewer dots per inch than magazines, which makes it possible for you to see the dot structure without a magnifying glass. Magazines, books, and advertisements use screens that are much finer.

Color adds life to anything that is printed. How do you suppose printers are able to print all the colors we see in the world

■ **Figure 7-6** ■ A halftone picture has continuous tone. This dog has dark areas, shaded areas, and very light areas that must be reproduced using one color ink.

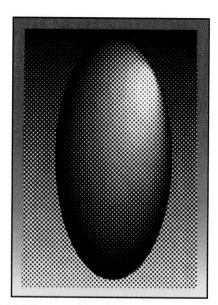

■ **Figure 7-7** ■ This egg was photographed through a halftone contact screen. The screen breaks the picture up into dots. At a far distance the dots that are all the same color will appear as black areas and continuous tones of gray.

194

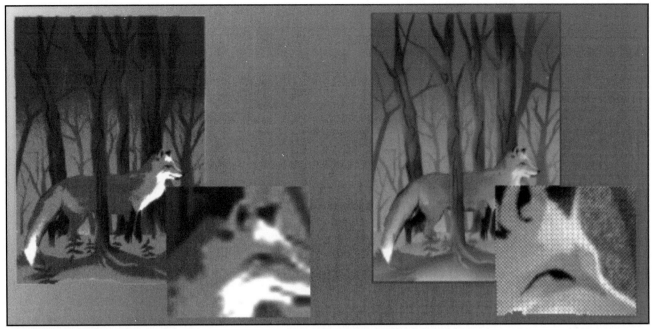

■ **Figure 7-8** ■ All the colors of the rainbow can be reproduced using magenta, cyan, yellow, and black ink.

around us? If color printing needed hundreds of ink colors and printing plates, not much would be printed in color. See Figure 7.8.

All colors in nature can be reproduced through the printing of three ink colors. These colors are magenta (similar to red), cyan (a blue), and yellow. A fourth color, black, is added in the printing process, creating what is called **four-color process printing**. Black is added because it deepens tones and is already being used to print the words on the page.

— Learning Time —

Recalling the Facts

1. How will a printing plate that contains a halftone image print differently than one that contains only line images?
2. What is the purpose of a halftone contact screen when making a halftone negative?
3. Name the four ink colors used by a printer during four-color process printing.
4. How is a halftone screen similar to a window screen?

Thinking for Yourself

1. When you look at a picture that was printed using four-color process printing, why don't you see the dots? Why can you see all the colors of the rainbow?
2. When you paint using a computer graphics program, are you using a lot of different colors or some form of three-color processing? Explain.

(Continued on next page)

— Learning Time —

Applying What You've Learned

1. In this activity you are going to see what happens when you mix different colors together. Combine color plastic sheets so that the light of an overhead projector or flashlight passes through them. On a sheet of paper, list the color plastics that you combine and what color light the combination creates. First use plastic sheets that are green, orange, and violet. After determining the colors created by combining these colors, do the same thing with plastic sheets that are the same colors as three of the inks used in color printing—yellow, cyan, and magenta. Can you find a relationship between these two sets of colors?
2. Construct a lucite wind chime that uses green, orange, violet, magenta, cyan, and yellow plastic pieces. When the wind causes your wind chime to move, the sunlight should cast a rainbow of color onto a nearby wall.

── Printing Processes ──

≡ *What are the three most popular methods of printing?*

A soda can and the pages of a book could not be printed on the same machine because they have different shapes and are made of different materials. However the basic printing process may be the same. All printing can be grouped into six basic printing processes:

1. Letterpress printing, also called relief printing
2. Gravure printing, also called intaglio printing
3. Lithography, also called planographic printing
4. Serigraphy, also called screen process printing
5. Xerography, also called electrostatic printing
6. Ink-jet printing

At this time most commercial printing is done by letterpress, gravure, and lithography. Computer developments and advances in ink-jet and electrostatic printing equipment have resulted in an increased use of these two systems. It is too early to tell if electrostatic and ink-jet printing will replace the other processes. The small, quick-print neighborhood shop is using these new printing processes. Most companies that use desktop publishing are also turning to the electrostatic and ink-jet processes as their processes of choice.

Printing process technology is developing very quickly. One of the new systems, which is called Scitex Digital on demand, allows the printer to take the entire halftone color printing job straight from the computer to a Xerox DocuColor printer.

Tech Talk

The word *relief* can mean to ease or remove stress, pain, or some other discomfort. A certain manufacturer of stomach antacid advertises that their product spells relief.

The word *relief* also refers to a surface that stands out from the rest of an object. Here the word is used to describe printing letters, coin surfaces, and maps that show the contour of the land.

Letterpress Printing, or Relief Printing

What part of a letterpress printing plate does the printing?

In **letterpress printing** the printing surface is raised above the rest of the plate. See Figure 7.9. Another name for letterpress printing is *relief printing* because a relief surface is raised above its surroundings just like the lines of your fingerprint or the letters on a rubber stamp.

Almost 40 percent of all the printing in the United States is done by letterpress printing. This process is used to print on a variety of materials using metal, rubber, and plastic plates.

When the printing plates are made of rubber, the process is called **flexography**. A rubber stamp is a flexographic printing plate. See Figure 7.10.

Gravure Printing, or Intaglio Printing

What part of a gravure printing plate does the printing?

In **gravure printing**, the printing plate has depressions etched into its surface. The hand-engraving or etching process creates thousands of pits in the copper or steel plates. The pits vary in depth and thickness and serve as miniature inkwells when the plate is printed on the printing press. The deeper the inkwell, the more ink it will hold and the darker it will print.

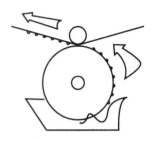

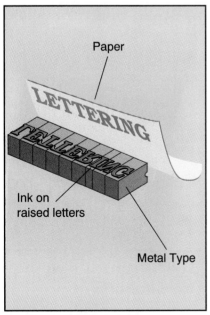

Figure 7-9 In letterpress printing the printing surface is raised above the rest of the printing plate.

Figure 7-10 This rubber stamp is an example of a flexographic printing plate.

Once placed on the printing press, the plate is dipped into a container of ink called the *ink fountain*. See Figure 7.11. A *doctor blade* scrapes the excess ink off the nonprinting surface of the plate. Printing is accomplished by pressing the paper under a great deal of pressure into the printing plate. Extremely fine lines can be accurately reproduced by the gravure process. This makes gravure printing the method of choice for printing U.S. currency, magazine sections of some newspapers, and many magazines.

Lithography

What is the difference between stone and offset lithography?

In **lithography**, sometimes known as planographic printing, the printing surface of the printing plate is perfectly flat. Printing is based on the principle that water and oil don't mix.

In *stone lithography*, the entire process is done by hand by an artist working on a limestone slab. The prints that are made are numbered and signed. If you see a lithograph that is numbered 25/100, it means that this print is the twenty-fifth print made out of a series of 100 prints.

This type of lithograph can be very valuable if the artist is famous. The value of the lithograph depends on the number of prints in the series and the specific number assigned to the print. After

lithography

a printing process that is based on the principle that oil and water don't mix. The image is created from oil-based ink. The printing surface is covered with water to prevent ink from sticking to it.

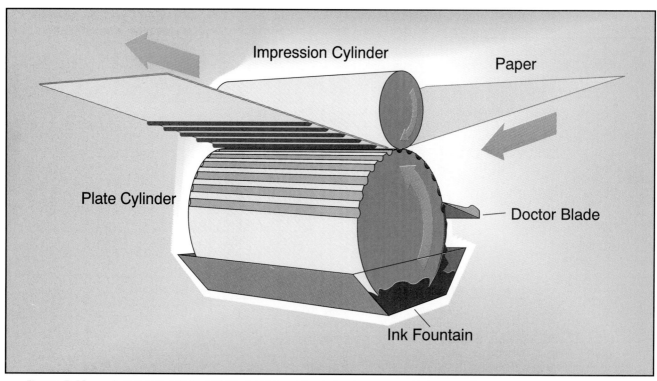

■ **Figure 7-11** ■ A gravure printing plate has ink wells etched or carved into its surface.

completing the prints in the series, the artist cleans the image off the stone so that no more prints can be made.

In *offset lithography,* the image is usually placed on a thin metal plate through a photographic process. The plate is mounted on an offset press, which prints the plate's image to a rubber blanket. Then the blanket transfers the image to the printing stock. See Figure 7.12.

Posters are often printed by offset lithography at over 9,000 copies per hour. Are all lithographs valuable? All items printed by lithography can be called lithographs, but not all lithographs are valuable.

Offset printing is used to print business forms, magazines, newspapers, and books. As the quality of offset lithography improved, more and more printing companies switched to this process. Today, approximately 45 percent of all commercial printing in the United States is done by this process. This textbook was printed on an offset lithographic press.

Spirit duplicators, also called *rexograph machines*, are office copying machines that print using a planographic printing plate.

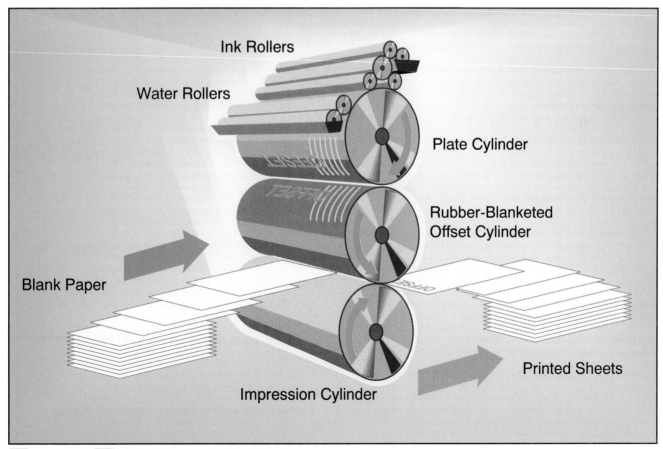

Figure 7-12 In offset lithography the flat printing plate prints on to a rubber blanket. The rubber blanket then transfers the ink to paper.

The image is typed directly onto a special paper plate. As you type, the letters are transferred onto the back of the plate by the impact pressure of the typing.

The prepared plate, called a *spirit master*, is placed onto the rexograph machine for printing. This machine coats a thin layer of duplicator fluid across the spirit master plate. The paper is then pressed against the plate, causing the image to transfer to the paper. This printing process will give you approximately seventy-five copies from one master sheet because they only print until the duplicator fluid has dissolved all of the special ink. This method of duplicating is being replaced by electrostatic printing.

■■■ Putting Knowledge To Work ■■■

The Inventor of Lithography

Alois Senefelder, a German inventor, developed lithography in 1796. He was trying to find an inexpensive way of printing sheet music.

When he was a teenager, he wasn't sure what he wanted to be. He tried to be an actor and then a writer. As a writer he got involved in printing his own books.

Senefelder went into business printing and selling sheet music. The printing plates that were needed to print the music sheets were expensive.

By accident he found that oil spilled on a limestone prevented water from being absorbed by the stone. However, when water covered the stone, oil crayons couldn't mark the stone. Senefelder was certain that this phenomenon (that oil and water won't mix) could be turned into a new printing process.

He experimented with thick slabs of limestone, grease inks, and water to perfect a new printing process that he called lithography. He developed the name for his new process by combining two Greek words: *lithos*, which means "stone," and *graphos*, which means "writing."

Senefelder developed a wax, animal fat, and lampblack grease crayon to draw the musical notes on the stone. You can compare his writing tool to the crayons that you now use in art. Remember, it wasn't the color of the crayon that determined the color of the printing, but rather the color of the ink that was used.

After drawing on the stone with his special greasy crayon, he coated the entire stone with water. The water was absorbed only by the parts of the stone that didn't have the drawing. He then coated the plate with ink. The ink coated the grease crayon areas while being repelled by the rest of the water-coated surface of the stone.

Using a press of his own design, he pressed paper to the stone to transfer the ink form the stone to the paper. To make the next print, he once again coated the stone with water and then ink.

Senefelder found that he could make as many copies as he wanted of his original drawing without any loss of quality. Once the printing was done, the stone image could be removed and a new image could be drawn on the stone.

Serigraphy, or Screen Process Printing

How is a serigraphy printing plate used?

In **serigraphy**, or screen process printing, the printing plate is an open screen of silk, nylon, or metal mesh. See Figure 7.13 on the next page. The openings in the screen are sealed in the nonprinting areas. The nonimage area can be sealed with a paper stencil, a special film stencil that is cut by hand, or a photographically prepared film stencil. Ink is then forced through the open areas of the screen by the movement of a tool called a *rubber squeegee*. Screen process printing can be used to print on paper, glass, metal, wood, and other materials. Objects of every size and shape have been printed by this process, including shirts, fabrics, toys, banners, and bottles. See Figure 7.14 on the next page.

Mimeograph machines are office copiers that print by screen process printing. The image is typed directly onto a mimeograph stencil. As you type, the letters cut openings in the stencil because of the pressure of your typing.

The Gestofax machine is an electronic scanner that makes mimeograph stencils electronically. The original and the special stencil are both mounted on a drum. As the drum turns, an electric eye scans the original. It directs a special needle to burn many tiny holes in the stencil in the exact shape as the letters. This method of duplicating is being replaced by electrostatic printing.

serigraphy
(sə-rĭg′rə-fē)
a printing process that uses a printing plate made of an open screen of silk, nylon, or metal mesh. The process is also known as *screen printing* or *silk screening*.

201

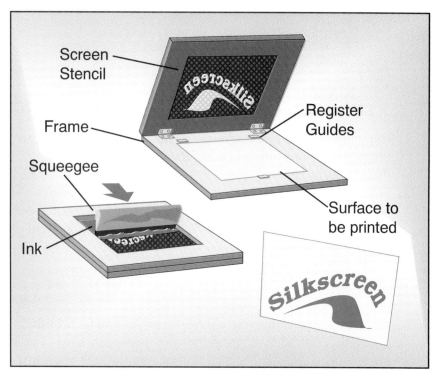

Figure 7-13 In serigraphy the ink is forced through openings in a screen.

Labels in figure: Screen Stencil, Frame, Squeegee, Ink, Register Guides, Surface to be printed, Silkscreen

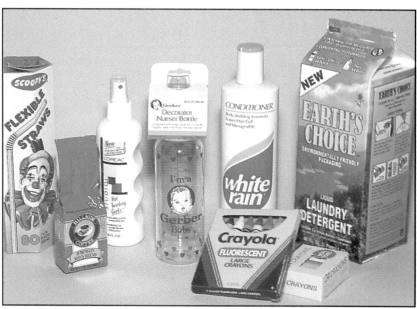

Figure 7-14 Screen process printing can be used to print on objects of different shapes, sizes, and materials.

Xerography, or Electrostatic Printing

How does electrostatic printing work?

Xerography, or **electrostatic printing** uses a flat printing plate, static electric charges, and a powder that is called a toner. The printing plate conducts electricity when exposed to light, but acts as an insulator when it is in the dark.

An image is projected onto this statically charged plate. The lit area of the plate conducts electricity and loses its static charge. The image area remains dark and therefore still charged. The image is made visible when the negatively charged toner sticks to the plate. The toner is transferred to paper when the paper is placed against the plate and then positively charged from the back. The powder is made permanent when the paper travels through a heater, causing the powder to melt onto the paper.

The best-known electrostatic printers are made by the Xerox Corp. See Figure 7.15. The equipment has been used for years as office duplicators of business records. New advances in the equipment have increased their speed to that of the small offset press.

Color electrostatic printing has also been introduced. A color copier operates exactly like a one-color copier except that it scans the original copy three times. During each scanning, a different filter covers the lens. After each exposure, the paper must go through the entire printing toner process. You are really making three prints on the same paper using three different color toners. Even though only three color toners have been used, all the colors of the rainbow can be seen, just like in four-color process printing.

Electrostatic printing is now of such high quality that government officials worry that U.S. currency can be easily counterfeited by these new color machines. Our $100 and $50 bills have already been changed to prevent this kind of counterfeiting. The special color-changing ink and watermark can't be easily counterfeited. Each year a new bill will be released until all our currency up to the dollar bill has been replaced.

Electrostatic printing lends itself to quick printing of newsletters, advertisements, and business forms. It is becoming the printing process that is preferred by the small printing shop.

Figure 7-15 Electrostatic printers have become very high tech printing presses.

Thermography: How to Print Without a Printing Plate

Wedding invitations are often printed with imitation gold letters. The process that is usually used to create this raised gold effect is called *thermography*.

The word *thermography* actually means "to print with heat." A special powder that sticks to wet areas is used to create the raised printing on greeting cards, business cards, and invitations.

This powder coats the card while it is still wet with ink. After the extra powder is removed, the print is heated, which causes the powder to fuse to the paper. Since thermography and xerography have a lot in common, you can use one to study the other. If you understand how thermography works, you understand the principles of xerography.

Ink-Jet Printing

How is ink applied to a surface in ink-jet printing?

In **ink-jet printing** there are no printing plates. The printer actually has very tiny ink spray guns that shoot their ink to the printing surface. Since the guns never touch the material being printed, this method can be used to print on very uneven or fragile surfaces. See Figure 7.16. To control printing accuracy, a static-electric charge is also used to help direct the movement of the ink.

With this process, a number of colors can be printed at the same time. In multicolor ink-jet printing, a separate ink gun is used to apply each color ink. Since the guns don't touch the paper, there is little chance of smearing. Because there is no direct contact between a plate and the *printing stock* (paper or other material to be printed), this type of printing is called nonimpact printing.

In the area of electronic publishing and printing, nonimpact printers account for 51 percent of the total printers sold in 1989. It is predicted that nonimpact printers will have 58 percent of the computer printer market by 1994. Ink-jet printers and electrostatic printers are expected to take over the short-run color reproduction market during the 1990s.

Did you ever wonder if advertising letters that seem to be personally addressed to you were typed by a person? Most of this mail is produced by computer-controlled printers that change the names and addresses as they print the same form letter over and over again. Other letters are printed by offset lithography, leaving space for your name to be added in the letter. Ink-jet printing is then used to customize your letter.

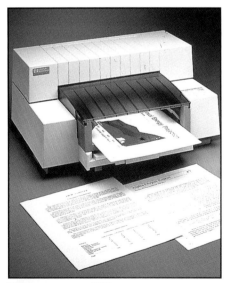

◼ **Figure 7-16** ◼ Ink-jet printers use tiny ink spray guns that shoot ink to the printing surface.

— Learning Time —

Recalling the Facts

1. Describe how a serigraphic printing plate prints.
2. What is a rubber squeegee used for?
3. Describe how printers make junk mail look like personal letters.

Thinking for Yourself

1. In what ways are thermography and xerography the same, and in what ways are they different?
2. Why can an ink-jet printer apply different color inks at the same time?

Applying What You've Learned

1. Develop an experiment that shows the basic principle of xerography. You can use static electricity with electrostatic toner or substitute thermographic powder. Prepare a large chart that shows how this process works. Videotape your demonstration for later presentation to the class.

   This presentation should be a sales pitch — as if you just developed this new printing process yourself. Carlson, the inventor of the xerographic process, did such a live presentation to Kodak to get their financial backing. They turned him down. Try to create the magic needed to sell your idea. Perhaps with your assistance, Carlson could have sold Kodak on developing xerography.

2. As a class project, set up a small business that will print T-shirts for your school's clubs and teams. Prepare sample shirts and sell them to the different school organizations. Plan an activity that will be funded from the profits of this business.

—— Desktop Publishing ——

What is desktop publishing?

Have you ever worked on a school newspaper or magazine that was duplicated and then passed out to your classmates? This kind of activity is now called **desktop publishing**—and it was actually being used in schools before making its way to the business world.

Computers and small, high-quality printers have made it possible for many companies to create their own "in-house" printing shops. Employees create sales reports, booklets, stationery, telephone pads, and interoffice memos without ever leaving their desks.

In business a desktop publishing system might include an IBM or Apple Macintosh computer, a hard disk drive, a mouse, desktop

desktop publishing
the process that allows people to create their own reports, booklets, and other items in their own school, office or home. Computers are a great help to the process.

publishing software, a laser printer, and an offset press or electro-static copying machine. In schools, a desktop publishing system can consist of a typewriter and a rexograph machine.

When you use desktop publishing software, you can set up your page exactly the way it will look when it is printed. See Figure 7.17. Your headings, text, and illustrations are all combined right on the computer screen. This is so much easier than using a regular word processing program or a typewriter, where you have to cut and paste your articles and pictures by hand.

Careers in the Printing Industry

What are some examples of printing careers you might enjoy?

Some of the careers in printing require only a high school diploma. Others require vocational or technical education at a vocational school. Some of these occupations require a technical education at a two- or four-year college. Still others are professional careers that will require a minimum of a bachelor's degree from a four-year college.

Some of these careers require a person who likes to be creative. If you are creative, you might consider a career as a designer, illustrator, artist or photographer.

All occupations require decision making. However, in certain careers, most of your time is spent evaluating the work of others and giving directions to the other members of your staff. Advertising executives, publishers, editors, and art directors all need good decision-making skills.

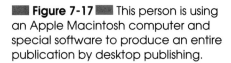 **Figure 7-17** This person is using an Apple Macintosh computer and special software to produce an entire publication by desktop publishing.

In other occupations, you operate the equipment that turns ideas and materials into books, posters, and other graphics. The people in these occupations include typesetters, camera operators (using large cameras used to rephotograph pictures), image assemblers, platemakers, press operators, and bookbinders.

— Learning Time —

Recalling the Facts

1. What is desktop publishing?
2. What kind of work is done in a company "in-house" printing shop?
3. If you use a regular typewriter to prepare material for desktop publishing, what tools will you have to use to lay out your pages?

Thinking for Yourself

1. Why were schools using desktop publishing before it was adopted by business and industry?
2. What steps would you follow to set up a school newspaper as a desktop publishing activity?

Applying What You've Learned

1. Develop a class newspaper that contains articles and illustrations about the work that you are doing in science, math, English, or social studies. Print your newspaper using your classroom's desktop publishing equipment.
2. Pick a printing occupation that interests you. Determine what a person performing this job would do and how much money the person would earn. Prepare a report, poster, or videotape that shows your findings.

Summary

Directions
The following sentences contain blanks. For each numbered blank, pick the answer that makes the most sense in the entire passage. Write your answer on a separate sheet of paper.

People often refer to graphic reproduction as printing. They don't limit printing to transferring images on a printing press. It is the eighth largest ___1___ in the United States.

All printing processes require the preparation of text, illustrations, and printing plates before you can start to ___2___ . Many

1.
a. place
b. industry
c. factory
d. manufacturer

2.
a. design
b. computerize
c. print
d. type

(Continued on next page)

Continued

of the same procedures are used to prepare the different plates, even though they might be printed by different ___3___ .

   Setting the letters of the text is called typesetting. Hot-type composition sets metal ___4___ that are made by pouring ___5___ metal into molds. In cold-type ___6___ letters are set photographically or by computer. When you set type in your technology lab, you set foundry type into a composing stick. When not in use, this ___7___ type is stored in a California job case.

   To make a printing plate out of a photograph, we must rephotograph the ___8___ through a screen. This ___9___ picture is called a halftone.

   The type of printing process used is determined by the part of the printing plate that does the ___10___ . Letterpress prints from the raised part of the ___11___ . Gravure ___12___ from inkwells etched into the plate. Lithography prints from a plate that is perfectly flat. Screen process printing prints right through little holes that exist in the plate. Electrostatic printing prints from a plate that is charged with ___13___ electricity. Ink-jet printing sprays the ___14___ directly on the material that you are printing. A computer directs the spray guns and tells them when to ___15___ .

   Desktop publishing is used to print school or company newspapers, stationery, booklets, and other needed items without having to go to a ___16___ . Small office computers and ___17___ have made this type of printing possible.

   Since printing is the ___18___ largest industry in the United States, you might want to consider a career in this industry. Openings exist for high school and college graduates.

3.
   a. people
   b. artists
   c. computers
   d. processes

4.
   a. letters
   b. things
   c. photographs
   d. presses

5.
   a. cold
   b. hot
   c. wet
   d. broken

6.
   a. composing sticks
   b. casters
   c. composition
   d. foundries

7.
   a. cold
   b. composing stick
   c. composition
   d. foundry

8.
   a. photo
   b. drawing
   c. text
   d. foundry type

9.
   a. copied
   b. screened
   c. color
   d. computerized

10.
   a. typesetting
   b. printing
   c. drawing
   d. planning

11.
   a. printing plate
   b. printing press
   c. copying machine
   d. negative

12.
   a. comes
   b. paints
   c. draws
   d. prints

13.
   a. static
   b. laser
   c. cold
   d. special

14.
   a. developer
   b. water
   c. ink
   d. fixer

15.
   a. clean
   b. revolve
   c. shoot
   d. lock

16.
   a. printer
   b. computer
   c. store
   d. art illustrator

(Continued on next page)

Summary

Continued

17.
   a. platemakers
   b. fax machines
   c. typewriters
   d. electrostatic
      printers

18.
   a. fifth
   b. sixth
   c. seventh
   d. eighth

Key Word Definitions

Directions

The column on the left contains the key words from this chapter. The column on the right contains a scrambled list of phrases that describe what these words mean. Match the correct meaning with each word. Write your answer on a separate sheet of paper.

Key Word		*Description*	
1.	**Graphic reproduction**	a)	Prints all the colors of the rainbow
2.	**Printing**	b)	Also called xerography
3.	**Halftone**	c)	Prints from indentations in the printing plate
4.	**Four-color process printing**	d)	In-house printing using computers and electrostatic printers
5.	**Letterpress printing**	e)	Prints because oil and water don't mix
6.	**Flexography**	f)	Often referred to as printing
7.	**Gravure printing**	g)	Reproduces continuous-tone pictures
8.	**Lithography**	h)	Prints from a rubber plate
9.	**Serigraphy**	i)	Prints with tiny ink spray guns
10.	**Xerography**	j)	Procedures performed to change artwork, photographs and text into printed material
11.	**Electrostatic printing**	k)	Also called electrostatic printing
12.	**Ink-jet printing**	l)	Screen process printing
13.	**Desktop publishing**	m)	Prints from a raised surface

PROBLEM-SOLVING ACTIVITY

Resources You Will Need

- Button machine
- Button parts
- Button print cutter
- Computer and printer
- Special computer ribbon used for heat transfers
- Heat transfer press or electric iron
- Heat transfer letters
- Shirts
- Paper
- Linoleum blocks
- Linoleum block cutters
- Proof press
- Printer's ink
- Heat transfer ink for letterpress printing
- Paper
- Padding cement
- Paper cutter
- Spiral binding machine
- Spirals
- Rubber stamp machine
- Type
- Rubber stamp supplies
- Engraver

Running a Printing Company

Did You Know?

A successful business has products or services that people want to buy. In the graphic arts industry, companies sell printed products. People design these products, prepare printing plates, run printing presses, and bind what they have printed into books, magazines, and note pads. Other people sell the products that were produced and take special orders for new ones.

Problems To Solve

You are now a part owner in the printing company *Great Impressions*. At this moment your company is on the verge of bankruptcy. The old management and employees have secretly sold their shares to you and the rest of your class.

You have many talented people that must be organized so that they can run your company. You must develop products and a sales network to make your company profitable.

To accomplish this task, you must pick company officers and managers. Since you all

A.

own equal shares, you should elect your leadership by voting the best people into each job. What offices do you need? How will you get things organized? Once your employees and managers are organized, you need to develop a product line and start printing.

A Procedure To Follow

1. Brainstorm to determine what officers your company will need. You might want to research what leadership jobs exist in a real company. You could then set up the same structure for your company, Great Impressions. Figure A.
2. Determine what you will do with the profits from the sale of your printed products. You could donate this money to a charitable organization, fund a special class trip, or agree on some other worthwhile use.
3. Determine what special talents your class members have, and hold elections for the different company positions.
4. Hire a consultant (your teacher) to determine the kinds of products that you can produce with your equipment.
5. Produce sample products using the equipment and materials of the shop. Figures B–E.
6. Determine a price structure based on the cost of your materials. This structure must be set up with the assistance of your consultant.
7. Print order forms, price sheets, and receipts.
8. Manufacture the needed items.
9. Deliver and collect payment for goods and services. If you are selling your products to students, you might want to collect charges in advance.
10. Distribute profits according to your plan.

. . . to Make Buttons

1. To make this product, you

(Continued on next page)

B.

must have a button machine. See Figure B on the previous page.

2. Draw designs or create them using a computer and a graphics program. The program *Print Shop* can be used.

3. Your school emblem and a senior button design will both be very popular.

4. Take orders for photographic buttons. Here people provide you with photographs, and you turn the photos into buttons.

5. Give people the opportunity to design their own buttons. Print a special order form that gives them a circle that is the exact size that the button design must fit.

6. Cut the design out using the button print cutter.

7. Place the design and button materials into the machine, and press the button.

8. Remove the completed button.

. . . to Make Printed Shirts

1. You must have a heat transfer press or clothing iron in the shop. See Figure C.

2. Place commercial letters and shop-produced transfers on the shirt, forming the desired design.

3. Transfers can be prepared using letterpress printing, offset lithography, or a computer with a graphics program. When using a computer, you must replace your printer's regular ribbon with one that is made for heat transfer printing. It is available at computer stores under the product name Underware.

4. Press the transfers and letters onto the shirt.

. . . to Make Notebooks

1. Prepare a linoleum block design for use as the cover for a small assignment pad or spiral notebook.

2. Transfer the design to the block, and carve it out using

C.

linoleum block cutters and a bench hook.

3. Print the design on cover stock at the proof press. Note that your linoleum block can be printed as a heat transfer by using special heat transfer inks.

4. Cut paper to size at the guillotine paper cutter.

5. Pad or spiral bind the covers and paper together. See Figure D.

. . . to Make Rubber Stamps

1. Set type or engrave the copy into a piece of lexan. The lexan mold must be engraved very deep into the plastic. See Figure E.

2. Lock up the type into the special tray or frame that goes with your rubber stamp machine.

3. Prepare a matrix (mold) from the type, following your teacher's instructions. Your engraved lexan is your matrix.

4. Place gum rubber on top of your mold.

5. Place these materials on the machine's tray and process them as per the requirements of your rubber stamp machine.

6. Once cooled, separate the mold from the vulcanized rubber and mount the stamp.

What Did You Learn?

1. What is the most difficult part about owning your own business?
2. What are some of the safety considerations that you must take into account when operating printing equipment?
3. What printing project did you enjoy the most? Why?

D.

E.

Chapter 8

Photography

How can you record visual events to share with others in the future? The title of this chapter gives you the answer. You use a camera and take a picture.

In chapter 5 you learned that whenever your ability to communicate is enhanced by technology, you are using communication technology. Photography is a very important tool of communication technology for individuals, business, and industry.

The camera is used to take both still and motion pictures. These pictures could end up in your photo album or be used by industry in a manufacturing process.

Photographic processing is the part of photography that develops these pictures. The chemistry and technology of photography is also used to develop printing plates for the printing industry and prepare motion picture films for the film industry. New photographic systems have recently been developed that combine chemical photography and digital photography. Our electronic industries create their tiny circuits through the use of photography. In this chapter you will learn about the technology that makes these many different uses of photography possible.

— **Key Words** —

In this chapter you will learn and study the meanings of the following important technology terms.

PHOTOGRAPHY
PINHOLE CAMERA
EMULSION
CELLULOID ROLL FILM
LENS
SHUTTER
APERTURE
FOCUS RANGE
LATENT IMAGE
NEGATIVE
DEVELOPER
STOP BATH
FIXER

The History of Photography

Can you identify some of the important developments that led to modern-day photography?

The process by which people create pictures using light energy is called **photography**. We make photographs using machines and materials that have been developed over the last 165 years.

Since light energy was always available, the principle of the camera only needed someone to discover it. The philosopher Aristotle, in the fourth century B.C., made the following observation: A small hole in a wall of a darkened room could cause an upside-down image to be projected on the opposite wall.

During the eleventh century, some Arabian scientists amused themselves with these images on the walls of their tents. No significant development of this principle took place until the late fifteenth century, however, when Leonardo da Vinci created large dark rooms for just such upside-down viewing. This kind of viewing room was called a *camera obscura*, which means "darkened chamber."

People would enter these darkened rooms and observe these naturally projected images. Artists would use these rooms to sketch pictures. See Figure 8.1. In a sense, the first photographs were done by artists who drew in the lines that were projected as shadows on a wall.

photography
the process that uses the energy of light to create pictures of objects

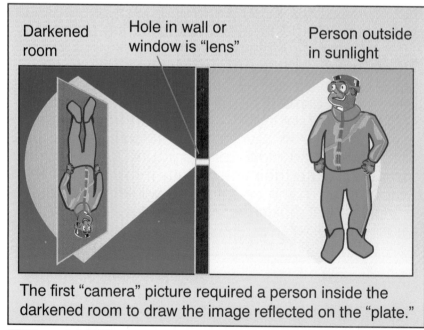

Darkened room — Hole in wall or window is "lens" — Person outside in sunlight

The first "camera" picture required a person inside the darkened room to draw the image reflected on the "plate."

■ **Figure 8-1** ■ The first cameras were room sized. Artists went inside and drew the image that was projected onto their canvas through an opening in the opposite wall.

216

Why couldn't the images, created in these room-size cameras of the sixteenth century, be recorded on photographic film or paper? The answer is simple. Photographic chemicals, film, and paper weren't invented until the middle of the nineteenth century.

In 1826, Joseph Niepce produced a permanent image in a small camera obscura that has become known as a **pinhole camera**. To create the world's first photograph, Niepce coated a metal plate with a light-sensitive chemical and then exposed the plate for 8 hours. See Figure 8.2. What would your friends say if you asked them to stand perfectly still for 8 hours while you took their picture?

In 1837, Louis Daguerre used a silver-coated sheet of copper, a 15- to-30-second exposure, a mercury vapor developer, and a table salt fixer to produce a highly detailed picture. See Figure 8.3. His process produced sharp, detailed pictures in a fraction of the time previously required by Niepce's process.

In 1839, William Fox Talbot invented light-sensitive paper negatives. From this paper negative he could produce many copies of the same picture. Talbot's pictures used a new fixing solution that stopped the photographic material from being sensitive to light.

■ **Figure 8-2** ■ This is a reprint of the world's first photograph.

■ **Figure 8-3** ■ Louis Daguerre, shown in the picture on the left, invented the Daguerreotype process that was used to produce the Paris street scene pictured above.

Although Talbot's prints weren't as sharp as Daguerre's metal photo plates, they could be included in books and magazines. Talbot's process quickly won acceptance.

Over the next 100 years, many improvements were made to the photographic process:

- Camera lenses that magnified the light and decreased the exposure time were added in the 1840s.
- Glass negative plates that were coated with a wet silver salt **emulsion**—a light-sensitive chemical—were developed in 1851.
- The rolling wagon photo studio (1850s) took your picture and then developed it on the spot. See Figure 8.4. In a sense, these wagons were the first quick-print photo centers. The wagon was

---

■ **Putting Knowledge To Work** ■

George Eastman—A Photography Pioneer

In 1877, George Eastman set out to change photography. His plan was to "make it foolproof for the lazy, casual millions" of people who would like to take pictures. He also said that he didn't develop a simple camera and film just because he loved being an inventor, but rather because "he smelled a money-maker."

At the age of twenty-three, George Eastman developed the emulsion for his new celluloid film on his mother's kitchen stove. See Figure 8.5. Twenty-three years later, his Eastman Kodak Company controlled photography.

When George was about eight years old, his father died. His desire to make money came from seeing his mother take in boarders in their family home, mop floors, and make other people's beds to support the family.

What was George like when he was your age? As a young teenager he liked making things. One day he showed a good friend a metal puzzle he had made out of knitting needles. His friend really liked the puzzle and wanted one. George wouldn't give it to him unless he paid 10 cents for it. His friend was mad but finally agreed to pay him for the puzzle.

George left school at age thirteen to earn money to help support himself and his mother. He worked his way up from an office boy earning $3 a week to a bookkeeper earning $1400 a year.

One of his friends got him interested in photography, and the rest is history. Once he developed his new film, he realized the importance of mass production, low prices, advertising, and domestic and foreign sales.

He purchased every photography patent from other inventors to guarantee that Eastman Kodak would control the market. He created a new mass market by selling a box camera that took 100 circular pictures. When you finished the film, you sent the whole camera off to Kodak for processing. His motto was, "you press the button, we do the rest." Figure 8.6 on page 220 shows the original products that made the Eastman Kodak Company famous.

needed because the wet emulsion film of the day required instant developing.

- A dry emulsion film was developed in 1871. This meant that the photographer could take your picture and develop it later. This gelatin emulsion allowed the picture to be printed larger than its negatives through the use of photo enlargers.
- In 1888, George Eastman developed **celluloid roll film** and the box camera. After the film was exposed, the film and camera were sent to Kodak—Eastman's company—for film processing. Kodak changed photography forever.
- The 35mm camera was introduced in the 1920s, the flashbulb in 1929, and the electronic flash in 1931.
- In 1935, color film was introduced, but it didn't surpass black-and-white photography until the 1950s.
- Polaroid introduced instant photography in 1947.
- Automatic computer-programmed cameras and sophisticated autofocus cameras were introduced in the 1980s.
- Computer cameras that use computer disks instead of film were introduced by Sony in 1987.
- The Advanced Photo System (APS), which combines chemical photography with digital, was introduced by Kodak, Canon, Fuji, Minolta, and Nikon in 1996.

■ **Figure 8-5** ■ George Eastman, the founder of the Eastman Kodak Company, is shown taking a picture using the camera and film he developed. This camera produced 100 circular pictures.

— Learning Time —

Recalling the Facts

1. What is a camera obscura, and how was it used?
2. Describe Joseph Niepce's contribution to the development of photography.
3. How did William Fox Talbot's photographic prints differ from Louis Daguerre's?

Thinking for Yourself

1. The development of technology moves from the simple to the complex. Using historical events, show if this is true or false for the development of photography.
2. What are the similarities and differences between the rolling wagon photo studio of the 1850s and the 1-hour photo stores of today?

Applying What You've Learned

1. Prepare a model of a camera obscura.
2. Construct a working model of the first Kodak camera, or pinhole camera. Develop a poster and report that explains how this camera worked.

lens
the part of a camera that magnifies light and the size of an object

The Modern Camera

Can you find the parts of a camera obscura in a modern camera?

The first camera was a shoe-box-size *camera obscura*. See Figure 8.7. The pinhole opening was set so that the photographer could place a cap over it to seal off the light when the negative was fully exposed.

The end of the camera opposite the pinhole had a place to put the light-sensitive plate (film). The entire box was placed on a *tripod* so that the camera could be held steady enough for long exposures. See Figure 8.8.

The average camera of today is still based on the camera obscura of the past. The basic parts of a camera still serve their original function, differing only in appearance and effectiveness.

Lens

What is the purpose of a camera lens?

The pinhole end of the camera now has a **lens** that magnifies light intensity. The lens also magnifies the size of the object. Special lenses called telephoto lenses make it possible for objects that are very far away to appear close in the photograph. Figure 8.9 shows the affect of using different camera lenses.

Figure 8-7 This 1895 Kodak camera featured daylight film loading. It bridged the gap between the pinhole camera and the modern camera.

Tech Talk

A shutter is something that you can open and close. You may have shutters on the windows of your house or apartment. How are those shutters similar to the ones on cameras?

Shutter

What does the camera shutter do when you take a picture?

The cap that the photographer used to place over the pinhole to shut out the light has been replaced by a **shutter**. The shutter blinks open for a very short period of time.

The shutter speed determines if moving objects will appear sharp or blurred. See Figure 8.10 on the next page. The shutter speed doesn't control your range of sharpness. The focus of close and distant objects is determined by the camera's aperture.

Aperture (Diaphragm)

What effect does the aperture of a camera have on the way you take pictures?

To control how much light can pass through the lens when the shutter is open, modern cameras use a device that goes by two names—**aperture** or *diaphragm*. The aperture is a round window that can be made larger or smaller. See Figure 8.11 on the next page. When the aperture is opened very wide, a lot of light can enter the camera.

The aperture works exactly like the iris of your eye, except that on many cameras it must be set by the photographer. The aperture on a computer-controlled camera automatically adjusts to lighting conditions just like the iris of your eye. Some very inexpensive cameras have a single aperture opening that can't be changed.

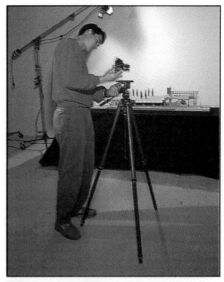

■ **Figure 8-8** ■ Photographers still rely on tripods to keep their cameras perfectly steady under special shooting conditions.

■ **Figure 8-9** ■ The pictures shown were shot from the same location by using a wide angle, normal, and telephoto lens.

■ Figure 8-10 ■ The shutter speed determines if moving objects will be in focus. In each picture the car was traveling at the same speed, but the shutter speed on the camera was changed.

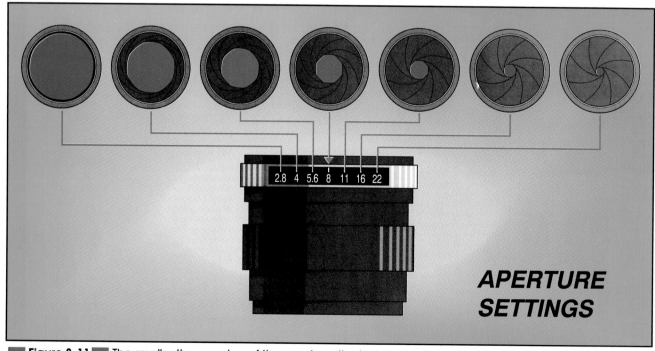

2.8 4 5.6 8 11 16 22

APERTURE SETTINGS

■ Figure 8-11 ■ The smaller the opening of the aperture, the larger your field of focus. This size is measured in units called f-stops. The relationship between the f-stop and the size of the opening is shown in this illustration.

If the aperture only reduces the amount of light that can reach the film, why can't it be replaced by a faster shutter? The aperture also affects how much of your picture will be in focus.

The smaller the opening of the aperture, the more the camera acts like a pinhole camera with everything in focus. By opening the aperture wider, you can take pictures with less light without the need of a flash. But a wide-opened aperture sacrifices the **focus range** of your picture. That is, less area inside your picture will be sharp. Since the aperture affects the focus range of your picture, the size that it is opened to is measured in units called **f**-stops.

A photographer often wants to control how much of the picture is in focus. In these photographs, the viewer is drawn to the area of the picture that the photographer wants to emphasize. Figure 8.12 shows the effect of different aperture settings on picture focus.

Viewfinder

What is the purpose of a viewfinder on a camera?

The viewfinder on a camera allows you to view your picture before you actually take it. If you don't carefully check your picture in the viewfinder, your photographs might end up missing the heads of your subjects.

In a single-lens reflex camera, your viewfinder sees directly through your camera's lens. When a picture is taken, the camera momentarily drops this mirror out of the way to expose the film. Figure 8.13 on the next page shows cameras with different types of viewfinders.

Film Winders

How do you get your camera ready for your next picture?

Film winders move the film forward to set the camera for the next picture. In some cameras this is done by pushing a lever. Many cameras today have motorized film winders.

Figure 8-12 The top picture was created using a large aperture opening, only the flowers are in focus. In the lower photograph much more is in focus because a very small aperture was used.

——— Photographic Film and Paper ———

Why is film made with different ISO numbers?

Photographic film is made for color or black-and-white photography. Color film can be purchased for slide or print photography. Today black-and-white film is only used by professional photographers, artists, and photo hobbyists. All of these films are produced in different sizes so that they can fit the different cameras that people use.

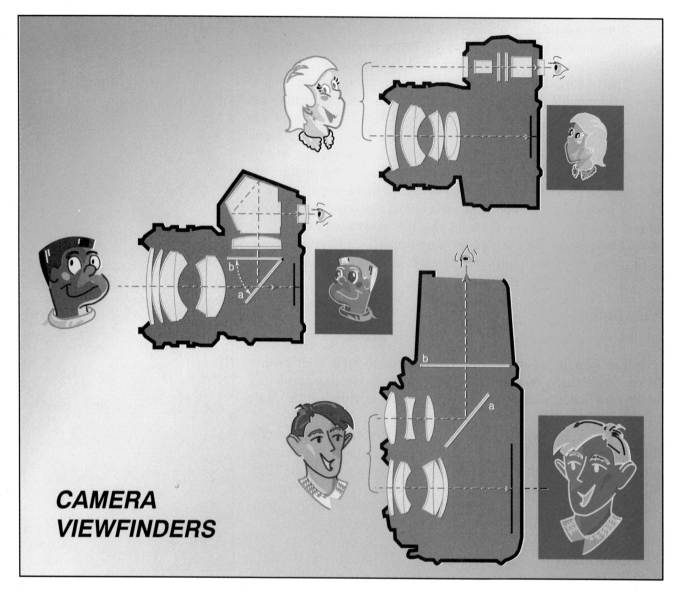

Figure 8-13 The viewfinder of a camera lets you compose your picture. This illustration shows you how some camera viewfinders operate.

CAMERA VIEWFINDERS

Photographic film is also produced with different levels of sensitivity to light. This light sensitivity is called *film speed*. The International Standards Organization (ISO) uses a number system to rate the speed of film manufactured by different companies throughout the world. Your camera must be set to the ISO number on your film. The number printed after the letters ISO is the speed of your film. The higher the number, the faster the film. If you know you will be taking photos in dark locations, you should select the faster (higher number) films.

Many new cameras automatically set the camera to match the speed of the film. With these cameras you must use film that is labeled DK. The camera reads the film speed using the same kind of technology used by a supermarket checkout scanner.

Photographic processing paper is sold with different contrast ratings. The quality of your prints will be improved if you use paper that matches the *density* of your negatives. The density of your negatives refers to how much light can pass through the clear and dark areas of the negative.

Digital Photography and Advanced Photo System (APS)

How does APS differ from digital photography?

In digital photography, images aren't recorded on light-sensitive film. The digital cameras have hundreds of thousands of microscopic resistors that convert each point of light from the subject into the individual pixels of the digital image. These pixels are stored as a series of 1's and 0's that a computer can change back into images. These images are stored in the camera's flash memory, chips, card, or even on a regular 3½" computer disk.

APS is a recently developed hybrid digital/chemical photographic technology. The APS film cartridge that is central to the system is interchangeable between all APS cameras.

With the APS, the image is stored as a standard photographic latent image along with digital information about the scene for use during photographic processing. The ease of downloading digital images from a digital camera to a computer will soon be matched by APS hardware, which will allow you to download photographs from the new APS cartridge directly into your computer.

All APS cameras have drop-in loading. Most of the cameras allow you to remove a roll of film before it is completed so that you can actually keep different rolls of film for specific topics. All APS cameras let you choose the shape of the final print at the time that you snap the picture. Your shape selection and other information about picture-taking conditions are recorded digitally on the data strip. On some camera models, the digital strip also can record captions, the number of prints you want for each individual shot, and the time and date when the picture was taken.

Once processed, the film is returned to you in its original film cartridge along with a special index print that contains postage-size prints of every picture on the roll. At this time, the digital recording for each picture is limited to about 80 bytes of information. In time, the digital part of the system will do more than just interface with the developing equipment. Kodak and Fuji have already demonstrated new display hardware that will let you scan your photos directly from the APS cartridge into your computer. With this equipment, you will be able to turn your T.V. or computer into a new type of slide projector with zoom, pan, fade, and even narration capabilities. As you can see in Figure 8.14, digital cameras don't look very different from APS cameras. However, the film cartridge design and available print sizes of APS cameras are very different from what was available in the past.

■ **Figure 8-14** ■ Digital cameras (top) record images as binary code. Film cameras record images as analog tones. The new APS camera (bottom) combines some digital recording with the analog picture.

— Learning Time —

Recalling the Facts

1. Describe the first camera, and explain how it worked.
2. What is the purpose of a camera aperture and shutter?
3. What is the meaning of a film's ISO number?

Thinking for Yourself

1. Describe how a modern camera and a pinhole camera are alike and how they are different.
2. Explain how the shutter speed, aperture setting, and ISO number of the film together affect your ability to take action photos.

Applying What You've Learned

1. Figure 8.12 on page 223 shows how the focus range is affected by the camera's aperture setting. Shoot a series of your own pictures that teach this same principle. Mount and label your pictures for display.
2. The Lego-Logo construction set contains all the parts needed to make a working model of a supermarket checkout scanner. This same technology is also used by new automatic cameras to read DK labeling on film. Using the Lego construction set, build a scanner that can read your own DK labels.

—— Film Developing ——

What is the purpose of film developing?

Most people develop their film by dropping it off at a store that sends the film to a photo laboratory. The same developing process can usually be done right in your technology laboratory.

Let's look closely at a roll of film and prepare to develop it. The film that is used today is very sensitive to light and must be protected from accidental exposure by special film *cartridges*. This cartridge is the plastic case that the film comes in.

Exposure to light chemically changes your film, but this change is invisible until after the film is developed. This invisible image on the film is called a **latent image**. When this latent image is developed, you have *negatives*. **Negatives** are reverse photographic images of your pictures on clear thin plastic. What appears as dark, shaded areas in your negative will appear as bright, lit areas in the photographic prints.

latent image
the image, or picture, on the film that is invisible until after the film is developed

negative
photographic film that contains an image in which the light parts of the image appear dark, and the dark parts of the image appear clear

Mixing Photo Chemicals

Why must you wear protective clothing and follow directions when mixing chemicals?

You will use a **developer, stop bath, fixer,** and *water wash* to bring out the latent image of your negatives. You should handle these chemicals only if you are wearing a protective plastic apron to protect your clothing and safety goggles to protect your eyes. Directions on how to mix and dilute the chemicals will be found on the package labels. If your chemicals are bought in powder form, it is a good idea to add water at least a day before you plan to use them.

The ideal temperature for photo developing is 68° F. The developing process will take place faster at higher temperatures and slower at lower temperatures. You will have problems with your negatives if your temperature is below 65° F or above 70° F.

Loading the Film into a Developing Tank

What is the procedure for loading film into a developing tank?

This operation can be done in a completely darkened room or in a special light-tight film-loading bag. There are no safe lights that can be used during film loading. You must practice the steps that you will follow to load the *developing tank* by using old film. Do this with the lights on until you can do the procedure blindfolded.

Developing the Film

What steps must be followed to develop a roll of film?

You are now ready to start developing your film into negatives. Since you are going to time the developing process, it is important that you have a watch or timer on hand before starting.

Pour enough of each solution into labeled measuring containers. Set these containers into large trays. The developing area is always set up so that the developing tray is the first tray on your left. If you reversed the trays, someone working in your darkroom could first pour the fixer into the developing tank and ruin the film.

The developer is added to the developing tank through a special opening in its top. See Figure 8.15 on page 229. The developer causes the exposed film emulsion to darken.

developer
the name given to all chemicals that are used to bring out the latent image in photographic processes

stop bath
a chemical that is used to stop the developing process and remove excess developer from the film or paper

fixer
a chemical that makes the photographic material no longer sensitive to light

Tech Talk

The word *film* means thin covering. Photographic film has a thin covering of a light sensitive material.

opaque
(ō-pāk′)

How opaque (dark) the film becomes depends on how long it is exposed to light and how long it is kept in contact with the developer. The tank should be carefully shaken for 15 seconds of every minute. This causes the chemicals that weaken from direct contact with the film to mix again with the rest of the solution. Since the developing tank is sitting in a large developing tray, any solution that might come out during shaking can be caught by the tray and poured back.

There must always be enough chemicals to cover the film in the tank. Once the developing time has been reached, the developer is poured back into the measuring container. It is replaced with a solution of stop bath that is shaken for 15 seconds and then poured out. The developer and the fixer have opposite chemical properties, and if they are poured in together, they neutralize each other.

After the stop bath is removed, the fixer is poured into the tank. The fixer will cause the unexposed emulsion to dissolve, leaving those areas of the film clear. Once fixing is completed—a process that takes approximately 10 minutes—the film is washed in running water for about 30 minutes to remove all chemicals from the film.

The last step is to hang the film up to dry. The chemicals can be used to develop a number of rolls of film during the same session. The film developer and stop bath cannot be saved from session to session.

Looking at your negatives, you will notice some interesting things. Light-colored areas of your original photo subject are now black on the film, and dark-colored areas of your original subject area are now clear on the film. For this reason, pictures seen as normal are called *positives* in photography, and pictures that are reversed are called *negatives*. Slides that are shown on a screen are examples of photographic positives. Developed film and dental X rays are examples of photographic negatives.

Printing Pictures from Your Negatives

What are contact prints and photo enlargements?

When your negatives have dried, they are ready to be printed onto photographic paper as *contact prints* or *photo enlargements*. The photographic paper that you will use can only be opened in a room that is equipped with proper safelights. Special red filters are used to block the light rays that will expose this paper.

A contact print is a photograph that is exactly the same size as its negative. Contact prints are made by placing the negative on top of the photographic paper and then exposing the paper to light. The

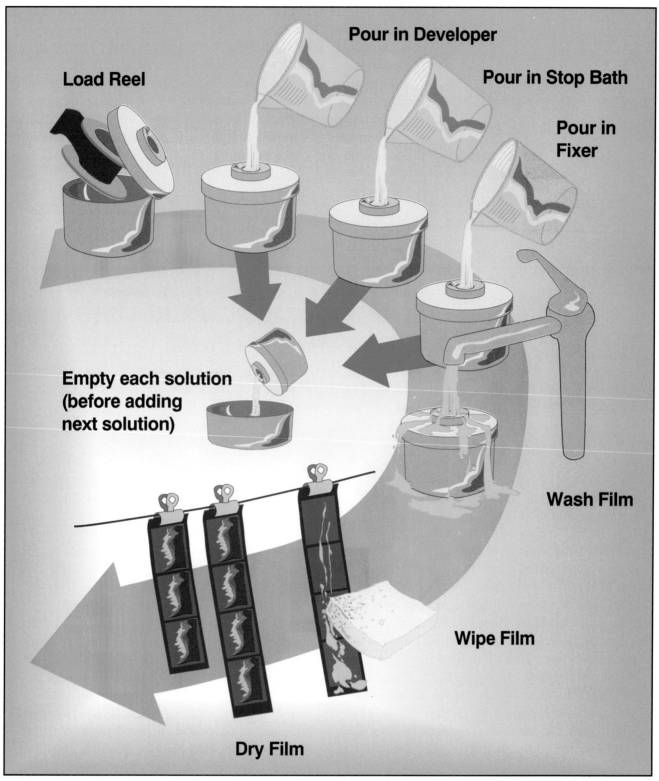

Load Reel

Pour in Developer

Pour in Stop Bath

Pour in Fixer

Empty each solution (before adding next solution)

Wash Film

Wipe Film

Dry Film

■ **Figure 8-15** ■ This flow chart shows you the step-by-step procedure you must follow to develop a roll of film.

amount of light that hits the paper is determined by the shaded areas of the negative. The paper is then developed using similar chemicals to the ones used in film developing. These chemicals are placed in trays, and the work is done under the safelights.

A photo enlarger is used to make photographic enlargements. The image on the negative is projected to the photographic paper. Your pictures can therefore be much larger than your negatives. After the paper is exposed to light, it is developed using the same procedure used in contact printing.

Edward Land and Instant Photography

When you take a picture you always wonder what it will look like. Did your friends have their eyes open? Was there enough light? Was the film loaded into the camera correctly? Did you remember to remove the lens cap? At one time you couldn't get the answers to these questions until your film came back from being processed—often two weeks after handing it in.

In 1947, Edward Land demonstrated a film and camera that made instant photography possible. With the Polaroid camera, you could see your pictures right on the spot. If you didn't like your picture, you could just shoot another one.

Land's original invention had an instant, peel-apart black-and-white film. This film contained the negative, chemicals, and print paper all packaged together. After taking your picture, you pulled the print from the camera. After a short wait, you stripped the negative from the print and coated the picture with a fixer using a sponge applicator.

From the moment his Polaroid Land cameras went on sale, Land began work on the development of an instant color film. Polaroid color film was finally marketed in 1963. This film has the chemicals, negative, and final print all packaged together. Once the picture is taken, you have nothing to peel or throw away. The chemicals and negative just stay hidden inside the picture.

To accomplish this, Land had to invent a sixteen-layer package that contained all the color dyes, developer, and fixer needed to make the picture. When the film is exposed to light, the rays pass through a clear plastic window that covers the film and strikes the different layers of the film. This creates a latent image ready for developing.

When the film rolls out from the camera, pressure rollers break the seal of the developing chemicals and spread them over the negative. These chemicals seep through the different layers of the film and activate the chemical dyes.

Once each chemical reaction is completed, another chemical causes that layer of the negative to become opaque. The final positive image covers the negative. The negative can't be seen because a white pigment layer has developed that hides the negative. Figure 8.16 shows a Polaroid color picture developing right before your eyes.

Figure 8-16 These photographs show a color polaroid picture at different stages of development.

— Learning Time —

Recalling the Facts

1. What is the purpose of a modern film cartridge?
2. What is a latent image?
3. What is the specific role of film developer, stop bath, and fixer when you develop a roll of film?

Thinking for Yourself

1. Why must film be developed in total darkness?
2. Why is it important that photographic chemicals be mixed with the exact amount of water called for by the manufacturer?

Applying What You've Learned

1. Photographs are often used to tell a story. Pick a technology theme, and then use a camera and black-and-white film to record your story.
2. Mix photo chemicals, load a roll of film into a developing tank, develop and print a roll of black-and-white film.

Summary

Directions

The following sentences contain blanks. For each numbered blank, pick the answer that makes the most sense in the entire passage. Write your answer on a separate sheet of paper.

The first photograph was produced only 165 years ago. The principle of the camera, however, was discovered over 2,030 years ago. It seems amazing that the ____1____ is older than the first ____2____ .

Originally, photographic film and paper were very fragile and needed to be processed immediately. ____3____ would take your picture and then ____4____ the picture for you. The emulsion is the part of the ____5____ that is sensitive to light. Photographic paper is also coated with ____6____ .

The modern camera is still a box with a lens on one end and a place to hold the ____7____ on the other. All the other parts of the camera just make sure that the amount of ____8____ that enters the camera won't overexpose the film. The shutter opens and closes very quickly so that you can take pictures of ____9____ objects that won't look blurred.

____10____ your film is developed, your pictures exist as latent images. The chemicals that are used change this latent image into negatives. To develop the film, you load it into a developing tank. This must be done ____11____ or the film will be ____12____ .

You use developer, stop bath, and fixer to develop your ____13____ and print your pictures. The developer causes the ____14____ to appear on the film and paper. The fixer makes the ____15____ no longer sensitive to light. A photo enlarger is used to make prints that are larger than your negatives.

1.
a. computer
b. photograph
c. camera
d. printing press

2.
a. computer
b. photograph
c. camera
d. printing press

3.
a. Illustrators
b. Drafters
c. Artists
d. Photographers

4.
a. pony express
b. mail
c. develop
d. buy

5.
a. film
b. picture
c. fixer
d. developer

6.
a. stop bath
b. fixer
c. developer
d. emulsion

7.
a. picture
b. object
c. film
d. photo paper

8.
a. time
b. X rays
c. light
d. heat

9.
a. moving
b. stationary
c. old
d. computerized

10.
a. After
b. While
c. Before
d. The instant

11.
a. in darkness
b. in water
c. quickly
d. slowly

12.
a. negatives
b. positives
c. ruined
d. stolen

13.
a. illustrations
b. positives
c. programs
d. negatives

(Continued on next page)

Summary

Continued

14.
   a. latent image
   b. computer image
   c. negative image
   d. positive image

15.
   a. darkroom
   b. developing tank
   c. developer
   d. film and paper

Key Word Definitions

Directions

The column on the left contains the key words from this chapter. The column on the right contains a scrambled list of phrases that describe what these words mean. Match the correct meaning with each word. Write your answer on a separate sheet of paper.

Key Word		*Description*	
1.	**Photography**	a)	Light-sensitive coating on film and photo paper
2.	**Pinhole camera**	b)	Camera part that opens and closes quickly
3.	**Emulsion**	c)	Stops the developing process
4.	**Celluloid roll film**	d)	Reverse images on clear, thin plastic
5.	**Lens**	e)	Photographic image before developing
6.	**Shutter**	f)	How much area inside your picture will be sharp
7.	**Aperture**	g)	Brings out the latent image
8.	**Focus range**	h)	Makes photo film and paper no longer sensitive to light
9.	**Latent image**	i)	Small camera obscura
10.	**Negative**	j)	The film you use in your camera
11.	**Developer**	k)	Works like the iris of your eye
12.	**Stop bath**	l)	Creating pictures using light energy
13.	**Fixer**	m)	Magnifies the size of the object

Developing and Printing Black-and-White Film

Did You Know?

Do you know what you looked like when you were a baby? Or who attended your first birthday party? Or what your parents looked like when they were teenagers? If you ask your mother about these things, she will probably respond by taking out a photo album.

Most people take pictures and get their pictures developed. Picture developing usually involves dropping off the roll of film at a place that does film processing.

Problems To Solve

In this activity you will develop and print a roll of black-and-white film. You will have to solve the problem of getting the film out of its plastic cassette and into the developing tank. You will learn to handle photographic chemicals and use photographic processing machinery. Figures A and B show all the equipment and materials necessary to develop your own film.

Resources You Will Need

- Exposed roll of film
- Darkroom
- Developing tank
- Photo thermometer
- Film developer
- Stop bath
- Fixer
- Timer
- Sink
- Spring clothespins
- Place to hang wet film to dry
- Safelights
- Contact printer
- Photographic paper
- Paper developer
- Photo enlarger
- Three 9 x 12-inch photographic trays
- Photo print dryer

A.

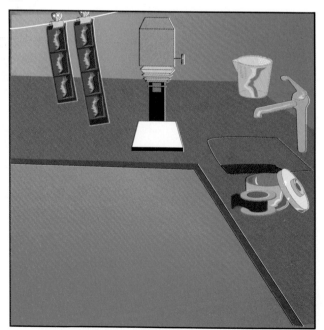

. . . to Develop Negatives

1. Mix photographic chemicals following manufacturers' directions. Make certain that these chemicals are 68°F before using.
2. In a darkroom with all lights out, load the black-and-white film into the developing tank. See Figure C.
3. Fill the developing tank with film developer.
4. Carefully shake the tank for 15 seconds per minute.
5. Carefully time your developing (approximately 8 minutes). Check the developer's instructions for recommended developing time.
6. Pour the developer back into a container for another student's use.
7. Pour in stop bath for 15 seconds.
8. Pour out the stop bath.
9. Pour in fixer for approximately 10 minutes.
10. Wash the film in running water for about 30 minutes.
11. Hang the negatives up to dry. Place a wood spring clothespin on top and bottom to prevent the negatives from curling.
12. The chemicals can be used to develop a number of rolls of film during the same session. The film developer and stop bath cannot be saved from session to session.

B.

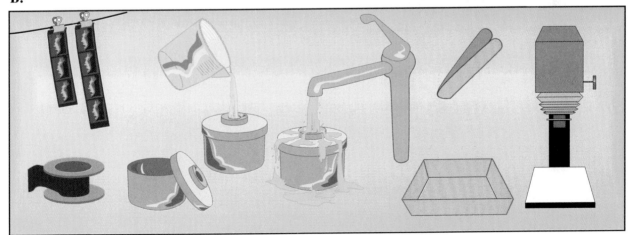

. . . to Make Contact Prints

1. Turn on the darkroom's safelights.
2. Fill your trays with 68° F paper developer, stop bath, and fixer.
3. Place your negatives onto the glass of the contact printing machine with the negatives' shiny side touching the glass.
4. Place your contact print paper directly on top of your negatives. The dull side of the negatives should now be in contact with the shiny side of the paper.
5. Close the cover and turn on the machine for the required exposure. Determine the length of exposure time by reading the instructions that came with the paper or by running timed test exposures.
6. Place the exposed paper into the developing tray with its shiny side facing up.
7. Rock the tray up and down to keep the developer in motion.
8. Watch as the latent image develops, and remove the paper when it appears to be slightly darker than desired. Developing time should be about 1 minute. If the picture develops very quickly, your exposure in the contact printer was too long.
9. Place the print in the stop bath for 15 seconds.
10. Place the print in the fixer for about 10 minutes.
11. Wash your prints for about a half hour in running water. Special chemicals can be used to cut your washing time and save water.
12. Place prints into a photo

C.

236

dryer until dry.

. . . to Make Photo Enlargements

1. Place your negatives into the enlarger, dull side facing down. See Figure D.
2. Turn on the enlarger and project the image onto the paper easel.
3. Raise or lower the enlarger to project the print size that you desire.
4. Focus the enlarger by turning the knob that moves the lens closer or farther away from your negatives.
5. Set the lens aperture so that a small amount of light shines through the film to the easel.
6. Turn off the enlarger.
7. Slide a piece of photographic paper, shiny side facing up, into the easel.
8. Turn on the enlarger and expose the paper for the desired time.
9. Develop, wash, and dry each print using the same chemicals and procedure used in contact printing.

What Did You Learn?

1. Why can't photographic films and paper be exposed to daylight before developing?
2. Why can film be developed with the regular light on once it has been placed in a developing tank?
3. What is the purpose of each chemical used in the photographic developing process?

D.

237

Electricity and Electronic Communication

Key Words

In this chapter you will learn and study the meanings of the following important technology terms.

VOLTAGE
ELECTRIC CIRCUIT
AMPERAGE
RESISTANCE
OHM'S LAW
WATTAGE
ELECTRON THEORY
CONDUCTOR
INSULATOR
SERIES CIRCUIT
PARALLEL CIRCUIT
SEMICONDUCTOR
SUPERCONDUCTOR
DIRECT CURRENT
ALTERNATING CURRENT
FREQUENCY
TRANSMITTER
FIBER OPTIC
LASER
MASS COMMUNICATION

You can see the static electricity jump and hear the crackle of the static discharge. Why are these shocks only annoying?

Some electrical devices require very little electricity to operate. Why can you touch the track of a Lionel train and not get shocked?

Your toaster, television, and stereo use a lot of electricity. Putting your hand inside these devices can kill you. Why?

When a storm is overhead, the electricity in a bolt of lightning can kill people outdoors if they are out in the open. See Figure 9.1 on the next page. Lightning can also kill people indoors if they are talking on a telephone. Lightning that strikes close to your home can destroy all electronic devices that are in your home. Why?

Every electric circuit contains voltage, amperage, and resistance. Together they determine what electricity can do and when electricity can be dangerous.

In this chapter you will learn what electricity is and how it is used. You will learn how modern communication devices operate through the use of electric circuits. You will also learn about new materials called *superconductors.*

Figure 9-1 Lightning is a very strong discharge of static electricity. It contains very high voltage and amperage.

Understanding Electricity

How can flowing water help us understand electricity?

Electricity flowing through a circuit can't be seen. It is difficult to visualize how this invisible force makes things work.

Although water and electricity don't look the same, they move through pipes and wires in very similar ways. You need water pressure to push water through a pipe, and you need electrical pressure to push electricity through a wire.

Voltage

What is voltage?

Voltage is the name given to the electrical pressure that pushes electricity through an electric circuit. An **electric circuit** is a complete electrical pathway. See Figure 9.2.

If all other things stay the same, the greater the electrical pressure, the more electricity will be pushed through the circuit. Voltage pressure is measured in units called *volts*.

voltage
the force or pressure needed to push electricity through a circuit

electric circuit
an electrical pathway that begins and ends at the same power source. Includes parts such as a wire for a path to conduct the flow and a device the electricity is being delivered to.

240

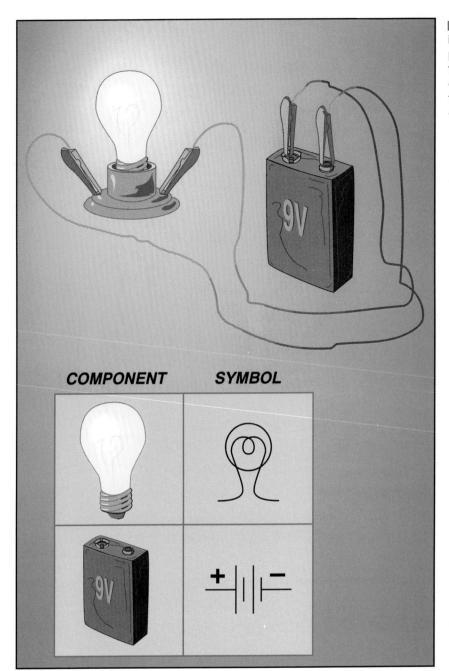

COMPONENT   **SYMBOL**

Tech Talk

The word *circuit* describes something that moves in a circular path. You have probably heard the expressions *theater circuit* and *circuit judge*. Electrical circuit is an important term in electricity.

Amperage

What is amperage?

The energy flowing along an electric wire is called the *current*. **Amperage** is the name used for the strength of that *current*. It is measured in units called *amperes*.

If a water pipe has a break in it, the water will flow out through the opening. You won't have the water where you want it because it will escape through the hole in the pipe. See Figure 9.3 on the next page. Electricity isn't like water in this respect. Electricity won't

amperage
the strength or power of the electrical energy through a circuit. It is also called the *current*.

ampere
(ăm′pîr′)

Figure 9-3 This illustration shows that water will flow out of a broken water pipe, while electricity will stop flowing if there is a break in the circuit.

Figure 9-3 This illustration shows that water will flow out of a broken water pipe, while electricity will stop flowing if there is a break in the circuit.

LAMP OFF

9V

SWITCH
OPEN

escape through a break in the circuit. Electricity can't flow at all when the circuit has a break or opening in it. It is as if someone shut off the energy. If, however, any other electrical pathway touches the opening, the electricity will start to flow through this new pathway.

Your body can carry electricity if the electrical pressure is high enough. That is why you can get a shock if you touch a broken wire or put your fingers inside a wall outlet.

resistance
the force that opposes or slows the flow of the electrical current

ohms
(ōmz)

Resistance

What is resistance?

In an electrical circuit, all the things that oppose the flow of electricity are called **resistance**. Resistance is measured in units called *ohms*.

242

Resistance in a water pipe is determined by the diameter, length, bends, and kinks in the pipe. The resistance in an electric circuit is determined by the electric wire's diameter, length, and temperature. See Figure 9.4. For example, the resistance of a wire increases as the wire gets longer. The resistance of a wire decreases as the wire gets fatter. Adding components in the circuit will also increase resistance.

If you have a water pipe with a very bad kink in it, no water will flow. If you have an electric circuit with too much resistance, the current will not flow. If you push the pressure high enough, you can get some water to flow through the kinked pipe and some electric current to flow through the high-resistance electric circuit.

Ohm's Law

☰ *What can you determine by using Ohm's law?*

If electrical devices are going to work properly, they must get enough electricity. When electric circuits are designed, the power needs of the components must be known.

A mathematical formula called **Ohm's law** is used to determine the voltage, amperage, or resistance of an electric circuit when only two of these three factors are known. The law gives the electrical designer a way of determining exactly what the circuit will need to make it run efficiently.

When you need to determine the current (amperage) of the electric circuit, use the following formula:

$$\text{Amperage} = \frac{\text{Voltage}}{\text{Resistance}}$$

When you want to determine the resistance of the electric circuit, the formula to use is:

$$\text{Resistance} = \frac{\text{Voltage}}{\text{Amperage}}$$

When you want to determine the electrical pressure (voltage) of the electric circuit, use:

$$\text{Voltage} = \text{Amperage} \times \text{Resistance}$$

According to Ohm's law, a pressure of 1 volt is needed to push 1 ampere of electric current through a circuit with a resistance of 1 ohm. By substituting the number 1 for resistance, voltage, and amperage in any of our three formulas, you will see this statement presented mathematically.

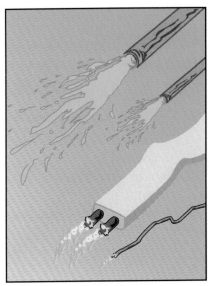

■ **Figure 9-4** ■ Resistance slows down the flow of water in a pipe or the flow of electrons in a circuit.

Ohm's Law
the relationship between amperage, voltage, and resistance

243

— Learning Time —

Recalling the Facts

1. What is voltage, and what unit is used to measure it?
2. What is amperage, and what unit is used to measure it?
3. What causes electrical resistance?

Thinking for Yourself

1. Compare water flowing through a pipe and electricity moving through an electric circuit. What are their similarities and differences?
2. Why would anyone need to use the formulas of Ohm's law?

Applying What You've Learned

1. Design and construct a battery-powered door alert (tells you when someone has entered the door). This device should contain one door contact (the switch), one buzzer, wire, and battery. (*Note:* Your teacher must approve all experimental plans before you hook up anything to any electric current.)
2. Construct a model or a chart that shows similarities and differences between water moving through a pipe and electricity moving through a circuit.

— Buying Electricity —

How is electricity sold?

The electricity that electrical appliances use is measured in watts. **Wattage** is the amount of amperage needed to run the electrical appliance multiplied by the voltage of the electric circuit:

$$\text{Wattage} = \text{Amperage} \times \text{Voltage}.$$

Your electric company meters (keeps track of) the electricity that your family uses. See Figure 9.5. When you turn on an electric appliance in your home, the wheels of your electric meter start to turn. The moving wheels are measuring the electricity that is being consumed in your home. The more appliances that you run, the faster the wheels turn and the bigger your electric bill.

Most electrical appliances tell you how much electricity will be consumed when they are in use. This information (the wattage) is usually printed on a metal plate on the bottom of the appliance.

If you watch the dials on your electric meter when your family is consuming very little electricity, the right-hand dial will move very slowly. This dial measures kilowatt-hours. A kilowatt-hour is

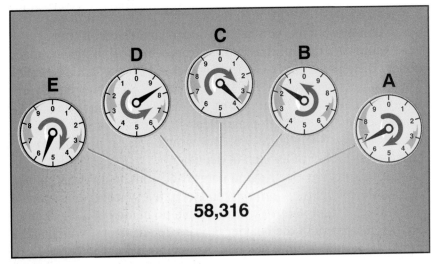

58,316

consumed electric power equaling 1,000 watts used for 1 hour's time. Each of the other wheels on the meter moves one number when the wheel to their right has made a full revolution. They are recording ten times the electric usage as the wheel to their right.

The way that one dial triggers the next dial's hand to turn can be compared to the second, minute, and hour hands on a watch. One full revolution of one hand makes a single-number movement of the next hand. Your home's electric meter has a separate dial for each hand, with each dial divided into ten parts.

To read your electric meter, notice between which two numbers the pointer is pointing in each dial. Choose the lower number in each case. Read the numbers from left to right.

When all the lights, televisions, and other electrical appliances are going, the dials will turn much faster. Each turn of the dial costs your parents more money. That is why your parents tell you to shut off the lights and the television when they are not in use.

Today, many air conditioners and refrigerators are sold with an efficiency rating. The higher this number, the cheaper it will be to run the appliance. Electric utility companies are encouraging people to use less electricity during high-demand periods of the day. A new electric meter has just been introduced that records the time that the electricity is used. When this meter is installed in your home, you will pay more money for the electricity that you use during high-demand times of the day and less money for the electricity that you use at low-demand times, like at night.

Conducting Electricity

How does electricity flow through a wire?

Our comparison of water and electricity should have helped you understand voltage, amperage, and resistance. To understand

Tech Talk

Did anyone ever tell you that he or she was electrocuted? If so, that person didn't know the meaning of the word *electrocuted*. Being electrocuted is much more serious than receiving an electric shock. Electrocution is a cause of death. It means someone dies of electrical shock.

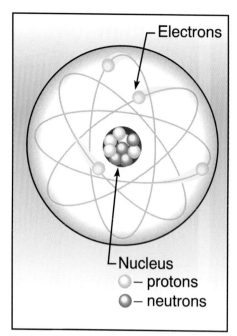

Figure 9-6 Atoms look like little solar systems. On average they are only a hundred-millionth of an inch in diameter.

how electricity flows through a solid wire and why it doesn't flow out of an open circuit, you need to understand **electron theory**.

Atoms are the building blocks from which all things are made. Ninety different atoms exist from which all objects are assembled. These ninety atoms are the ninety natural elements listed on a *periodic table* (a chart that lists all the natural elements you study in science). Under very high magnification, atoms look like very little solar systems. See Figure 9.6. But atoms are, on average, only a hundred-millionth of an inch in diameter.

All atoms are electrical in nature and contain an equal number of positively charged *protons* and negatively charged *electrons*. They also contain *neutrons*, which have no charge at all. All atoms have tight control of the heavier protons and neutrons, which are located within the atom's nucleus. Atoms have weaker control over the lighter electrons, which circle the nucleus like planets. Some atoms have less control than others over these electrons.

The movement of electrons causes atoms to lose their electrical neutrality. When an atom picks up extra electrons, it becomes negatively charged. When an atom loses electrons, it becomes positively charged. See Figure 9.7.

When atoms approach each other with the same electrical charge, they repel each other. If the charges of the atoms are different, they attract each other. Atoms obey the same rules of repulsion (pushing away) and attraction (moving together) as magnets. When atoms aren't under stress, they will pick up or lose electrons until they are in a neutral state.

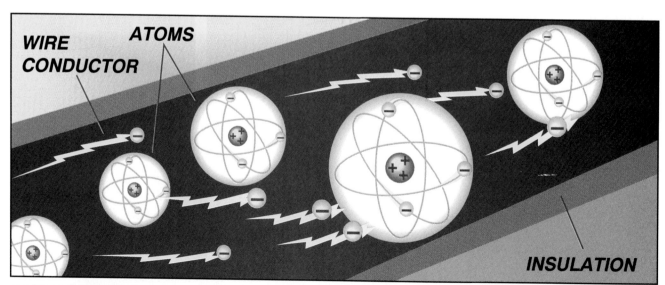

Figure 9-7 Electrons in an electrical conductor can be pushed from atom to atom. When an atom loses electrons it becomes positively charged.

Electrical Conductors

 ⬚ *What is an electrical conductor?*

Certain atoms have a very weak hold on their electrons. Materials that are made up of these atoms are called **conductors** because electricity will flow through these materials easily. Copper and silver are excellent conductors of electricity. Think of the electrical conductors as highways on which electrons travel. Electricity will push electrons along a conductor near the speed of light.

Electrical Insulators

 ⬚ *What are electrical insulators?*

Other materials contain atoms that hold onto their electrons. These materials are used to insulate electrical conductors. Rubber, plastic, and ceramics make good **insulators**. See Figure 9.8. This insulation prevents the electrons from getting out of their intended path.

If an electrical extension cord has a break in its insulation and you touch the broken area, the electron flow can move into your body, giving you an electric shock. If your body is wet, you will be a good conductor of electricity and could be electrocuted.

Semiconductors

 ⬚ *What are semiconductors?*

Some materials can act as conductors and other times as insulators. These materials are called **semiconductors**.

The vacuum tube is a relatively large device that is shaped like a light bulb. Vacuum tubes use a lot of electricity and give off a lot of heat. Vacuum tubes act as electrical conductors and insulators. They were the first device to work as an electric switch and opened up the entire field of electronics (see Chapter 3). The picture tube of your television set and computer monitor, unless they are of the portable flat screen variety, are vacuum tubes. See Figure 9.9 on the next page.

With the exception of the television tube, vacuum tubes have been replaced by the transistor, which gets its semiconducting properties from the material it is made of. By adding impurities to the white rare metal germanium or the second most abundant element silicon, we get a composite material that has semiconducting properties.

Superconductors

 ⬚ *What are superconductors?*

Superconductors are to electricity what diet foods are to people. With diet foods you shed calories. With superconductors you shed resistance. Superconductors are important because they make

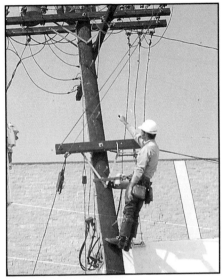

■ Figure 9-8 ■ Electrical power lines use insulators at pole locations to prevent the electrons from getting out of their intended path.

semiconductors
materials, such as the transistor, that can be used as either conductors or insulators

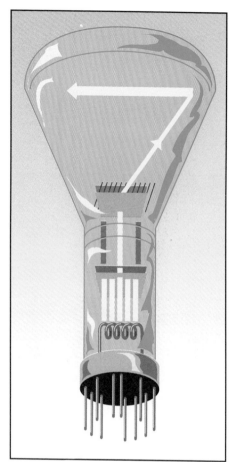

Figure 9-9 Television picture tubes are vacuum tubes. This illustration shows what a full television picture tube looks like.

Kamerlingh Onnes
(käm-ər-lin-'ón-əs)

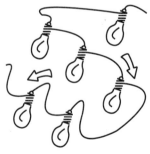

series circuit
a single pathway that current flows through to more than one electrical device. If one device in the path stops working, they all stop. Old Christmas tree lights are a good example.

it possible to generate more productive and efficient electrical power.

In 1911, Dutch physicist Heike Kamerlingh Onnes discovered that when certain materials were made very cold (−459° F), they could carry electricity without any resistance. The difficulty and cost of creating such a temperature put superconductivity out of technology's reach—for a while.

In 1986, two IBM physicists, Alex Mueller and George Bednorz, created a superconductor that worked at a temperature of −397° F. This proved that superconductors were possible in temperatures warmer than −459° F.

Scientists continue to look for easier-to-produce, more economical superconductors. Meanwhile, superconductor technology is in use today. For example, superconductors are used in a new type of machine that lets doctors see what is happening inside the human body. These MRI machines use super magnets and computer imaging to create detailed pictures for use in medical diagnosis and research (see chapter 16).

Electrical Circuits

What does an electrical circuit consist of?

An electrical path along a conductor that begins at a power source and ends back at the power source is called an *electric circuit*. If the circuit isn't complete, electrons will stop flowing, which turns off the electricity. A device that opens and closes the circuit to turn things on and off is called an *electric switch*. See Figure 9.10. There are different kinds of electric circuits.

Series Circuits. When you wire batteries and energy-using devices in a line, you have a **series circuit**. If any item in the circuit breaks, power is lost to the entire circuit.

When batteries are connected in a series, each one increases the pressure (voltage) of the circuit. See Figure 9.11 on page 250. You can compare this to attaching water guns together. The pressure of the first gun is added to the shot of the second gun, so that when the second gun fires, the water goes farther. To increase power, most battery-powered electric devices connect batteries in series. When you place three 1.5-volt batteries together in a series, the circuit will be powered by 4.5 volts. Do you understand why?

Light bulbs and other electrical devices can also be wired together in series. See Figure 9.11 again. In this type of electric circuit, the electricity must pass through each device on its way to the next. A major disadvantage of series wiring is that if one device burns out, all the components in the circuit stop working. Are the circuits in your home wired as series circuits? Do all the lights go out when one bulb burns out? Most of the circuits in your home are not wired as series circuits. But each circuit has a fuse or circuit breaker that is wired in series to the location where electricity enters your home. Why?

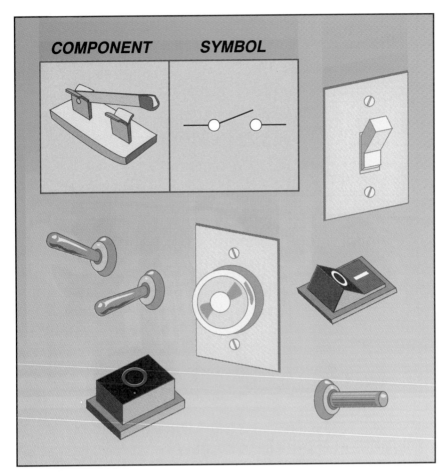

COMPONENT SYMBOL

Old Christmas lights were wired as a series circuit. Here a single broken bulb stopped electricity from flowing throughout the entire string of lights. When lamps or other electrical devices are connected in a series, each one decreases the pressure (voltage) of the circuit. In this case the series circuit, by dividing the available voltage equally between all the lamps, reduced the possibility that a string of light would give you a shock or start an electrical fire. Modern electrical methods have led to strings of low-voltage bulbs for which a burnt-out bulb is easily replaced.

If you place too many energy-using devices into a series circuit, the voltage will be so low that the devices won't operate.

Parallel Circuits. In a **parallel circuit**, an electrical path exists around each energy-using device. If one device burns out, the electricity doesn't stop flowing to the other items in the circuit.

When you connect batteries together using a parallel circuit, the available current increases while voltage remains the same. See Figure 9.11 again. Designers of battery-operated electrical devices take advantage of parallel and series circuits to make their devices run for a reasonable length of time on a single set of batteries.

Most of the circuits in your home are parallel circuits. In all parallel circuits, the voltage remains the same throughout the circuit

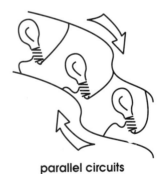

parallel circuits
multiple pathways that electrical current flows through to individual electrical devices. If one device stops working, the others are not affected. Homes are wired with parallel circuits.

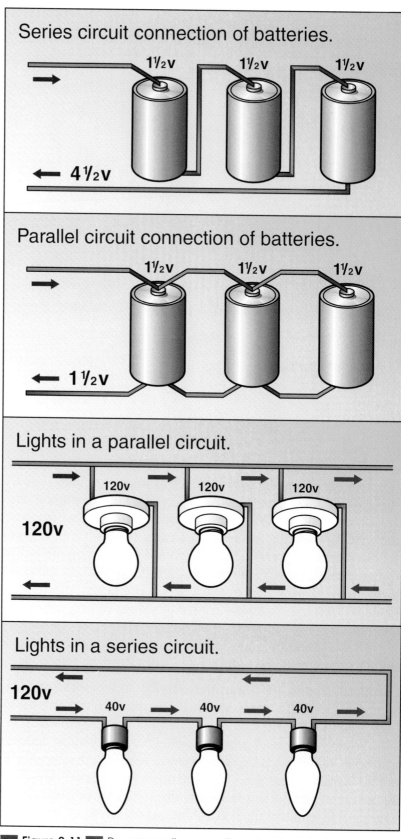

Figure 9-11 Power supplies, as well as electrical devices, can be wired as parallel or series circuits.

regardless of the number of devices connected. See Figure 9.11 again. Each electrical device connected to a parallel circuit draws the current it needs to operate. When the current demand of the electrical devices exceeds the rating of the fuse or circuit breaker, power to the circuit is cut off. Current that exceeds the capacity of the wiring in your home can start an electrical fire.

Electricity and Magnetism

How are electricity and magnetism similar?

Electricity is related to magnetism in many ways. Electricity flowing through a wire creates a magnetic field around the wire. See Figure 9.12. A magnetic field is used to generate electricity. If a wire is wrapped around an iron bar, the bar will act like a magnet when electricity flows through the wire. This type of magnet is called an electromagnet because electricity causes the metal to become a temporary magnet. See Figure 9.13 on the next page.

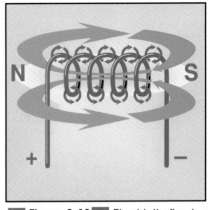

Figure 9-12 Electricity flowing through a coil of wire will generate a magnetic field.

— Learning Time —

Recalling the Facts

1. Describe what causes atoms to become negatively and positively charged.
2. What is the difference between an electrical insulator and an electrical conductor?
3. What are superconductors and what are they used for?
4. What is the difference between a series circuit and a parallel circuit?
5. Which wheel on your electric meter shows kilowatt-hours?

Thinking for Yourself

1. Would you buy an air conditioner that cost $100 more if it was rated twice as efficient as the cheaper model? Why?
2. For one week, read the electric meter in your home. Compare your family's use of electricity for each day of the week. On which day did your family consume the most and least electricity? What reason can you give?
3. Why can a semiconductor act as an electric switch?

Applying What You've Learned

1. Working as a team, use wire, a magnet, tape, cardboard tubing, and an electric meter to construct an electric generator that produces enough electricity to get a reading on the meter.
2. Wrap forty turns of wire around a 3-inch nail, and then connect one end of the wire to two batteries that are connected in series. When you touch the other terminal, you create an electromagnet. While holding this last connection, see how many paper clips the magnet will lift. Now switch your batteries so that they are wired in parallel, and determine your electromagnet's power. Record your results.

What are some different types of electricity?

Is the electricity that comes out of the wall sockets in your home the same as the electricity that comes out of the wall sockets in other countries? Are static electricity, lightning, and battery-pow-

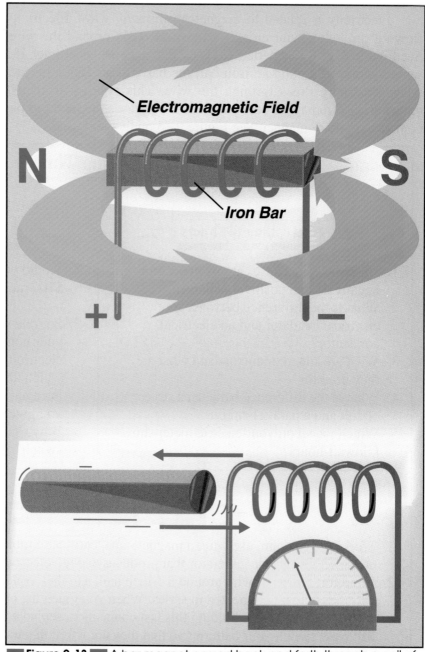

Figure 9-13 A bar magnet passed back and forth through a coil of wire will generate electricity. This is the exact reverse of electricity flowing through a coil of wire which will turn a steel bar into an electromagnet.

ered electricity different from the electricity that powers your home?

All electricity consists of electrons in motion. All electricity has voltage and amperage. All electricity is affected by the resistance of the electrical circuits. But except for these similarities, the types of electricity are so different that powering a device with the wrong type can burn it out.

Static Electricity

What causes static electricity?

A *static-electric* charge is a buildup of electrons that aren't in motion. Static electricity is electricity looking for a circuit. When the static discharges, the electrons are set in motion across a poor electrical conductor. When you see the spark, that means that enough voltage (electrical pressure) has built up to allow the electrons to move. Given enough electrical pressure (voltage), even nonconductors can be forced to become poor conductors of electricity and allow electricity to jump small distances.

When you receive a shock from static electricity, you aren't hurt because the amperage is low. Static electricity does, however, carry enough amperage to destroy electronic circuits that need very little electricity to operate. To ignite flammable liquids, all you need is a little spark from a static discharge.

Lightning is a very strong discharge of static electricity. The electron buildup is so big that the electrons can actually create an electrical path through our atmosphere.

The lightning bolt is looking for the best conductor and the shortest route to ground. Buildings, trees, and people conduct electricity better than air. If you are out in a thunderstorm, stay clear of trees and open fields.

When lightning hits a power line, the electricity from the bolt surges (sends high voltage and amperage) through the line. This power surge can destroy the electronic appliances in your home and kill you if you are using a telephone connected to a phone line that has been hit.

> ### Tech Talk
>
> The word *static* has many meanings in addition to its use in the term *static electricity*. When the sound on your radio, television or telephone "breaks up," you call the interference *static*. Did you ever get "*static*" from your parents or friends?

Batteries

What is an advantage and a disadvantage of batteries as a source of electricity?

Batteries contain stored electricity. Batteries are useful because they provide a source of electricity that can be easily moved from one place to another. You can take them to places that aren't hooked up to a permanent source of electricity. Of course batteries do not contain a limitless supply of electricity. Sooner or later they need to be replaced.

When batteries are in use, electrons flow out from the negative terminal of the battery, around the circuit, and then back through the battery's positive terminal. At one time, scientists believed that electrons moved in the opposite direction. When your parents were in school, they were probably taught that electrons flowed from the positive terminal to the negative terminal.

When the electrons flow only in one direction in an electric circuit, the electricity is called **direct current**, abbreviated DC. All batteries and special DC generators use direct current.

direct current
the one-directional flow of electrons in an electrical circuit. All batteries have **D**irect **C**urrent (DC) flow.

Household Electricity

Why are circuit breakers important?

All the electric outlets in your home are connected to a fuse box or circuit breaker panel. These safety fuses or electric breakers will shut off the electricity in case of a power overload. A power overload will happen if the toaster, lights, hairdryer, and other energy-using devices "pull" more electrons than the wire in the circuit can safely carry. See Figure 9.14. If an electric wire is not properly protected by a fuse or a circuit breaker, an overload can superheat the wire and start a fire.

The wires that come into your home can be traced all the way back to a power-generating station. The electrons that flow through these wires change direction 120 times per second. This constant directional change is the reason that this electricity is called **alternating current**, abbreviated AC.

alternating current
electrical flow that constantly changes direction. The electrical flow that powers your home is **A**lternating **C**urrent (AC).

Each back-and-forth motion is called a *cycle*. This means that the electricity that powers your home is measured at 60 cycles per second. The number of cycles per second is called the **frequency** of the alternating current. Another name for cycles per second is *hertz*.

frequency
the number of cycles or changes in direction of the **A**lternating **C**urrent. Frequency is measured in hertz, or cycles per second.

Most countries generate alternating current. But many countries have their current cycle and voltage at a different rate than ours. Many appliances sold in the United States can't be used in other parts of the world.

Kits are sold that let us use our electrical devices on trips to other countries. Don't buy any electrical device in a foreign country without making sure that it is built to operate in the United States.

Some radios and other devices are built to run on batteries or household current. When these devices are plugged into a wall outlet, the electricity is transformed (changed) to match the requirements of the device. See Figure 9.15 on page 256. Electricity is changed from one voltage to another voltage by an electrical device

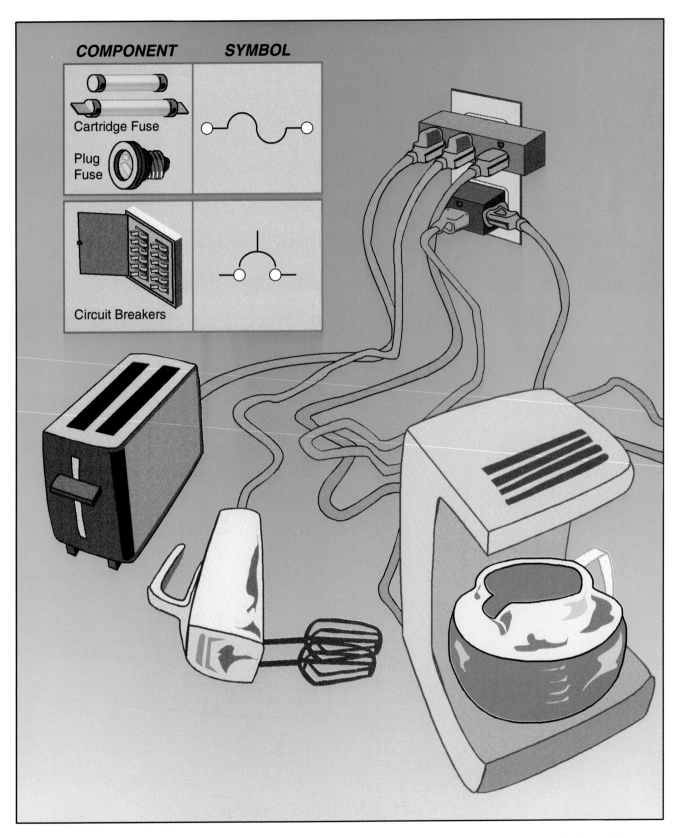

COMPONENT	SYMBOL
Cartridge Fuse	
Plug Fuse	
Circuit Breakers	

Figure 9-14 It is dangerous to overload an electrical outlet. Fuses and circuit breakers are designed to protect against such electrical overloads.

Figure 9-15 These household electrical devices have built in transformers that convert 110 electricity to the voltage that they require.

called a *transformer*. See Figure 9.16. The electricity has to be changed from alternating current to direct current because all battery-operated devices work on direct current.

—— Electricity at Work ——

What are some of the uses of electricity?

The different types of electricity are used in five basic ways:

- To turn electric motors
- To heat ovens and other appliances, and homes
- To create artificial light
- To power electromagnets
- To carry electronic communication signals

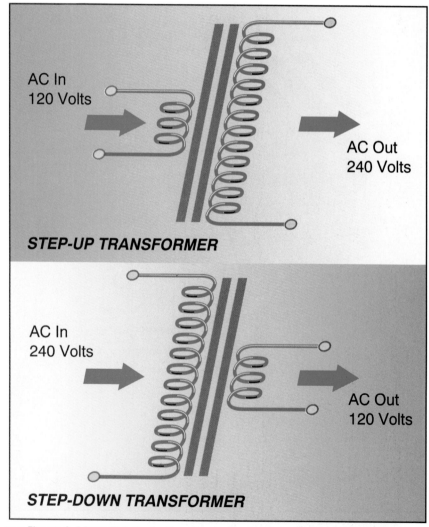

AC In
120 Volts

AC Out
240 Volts

STEP-UP TRANSFORMER

AC In
240 Volts

AC Out
120 Volts

STEP-DOWN TRANSFORMER

Figure 9-16 This illustration shows the basic principle behind the operation of a transformer. The magnetic field from one coil generates a magnetic field in the other coil.

Turning Electric Motors. Electricity is used to power the motors in power tools, office machines, industrial machines, kitchen appliances, and consumer electronic devices. If a device has parts that spin, it uses electric motors to get things spinning. Your CD player, computer disk drive, record player, electric drill, kitchen blender, electric can opener, electric pencil sharpener, and video player all have motors.

Heating Electrical Appliances. Electricity flowing through a high-resistance wire generates heat. This principle is used in toasters, ovens, heaters, hair blowers, curling irons, coffee pots, and electric dryers.

Creating Artificial Light. All light bulbs are not alike. In the most popular variety, electricity heats the *filament* to a temperature of about 4,500° F. See Figure 9.17. The filament is the part of the bulb that glows and gives off light. The bulb is filled with a gas that contains no oxygen, and this prevents the filament from burning.

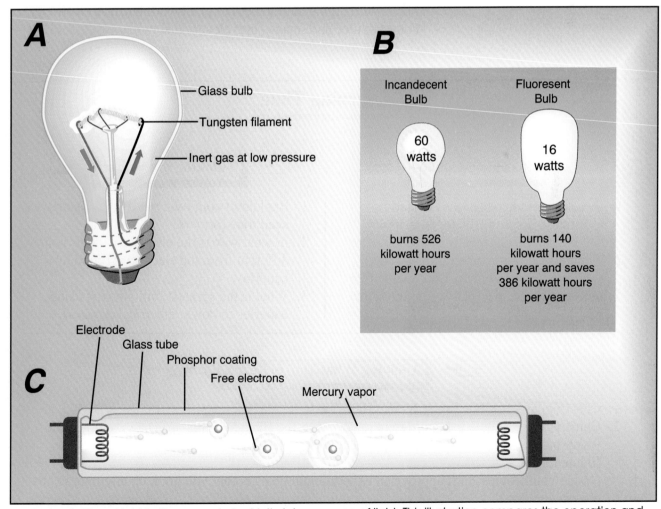

A

Glass bulb

Tungsten filament

Inert gas at low pressure

B

Incandecent Bulb

60 watts

burns 526 kilowatt hours per year

Fluoresent Bulb

16 watts

burns 140 kilowatt hours per year and saves 386 kilowatt hours per year

C

Electrode

Glass tube

Phosphor coating

Free electrons

Mercury vapor

■ **Figure 9-17** ■ Light bulbs convert electricity into a source of light. This illustration compares the operation and efficiency of incandescent and fluorescent light bulbs.

The fluorescent light has a tube that glows white when electricity is flowing. Electrodes at each end of the tube send electrons to crash into the mercury vapor that fills the tube. When the mercury vapor atoms are struck, they give off ultraviolet light. The ultraviolet light is converted to white light when it causes the phosphor coating of the tube to glow. See Figure 9.17 again.

Powering Electromagnets. Electromagnets are used to separate iron particles from other material, lift heavy iron and steel weights, turn motors and lock doors. Electromagnets are responsible for the sound from telephones, buzzers, speakers, and bells. They are used to lift and move maglev trains (see Chapter 16), create images in MRI diagnostic machines (see Chapter 15), and align the electron gun in your T.V. picture tube.

Carrying Electronic Communication Signals. Electronic communication devices use electricity to convert voice and picture images into signals that can be transmitted to remote locations. The telephone, radio, television, tape recorder, VCR, video camera, CD player, and digital audiotape (DAT) all work because electricity can convert voice and images into electrical signals.

— Learning Time —

Recalling the Facts

1. Describe the characteristics of static electricity, lightning, battery electricity, and household electricity.
2. Name five different uses for electricity. Count motors only once. Don't name five objects that use electricity to do the same job.

Thinking for Yourself

1. Explain in your own words the reason we need transformers.
2. In what way is the electricity in our homes different from the electricity found in batteries?
3. What is the greatest limitation of using batteries to power electrical devices?

Applying What You've Learned

1. Construct a small DC electric motor.
2. Construct a light bulb using different materials as the filament. This project should include a 9-volt battery, electric wire, two paper fasteners, a plastic jar, and poor conductors for the filament.
   (*Note:* Your teacher must approve all experimental plans before you hook up anything to any electric current.)

Electronic Communication

What is the main difference between the telephone and the telegraph?

Probably the most important electronic communication device is the telephone. When Alexander Graham Bell invented the telephone in 1876, he created a new communication device that was based on many of the principles of Samuel Morse's 1844 invention, the telegraph.

▰▰▰ Putting Knowledge To Work ▰▰▰

The Video Telephone

On January 6, 1992, AT&T unveiled the video telephone. When it was first introduced, a desk set cost $1,500. If you wanted to have a videophone conversation, each person needed his or her own phone. The picture part of the videophone conversation changed 10 frames per second, which made the video movement disjointed or jerky.

By the middle of 1997, advances in the speed of analog modems, video compaction, and computer microprocessors were enough for two companies to introduce two new videophone concepts at a reasonable cost. The ViaTV videophone is manufactured by 8x8, Inc., and current models range in price from $329 to $499. The VC105 model combines your TV set, touch-tone phone, and phone line with its TV-top box. This box contains a tiny digital camera, a high-speed microprocessor, and a fast analog modem. The nice part is that the hookup is as easy as attaching a VCR to your TV.

You don't need a computer to use ViaTV's videophone. It will also work with other manufacturers' videophone products. This videophone gives you standard phone call pricing, zoom, and the ability to change picture size and frame speed to a maximum of 20 frames per second. This is still 4 frames less than full-motion video.

The Intel Create and Share Camera Pack is a videophone for your computer. The North America version includes a digital camera, video capture card, and a fast modem. The upgrade kit for your computer contains software packages that include the voice picture hookup and a photo enhancer. To use their most enhanced model (USB), you must have a 166 MHz Pentium computer, Windows 95, a minimum of 16 MG RAM, 75 MB free on your hard drive, and a 4x or faster CD-ROM drive.

Once all the hardware is installed, your computer will be ready to serve also as a videophone. The installation is not as simple as hooking up a TV-top box. You may want the installation to be handled by a professional computer technician. This technology should become a standard feature on many new computer models that will be released during 1998.

The Intel videophone is also compatible with regular phone lines. This means that an Intel videophone user could have a videophone conversation with the owner of a ViaTV unit.

259

Figure 9-18 Communication devices can be combined to make new products. For example, hardware and software are available to turn your computer into a videophone.

The crucial difference between the telephone and the telegraph is that the telephone sends speech, while the telegraph sends code. Each new technological invention is built on the knowledge of earlier inventors.

Let's find out how our old and new electronic communication devices operate. Devices based on similar principles of operation will be studied near each other to help you see the similarity of their construction. See Figure 9.18 above.

The Telegraph

How does the telegraph work?

Of all the communication devices, the telegraph is the easiest to build and the hardest to use. It is the hardest to use because you must learn to communicate by sending and receiving dots and dashes (Morse-coded messages).

The telegraph circuit consists of a power source, electric wire, an electromagnet with metal clicker, and a key switch. The power source creates the magnetic field for the electromagnet to attract the metal clicker. The key switch opens and closes the circuit to control the length and spacing of the clicking sound. The wire is the pathway that connects all of the parts.

The telegraph in Figure 9.19 could be used to communicate with someone in the same room. What you want is a telegraph that can be used to communicate over distances. What would you have

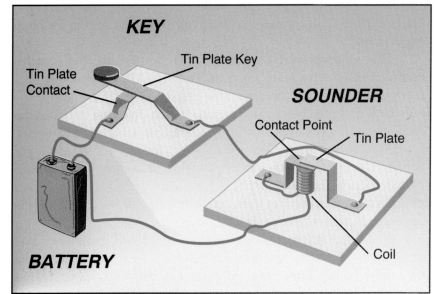

KEY

Tin Plate Contact

Tin Plate Key

SOUNDER

Contact Point

Tin Plate

BATTERY

Coil

Figure 9-19 The telegraph allows people to send and receive information by code. When you build this model of the telegraph, the two tin plates of the sounder must touch each other or your electric circuit will not work when the key is pressed.

to do to convert the electric circuits shown so that a sounder and key at one location could communicate with a sounder and key at another location?

The Telephone

What did Alexander Graham Bell have to invent in order to build his first telephone?

The first telephone to carry the human voice used a liquid, voice-activated battery. When Alexander Graham Bell spoke into the mouthpiece, a platinum wire was pushed up and down into battery acid. The farther the wire went into the acid, the greater the electric current sent into the circuit. The varying current created by a microphone combined with the battery traveled down the electric circuit and then caused a varying vibration at the receiving phone. This mouthpiece battery was the first **transmitter**, converting sound waves into electrical impulses. See Figure 9.20.

The circuit on the telegraph had the electromagnet clicking a metal plate. Bell's first telephone replaced the metal clicker with a diaphragm (rubber drum) attached to the electromagnet. Instead of getting a clicking noise, the diaphragm, when it vibrated, reproduced the sound of the human voice.

A little voice pressure on the microphone of the telephone caused a slight dipping of the platinum wire into the battery acid and the generation of a weak current. This in turn caused a weak vibration of the diaphragm at the receiving phone. The greater the current, the greater the vibration of the diaphragm. To build the telephone, Bell had to invent the microphone as well as the speaker.

It was necessary to place the battery in the microphone since the battery increased and decreased just how much electricity was sent to the circuit. This design worked, but it meant that each person's phone was going to have acid inside it.

transmitter
the part of a communication system, such as a radio, that changes the sender's message into electrical impulses and sends it through the channel to the receiving end

Figure 9-20 Alexander Graham Bell speaks into an early model of the telephone.

The first statement ever heard on a phone was Alexander Graham Bell's call for help. He was burned by the acid in his phone experiment.

This voice-powered telephone was the first phone to be used commercially. Alexander Graham Bell abandoned this design in favor of a battery-powered variable resistance design.

Bell designed a new microphone with a diaphragm that pushed carbon granules together. When the diaphragm vibrated, it caused the granules to vary the resistance of the electric circuit. See Figure 9.21A. The current in the circuit was now increased and decreased by electrical resistance. Just as before, the earpiece at the receiving phone vibrated according to the varying strength of the electric current. See Figure 9.21A again.

Condenser microphones such as the electret microphone have a better sound, are more sensitive, can be produced in tiny sizes, and can take more abuse. They are used in cellular phones, regular phones, hearing aids, lapel microphones, and listening devices (bugs). See Figure 9.21B.

■ **Figure 9-21A** ■ This illustration shows how the microphone and earpiece of a telephone operate.

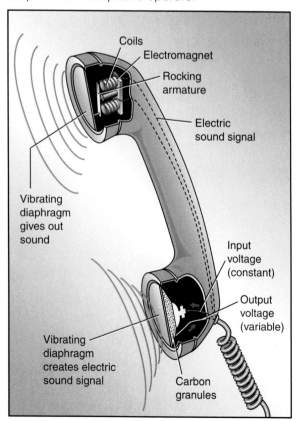

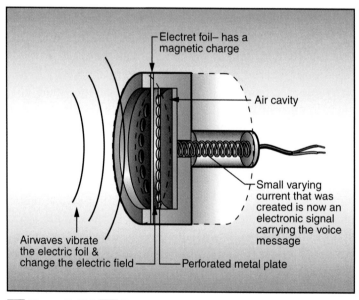

■ **Figure 9-21B** ■ The parts of an electret microphone are shown above.

262

Telephones of today still use the variable resistance principle that was developed by Bell over 100 years ago. Did you ever use a phone during a power failure? You can, since phones are not connected to your home's electrical service. The phone system is still powered by low-voltage direct current that comes to your home straight from the phone company.

Your phone signal leaves the world of Alexander Graham Bell when it reaches the computer exchange network, our modern equivalent of the old-fashioned switchboard. See Figure 9.22. Here your call is likely to be changed from an *analog* signal to a digital signal. It left your home on electric wires, but it might now be sent on its way as waves of light traveling along fiber-optic cables. Your call might even find its way to your caller as radio waves transmitted to someone's cellular telephone.

Fiber Optics. Fiber optics are cables made out of very pure glass. Each glass strand in a cable can be as thin as half the thickness of a human hair. See Figure 9.23.

The glass strands have an outer coating of glass with different reflective characteristics. This outer layer causes the rays of light traveling at the core to stay inside the filament (glass strand), regardless of twists and turns. These cables can carry a signal, without loss of power, for very long distances. When necessary, their power can easily be boosted so that messages will not be lost.

Just like with copper wires, it is possible to have each filament carry a number of signals at the same time. A fiber optic cable that is one-sixth as thick as an old copper cable can carry over four times as many calls.

fiber optic
a type of system that uses thin, flexible glass strands or cables to transmit light over great distances

■ **Figure 9-23** ■ Fiber optic cables are made up of very fine strands of very pure glass.

■ **Figure 9-22** ■ Switchboard operators are still used to complete telephone connections in some companies.

New fiber optic cables that are under our oceans are so efficient that many of the older copper cables are no longer in use. The signals that are sent along fiber optic cables are powered by lasers.

Lasers. Lasers produce very narrow beams of intense light. All the light rays in a laser beam have exactly the same wavelength. The beam that is sent by a laser can be visible light or invisible infrared light.

Lasers that send out visible light are used in digital communication. See Figure 9.24. Messages are sent as pulses of light along fiber optic cables.

By transmitting signal beams that are on different wavelengths, it is possible to have many messages travel along the same

laser
an instrument that produces a very powerful, narrow beam of light. Used to cut material, carry communication signals, build and guide weapons, and produce 3D pictures.

■■■ Putting Knowledge To Work ■■■

Laser Cleaning

Do you know of a dirty, ancient statue that is in need of a good scrubbing? A physicist by the name of John Asmus travels around the world with his million-watt laser to clean irreplaceable works of art.

His laser looks like a yellow, wheeled dishwasher with a vacuum hose attachment. On the other end of the hose is an aluminum light fixture with what looks like a short fluorescent light.

Wearing protective eye goggles, Asmus places the tube against the statue and presses a trigger. A light flashes followed by a loud pop. Smoke now drifts up from the area where the laser has struck the statue.

It takes thousands of blasts from the laser to clean a statue and restore it. Why not just wash it or blast it with something not so high tech? Other methods of cleaning damage the stone.

Once he has taught the curator how to operate the machine, this physicist is ready to take on the challenge of another art treasure. He believes that it is possible to fine-tune the fre-

quency of a laser so that it can restore the color pigment in a painting. This means that the original color—which has long ago changed because of oxidation—can be turned into fresh-looking paint in the most delicate of ancient colors.

The cleaning career of the laser began years ago when an art curator asked if the laser could zap their statues clean. She explained to Asmus and his group of scientists that air pollution from oil refineries were turning their marble statues into crumbling soft plaster that was a few hundredths of an inch thick. This crust, combined with black dust, made the statues look terrible.

After the group finished using the lasers for photographing certain statues, Asmus, with the curator's permission, turned the laser up to full blast and shot an unimportant statue to see what effect the laser would have. His result was a perfectly clean spot without any damage to the stone. All the soft material was removed evenly, so the features of the statue weren't affected at all.

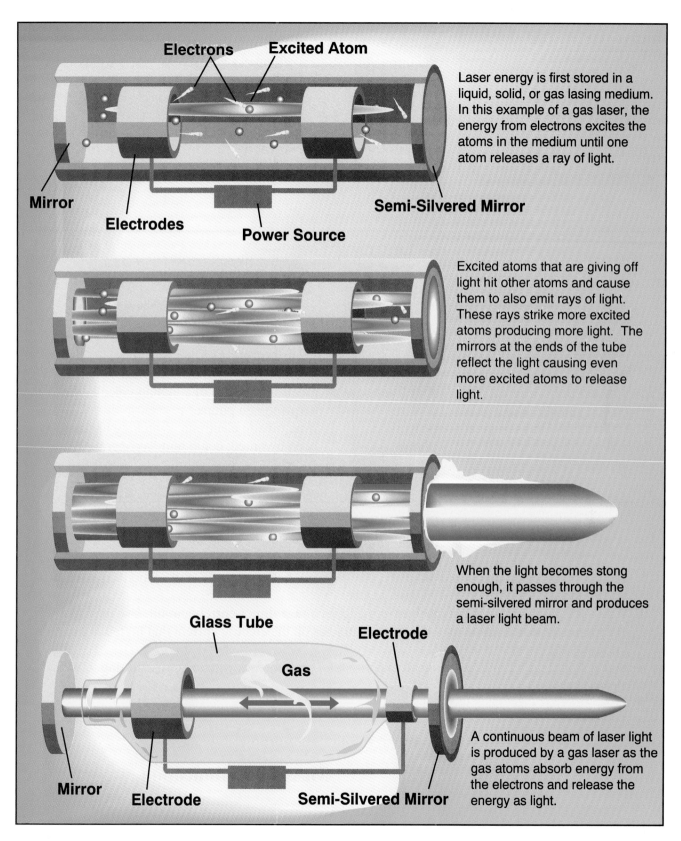

Electrons **Excited Atom**

Laser energy is first stored in a liquid, solid, or gas lasing medium. In this example of a gas laser, the energy from electrons excites the atoms in the medium until one atom releases a ray of light.

Mirror

Semi-Silvered Mirror

Electrodes

Power Source

Excited atoms that are giving off light hit other atoms and cause them to also emit rays of light. These rays strike more excited atoms producing more light. The mirrors at the ends of the tube reflect the light causing even more excited atoms to release light.

When the light becomes stong enough, it passes through the semi-silvered mirror and produces a laser light beam.

Glass Tube

Electrode

Gas

Mirror

Electrode

Semi-Silvered Mirror

A continuous beam of laser light is produced by a gas laser as the gas atoms absorb energy from the electrons and release the energy as light.

Figure 9-24 This diagram shows the operation of a laser.

fiber optic strand at the same time. At the end of travel, the different wavelengths are separated, just like in radio communication.

Infrared lasers give off intense heat. They are used in industry to cut and process materials. Lasers also have military applications. At this time they are being used to guide "smart" weapons to their targets. The particle beams that were developed by the Star Wars Defense program are very high-powered laser beams capable of destroying incoming missiles and earth satellites. In 1997, the MIRACL (Mid-Infrared Advanced Chemical Laser) was successfully test fired at a dying U.S. satellite from White Sands Missile Range.

— Learning Time —

Recalling the Facts

1. What is the most significant difference between the telegraph and the telephone?
2. How did Bell's original phone differ from the one you use today?
3. Explain how Bell's original phone worked.
4. How do fiber optic cables carry phone signals?

Thinking for Yourself

1. Describe how a carbon granule microphone forms the electrical signal that creates the sound of a human voice in the earpiece of a phone.
2. If automatic electronic switching had not been invented, could humans operate a switchboard system today? Why or why not?

Applying What You've Learned

1. Build a small telegraph out of a tin plate, electric wire, batteries, nails, and a wood base. All project ideas must be approved by your teacher.
2. Wire a telephone speaker and receiver into a low-voltage direct-current circuit. Use the electric meter to determine the effect that sound has on the electric current that is flowing through your circuit.
   (*Note:* Your teacher must approve all experimental plans before you hook up anything to any electric current.)

—— Mass Communication ——

What is mass communication?

The telegraph and telephone were like circus performers who were fantastic at what they could do, except they couldn't work

Searching for E.T.

SETI—or Search for Extraterrestrial Intelligence—has listened to millions of radio signals from space for a chance message from intelligent life. Radio-wave signals are natural occurrences that cut across the solar system without purpose or destination.

Radio and television stray signals have been transmitted into space by people on our planet for at least fifty years. They left our planet traveling at the speed of light, and these signals have already passed 100 suns on their travels through the universe. They will continue to travel on through time unless an intelligent life-form with very sensitive equipment receives the signal. Perhaps a thousand years from now, an intelligent being will see "I Love Lucy" and wonder where the signal came from.

Many researchers feel that intelligent life in space predates our own by perhaps millions of years. They hope that signals from these ancient worlds will by chance reach our planet. To receive such a signal would forever end the question, Are we the only planet that has intelligent life?

Many other people feel that SETI is looking for a needle in a haystack. Would intelligent life-forms have had to invent radio as part of their technology? Would they have learned to use radio communication signals that we can listen to? What's your opinion? Do you think that E.T. will ever phone home, or do you feel that SETI is looking for an unlisted number?

---

without a safety net. The telephone and telegraph couldn't, at the time of the introduction, work without wires.

At first, radio transmissions only carried coded telegraph signals rather than voice communication. With the development of voice transmission, radio allowed many people to listen to the same conversation at the same time without being wired to the original source. The name given to this sharing of a signal is **mass communication**.

Television is the perfect method of mass communication because it provides the audience with the ability to hear and see things in the comfort of their own homes.

mass communication
the systems that make it possible for many people to hear the same message at the same time. Radio and television are examples of mass communication systems.

Radio

How does a radio work?

To transmit a voice signal by radio, a microphone at the radio station converts speech into a variable current flow signal. This type of microphone works just like the one in your telephone.

The variable signal attaches itself to a radio carrier wave. You can compare the joining of these two signals to a class trip on a bus.

The voice signals are like passengers on the bus. The bus is the carrier wave. Just as the bus carries you to your destination, the carrier wave carries the signal to its destination (a radio).

When you tune your radio, you are adjusting which radio carrier wave will make it to the circuits of your radio. The chosen frequency creates a weak but exact duplicate of the signal that left the radio station. This signal is then amplified by vacuum tubes, transistors or integrated circuits. The signal is then sent to the speaker of your radio.

Putting Knowledge To Work

HDTV

By Christmas 1998, the first digital high-definition television (HDTV) sets will go on sale in the United States. To start building HDTV systems, television manufacturers and broadcasting companies had to wait for the Federal Communications Commission (FCC) to decide on one standard for all HDTV broadcast systems.

HDTV isn't compatible with our current analog TVs. TV broadcasting companies are required to transmit both analog and HDTV broadcasts until the year 2006. Even though analog transmissions will stop around that time, special TV boxes will exist that will convert the digital signals to analog signals.

The significant difference between HDTV and analog TV isn't the shape of the picture tube. Rather, the big difference is how the picture is sent from the broadcasting company to your TV set.

Let's look at how analog and digital programs are transmitted. In both cases, they start with an analog video camera that produces a variable-strength electric signal as it scans the scene in front of it. In an analog system, a bright spot in front of the camera becomes a strong signal; a dark spot becomes a weak signal. These signals are then joined to a radio carrier wave and transmitted to your TV receiver.

The entire transmission process of the analog system keeps the analog energy wave, which is constantly changing in strength, from getting distorted. When the signal is received by your television, its varying strength determines how bright the areas of your picture tube glow.

In digital HDTV, the camera's analog image is converted into compressed digital code. This stream of digital code is then joined with a carrier wave and transmitted to your TV receiver. This digital stream is made up of 1's and 0's that the computer in your HDTV converts into motion pictures.

The main difference between the two systems is the conversion to a digital signal and its compression so that it can be transmitted to your home at 19.39 million bits per second. If you have a DBS (Direct Broadcast Satellite) system, such as PrimeStar, or a DSS (Digital Satellite System), such as RCA's system, your broadcast is received from a satellite in a geosynchronous orbit and then transmitted directly into your HDTV set or a TV box if your set is analog.

HDTV will give you 3-D computer-monitor-quality TV pictures from a digital television. You will therefore be able to surf the Web, play video games, and add on the necessary computer hardware to do anything that your regular computer can do.

Television

How does a television work?

In your television the picture transmission and the sound transmission are handled separately. The picture and sound come to your home through completely different and separate systems that function together.

The *audio* system includes the television station's microphones and transmitter and your home television set's built-in radio. The *video* system includes the television station's cameras and transmitter and your home television set's tuner, video decoder, and picture tube. Your television picture tube reverses the electronic signal created by the television camera. See Figure 9.25.

The Recording of Sound

How do tape recorders, video recorders, compact disc players, and DVDs work?

All recording and playback systems have several things in common. They all need to convert sound into a variable strength electric signal. These signals must be stored on some material. Each system must have a way of converting the stored signal back into a variable electric current to reproduce the sound.

Tape Recorders. You are probably familiar with cassette recorders. You might have wondered how they record and play back sound.

The electric signal exists on the surface of the recording tape. As the tape passes the playback head, two small coils of electric wire pick up variable currents from the magnetic tape. These two variable signals (two signals give you stereo) are sent to the amplifiers and then on to the player's speakers. See Figure 9.26.

When you press the record button, the variable signal is sent into the coils of the recording head. The magnetic fields of these coils cause the magnetic tape to become slightly, moderately, or fully magnetized according to the sound strength of the music.

Figure 9-25 The circuits of your television set are designed to receive the signal from the television station and then convert it into picture and sound.

Figure 9-26 An audiotape is ready to be played on a cassette tape deck.

Figure 9-27 The newest models of video recorders are digital. They record onto small tapes. This model has a 3.8" color LCD monitor, 100X digital zoom, a remote control, and a color viewfinder.

Figure 9-28 This personal DVD player includes a display screen. The unit is battery operated, so you can take it anywhere. DVD drives are also now available to replace CD-ROM drives. DVDs are backward compatible and can play audio CDs, CD-ROMs, CD-Rs, and CD-RWs.

Video Recorders. Video recorders record video pictures in the same way that cassette recorders record music. The only difference between the two is that video pictures have huge amounts of information to record.

To handle the amount of information, videotape is much wider than audiotape. The video head records on an angle, which further increases the tape surface area for storing information. See Figure 9.27.

Compact Disc Players. Compact disc players use a laser beam to read the recorded signal. The laser light is reflected off the disc to a photoelectric cell. The signal is created because the entire disc is made up of bright spots and little, dark crevices. The signal from the photoelectric cell is then sent through the machine's computer processor to the speaker system.

DVD. DVD is an abbreviation for Digital Video Disc or Digital Versatile Disc. Different manufacturers' products have different features. See Figure 9.28. On all units, you can choose the shape of the picture, change languages, switch camera angles, run subtitles, and access any point that you want in the movie. Some units allow you to change R-rated movies into PG or just lock out certain channels completely. On some units, you can zoom in or out for close-ups or long shots of particular parts of a scene.

The disc is the physical size of a CD-ROM, yet it can hold 17 gigabytes of data, which is the equivalent of a full-length movie. Its storage capacity makes many believe that it will eventually replace the compact disc player, CD-ROM drive, laser disc player, and VCR.

DVD's principle method of operation is the same as the compact disc player described above. It does have backward compatibility with today's video and audio CDs. Some manufacturers already are placing DVD drives onto their computers.

— Learning Time —

Recalling the Facts

1. What is mass communication?
2. What kind of signals were carried by the first radio transmissions?
3. What systems exist at the television studio and in your home that make television possible?

Thinking for Yourself

1. What is the main advantage of HDTV over our current system?
2. Why can't many different companies market their own system for HDTV?

Applying What You've Learned

1. As a group activity, design a model or chart that shows what your group considers the ideal mass communication machine.
2. Pick your favorite electronic communication device (such as the telephone, radio, TV, tape recorder, or video recorder). Using research books in your technology laboratory, school library, or public library, prepare a chart or model that shows how this device operates.

Summary

Directions

The following sentences contain blanks. For each numbered blank, pick the answer that makes the most sense in the entire passage. Write your answer on a separate sheet of paper.

An invisible force called electricity is used to power the electrical devices found in our homes, schools, and factories. These devices include ___1___ motors, heaters, artificial lights, electromagnets, and electronic communication transmitters and receivers.

Electricity will only flow through an electric circuit that has no openings. Such a ___2___ consists of wire made out of an electric conducting material, a power supply, and a lamp or other electrical device. ___3___ are used to ___4___ an electric circuit. They start and stop the electricity from flowing.

Electricity is forced through the wire by its electrical pressure, called voltage. This ___5___ pushes the current through the circuit.

1.
a. electric
b. new
c. mechanical
d. solar

c. Wire cutters
d. Switches

4.
a. convert
b. increase
c. open and close
d. eliminate

2.
a. circuit
b. machine
c. carriage
d. device

5.
a. solar energy
b. pressure
c. laser
d. resistance

3.
a. Electronic circuits
b. Fuses

(Continued on next page)

Continued

The current is the ___6___ of the electricity that makes your device ___7___ . The circuit wire's thickness, length, and ability to conduct electricity will determine how much opposition there is to the flow of electricity. This ___8___ is called resistance.

Electrical insulators have very high resistance to the flow of electricity. They are used to cover electrical ___9___ . They prevent us from touching the wire and getting ___10___ . Some materials under special conditions can act as insulators or conductors. These materials are called semiconductors. New special materials are now being developed that have no electrical resistance. They are called ___11___ .

Electric utility companies ___12___ electricity to people who want to run electrical devices. An electric meter tells the utility company how much electricity has been ___13___ . You are charged by the kilowatt hour which is 1000 watts of electricity used for one hour. A 100 watt light bulb that is burning for ___14___ will consume one kilowatt hour of electricity.

Over time, an inexpensive air conditioner that uses a lot of electricity will cost more money to own than ___15___ air conditioner. Many new appliances list an efficiency rating. A very efficient machine will use less electricity.

In a series electric circuit, electricity must pass through ___16___ device in the circuit. If one device burns out, the electricity will stop flowing. In parallel circuits, the electricity has ___17___ pathways to each device in the circuit. If one device burns out, the electricity isn't prevented from reaching the other devices in the circuit.

Battery electricity is called direct current. Your telephone is powered by low voltage direct current that comes to your house direct from the telephone company. That is why you might have experienced a ___18___ and found that your phones were still working.

Your home is powered by a type of electricity called alternating current.

6.
  a. part
  b. core
  c. resistance
  d. voltage

7.
  a. break
  b. energetic
  c. worthy
  d. work

8.
  a. voltage
  b. opposition
  c. amperage
  d. wattage

9.
  a. insulators
  b. fiber-optic cables
  c. conductors
  d. motors

10.
  a. an electric shock
  b. extra energy
  c. free power
  d. battery power

11.
  a. superinsulators
  b. superconductors
  c. voltage meters
  d. electrical conductors

12.
  a. give
  b. rent
  c. sell
  d. donate

13.
  a. wasted
  b. found
  c. squandered
  d. used

14.
  a. one hour
  b. ten hours
  c. one day
  d. one week

15.
  a. an efficient
  b. an inefficient
  c. a foreign
  d. a modular

16.
  a. one
  b. not one
  c. every other
  d. each

17.
  a. two
  b. three
  c. four
  d. separate

18.
  a. bill reduction
  b. power failure
  c. broken light bulb
  d. power reduction

Key Word Definitions

Directions

The column on the left contains the key words from this chapter. The column on the right contains a scrambled list of phrases that describe what these words mean. Match the correct meaning with each word. Write your answer on a separate sheet of paper.

Key Word

1. **Voltage**
2. **Electric circuit**
3. **Amperage**
4. **Resistance**
5. **Ohm's law**
6. **Wattage**
7. **Electron theory**
8. **Conductor**
9. **Insulator**
10. **Series circuit**
11. **Parallel circuit**
12. **Semiconductor**
13. **Superconductor**
14. **Direct current**
15. **Alternating current**
16. **Frequency**
17. **Transmitter**
18. **Fiber optics**
19. **Laser**
20. **Mass communication**

Description

a) How electrons flow

b) Strength of electricity

c) A separate electrical path to each energy-using device

d) Electrons flow in one direction in the electric circuit

e) Voltage = Amperage × Resistance

f) A complete electrical pathway

g) Number of cycles per second

h) The rubber or plastic coating on an electric wire

i) Large audiences receiving the same communication

j) Can act as a conductor or insulator

k) Electrons constantly change direction in the electric circuit

l) Cables of very pure glass

m) Electrical pressure

n) Electricity flows through easily

o) Only one electrical path

p) Very narrow beam of intense light

q) Can conduct electricity without resistance

r) Opposes electrical flow

s) Amount of electric power needed to run an electrical appliance

t) Converts sound waves to electrical impulses

Resources You Will Need

- One clothespin
- 6 inches of no. 20 bell wire
- Wire stripper
- Metal punch
- Drill
- ⁹⁄₆₄ drill bit
- 3½ x 4½ wood base
- 9-volt battery
- 9-volt battery snap (Radio Shack no. 270-325 or equal)
- Piezo buzzer (Radio Shack no. 273-060 or equal)
- Two 1-inch ⁶⁄₃₂ machine bolts
- Four ⁶⁄₃₂ nuts
- Six small washers
- Four ½-inch no. 6 wood screws
- One 1-inch no. 7 wood screw
- Electrical tape
- Tongue depressor
- String
- Items needed to individualize your alarm system
- 3½ x ¾ tin plate strip

The Simple Alarm System

Did You Know?

The principle of the telegraph led to the development of the telephone. It also led to the development of the alarm system.

When you think about an alarm system, do you only think about the type of system that protects against theft? Less noisy alarms also warn people when technological devices need servicing, when messages have arrived, when medical emergencies exist, and when it is time to get up in the morning.

All alarm systems include a circuit, switch, and bell. In an alarm clock the switch is connected to a clock, and the bell is the ringer or radio that wakes you up in the morning. Can you name some other alarm systems and identify their switches and bells?

The modern burglar alarm system still includes the simple switches of the past along with sensors that react to physical movement and changes in temperature. The burglar alarm bell of the past has been replaced by ear-splitting sirens and silent calls to the police. The parallel and series circuits that are used in these alarms have been enhanced by computer monitoring that reads the resistance in the alarm's wire. These circuits constantly check for breaks or attempted bypass by burglars.

A.

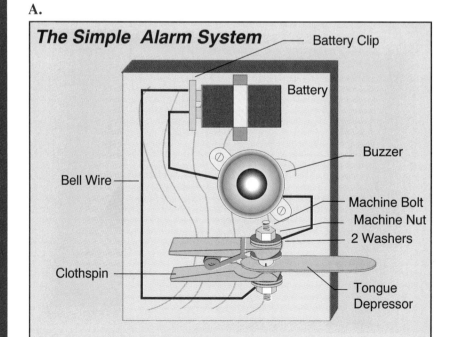

The Simple Alarm System

- Battery Clip
- Battery
- Buzzer
- Machine Bolt
- Machine Nut
- 2 Washers
- Tongue Depressor
- Clothspin
- Bell Wire

Problems To Solve

The simple alarm shown in Figure A can be triggered by all kinds of situations. The alarm will sound when the insulation is pulled from the clothespin. If this insulation is connected to a Ping-Pong ball, it can be rigged so that rising water could float the ball and set off the alarm. How would you accomplish this task and keep your alarm above water?

If this alarm is attached to the handle of a shopping bag, it could be wired to go off if someone tried to pick up the bag.

Your problem is to build the alarm and then find a useful or mischievous place to install it. Demonstrate your alarm and how it will be used to the class. You could videotape your "alarming situation" showing classmates or family members caught by surprise. If your video is funny, you might send it in to the T.V. show *Funniest Home Videos*.

A Procedure To Follow

1. Prepare the wood base for your alarm. Figure B.
2. Cut and bend the metal strip that will hold your battery to the base. Figure C.
3. Use the metal punch to cut the holes in your battery holder. Figure D on the next page.
4. Use two no. 6 wood screws to attach your battery

B.

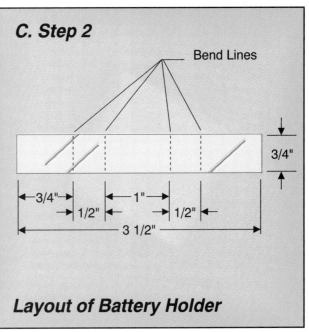

B. Step 1

4 1/2"

3/4"

3 1/2"

Wood Base

C.

C. Step 2

Bend Lines

3/4"

3/4" 1/2" 1" 1/2"

3 1/2"

Layout of Battery Holder

holder to the base. Do not attach the battery clip to the battery until all work on your alarm is completed.

5. Use two no. 6 wood screws to attach the piezo buzzer to your base.
6. Drill a ⁹⁄₆₄ inch hole through the two closed ends of your clothespin. Figure E.
7. Through each of these holes, slip a machine bolt and washer. *Note:* Bolt heads are on the inside of the clothespin. Figure F.
8. Secure each bolt with a nut.
9. Strip ½ inch of insulation from the ends of your wire.
10. Slip two washers onto the ends of each bolt. Hook one end of your bell wire between these washers and secure it in place with a nut.
11. Follow this same procedure to hook the black wire from the piezo buzzer to the other side of the clothespin.
12. Secure the clothespin to the base by screwing the no. 7 wood screw through its spring.
13. Splice the red wire from the buzzer to the red wire of the battery snap. Tape your connection with electrical tape. Figure G.

D.

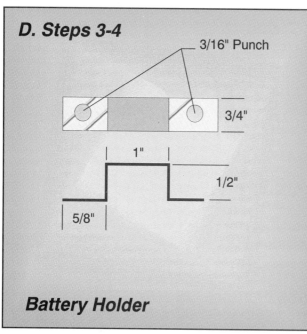

D. Steps 3-4

3/16" Punch

3/4"

1"

1/2"

5/8"

Battery Holder

E.

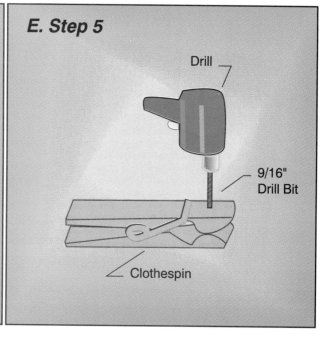

E. Step 5

Drill

9/16" Drill Bit

Clothespin

14. Splice the free end of your bell wire to the black wire of the battery snap. Tape your connection.
15. Check that all components have been connected correctly.
16. Place a tongue depressor between clothespin contacts.
17. Connect the battery clip to the battery.
18. Test the alarm by pulling the tongue depressor out from between the contacts.
19. Shorten the depressor, and drill a small hole into its end for a pull string.
20. Adapt your alarm to work where you plan to install it.

What Did You Learn?

1. Why doesn't the buzzer ring before you pull the tongue depressor?
2. Why can't a larger model of this alarm protect your home?
3. Is there a relationship between the voltage of the battery and the size of the buzzer?

F.

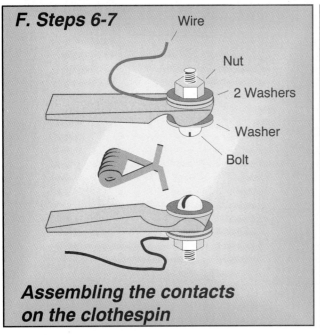

F. Steps 6-7

Wire
Nut
2 Washers
Washer
Bolt

Assembling the contacts on the clothespin

G.

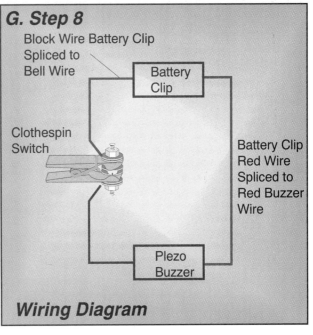

G. Step 8

Block Wire Battery Clip Spliced to Bell Wire

Battery Clip

Clothespin Switch

Battery Clip Red Wire Spliced to Red Buzzer Wire

Plezo Buzzer

Wiring Diagram

Clean Room

Old-time factories were often dirty, noisy places. Today, manufacturing may take place in a "clean room"—a dust-free room where temperature and humidity are strictly controlled. Computer chips and other delicate electronic parts are made in clean rooms.

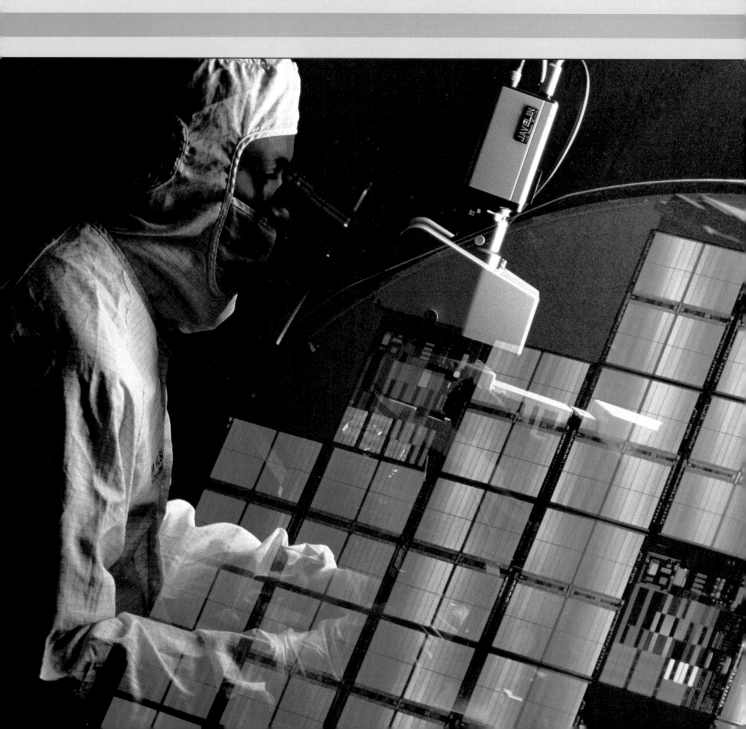

Chapter 10

The Story of Manufacturing

— Key Words —

In this chapter you will learn and study the meanings of the following important technology terms.

MANUFACTURED PRODUCTS
FACTORY
CONSUMERS
CRAFT
CRAFTSPERSON
INDUSTRIAL REVOLUTION
DIVISION OF LABOR
SCIENTIFIC MANAGEMENT
TIME AND MOTION STUDY
MOVING ASSEMBLY LINE
DESIGNING
MARKET RESEARCH
SUPPLIERS
CAPITAL
JUST IN TIME
PRODUCTION MATERIALS
ADDED VALUE
QUALITY CONTROL
ASSEMBLY LINE
COMMISSION
ORGANIZATION CHART

Look around at the objects in your classroom. What do you see? Doors? Tables? Clocks? Desks? These items, and probably everything else in your classroom, are **manufactured products**. See Figure 10.1 on the next page. They are products made in special buildings called **factories**.

Most of the things you use every day were manufactured in factories. In fact, you may find it difficult to name things you use that were never inside a factory. Factory production is a manufacturing system that is essential to our modern way of life. Without factories, we would not have cars, contact lenses, televisions, electric drills, wrist watches, books, microwave ovens, or board games like Monopoly.

Manufactured items are sent to stores, where we can purchase them. We are *consumers* of manufactured products. **Consumers** are people who buy and use the items made in factories. We are consuming products just by using soap, wearing clothes, and reading books.

Manufacturing is a pretty remarkable process when you think about it. Imagine being told, for example, that you had to manufacture T.V. sets. Where would you start? Think of all the parts you'd need. How would you ever make them? How would you fasten them all together? Sounds impossible, doesn't it? In this chapter and the next, you'll learn to appreciate this remarkable manufacturing process that we all take for granted.

Figure 10-1 Most of the items you buy in a shoe store, electronics store, drug store, and every other store, were made in factories.

The History of Manufacturing

How were some early products manufactured?

The products that people used in the past weren't always made in factories. Many years ago, each family had to make almost everything they needed. They made their own cloth for clothing. They made their own farming tools, their candles for lighting, their wagons for transportation, their furniture, and even their children's toys. It took a long time to make each item.

Not everyone could do a good job at making so many different things. Some people could make good wagons but not good tools. Others were better at making cloth. People soon began to specialize in just one type of job. The job became known as a **craft**. Blacksmiths, wagon makers, and shoemakers, for example, were known as **craftspeople**. See Figure 10.2.

craft
a skilled occupation or job, usually done with the hands, such as carpentry or sewing

craftsperson
a person who specializes in a particular job or craft

Figure 10-2 John Deere (left), inventor of the steel plow, watched his 1874 horse-drawn Gilpin Sulky that could plow over three acres of soil in 12 hours.

The Industrial Revolution: The Early Years

Why was the Industrial Revolution so important to technology?

During the late 1700s, great changes started to take place around the world because of new methods of manufacturing. Because the new methods were so important, that period is called the **Industrial Revolution**. Goods once produced by hand were now produced by machines and power tools. Cities boomed when factories drew people from the farms to the cities. The Industrial Revolution was a great turning point in the history of the entire world.

The first factories in the world made textiles in England and Scotland between 1750 and 1800. *Textiles* are woven fabrics like wool and cotton. The factory's power usually came from rivers that turned waterwheels. The spinning waterwheels operated machinery inside a building. The factory system was so important to England and Scotland that it was illegal to leave either country with drawings or books about the new factories.

The machinery in the first U.S. factory was put together by Samuel Slater in Pawtucket, Rhode Island. Slater came from England and had memorized everything he could about England's textile factories. Using water power from the Blackstone River, the factory he set up produced its first cotton thread in 1790. It later produced cloth. See Figure 10.3 on the next page. Cotton manufacturing was the first industry in America.

In 1798 Eli Whitney received a contract from the U.S. government to manufacture 10,000 rifles. Whitney felt that his idea of division of labor would help him meet the goal. See Figure 10.4 on page 285. **Division of labor** means dividing the work into many simple steps. Whitney did not have each worker making one rifle from start to finish as did other manufacturers. He had fifty workers making or

Tech Talk

The Industrial Revolution caused many social changes. It changed America from a country based on farming to a country based on industry. That's why it's called an Industrial *Revolution*. It had no precise beginning date, but 1760 is sometimes used.

Industrial Revolution
a period during the late 1700s when machines were used to complete many of the tasks previously done by hand. Factory production replaced much of the home manufacturing.

division of labor
the idea of dividing work into many steps. A different person is responsible for each different step. This allows the job to be completed quickly and accurately.

Figure 10-3 Samuel Slater's 1790 factory is a registered historic site in Pawtucket, Rhode Island and is open to the public.

scientific management
the system of developing standard ways of doing particular jobs. One goal of the system is to cut out all wasted time and motion.

time and motion study
an investigation into the best ways of doing a job. Things such as working conditions, wasted time, and unnecessary movement are looked at.

assembling individual parts. Each job was simple and could be completed quickly and accurately.

Division of labor proved so successful that it is still used today. Still, Whitney was the first to use the method, and he had many problems getting it established. His contract with the government took ten years to complete.

James Watt designed a steam engine that could operate machines in factories. Watt offered his first engines for sale in 1774. However, it took many years before factories completely switched from water power to steam engines.

The Industrial Revolution: The Later Years

Who were some people who improved working conditions in factories?

As factory production increased, people began to look for better ways to do things. Factories had become wasteful of the worker's efforts. Frederick Taylor worked for a Philadelphia steel mill in 1880. See Figure 10.5 on page 286. He was the first person to use a stopwatch to measure how long certain factory operations took to complete. He developed standard ways of doing things, making it easier for workers to manufacture products. His method is called **scientific management** or the *Taylor system of manufacture* and is an important part of modern factories. Taylor earned over 100 patents.

The husband-and-wife team of Frank and Lillian Gilbreth also worked on better ways of doing things. Like Taylor, they worked in the new field of **time and motion study**. Time and motion study is a way of finding the best way to do a job. The Gilbreths improved factory working conditions and developed charts and motion picture techniques to measure factory efficiency. See Figure 10.6 on page 286.

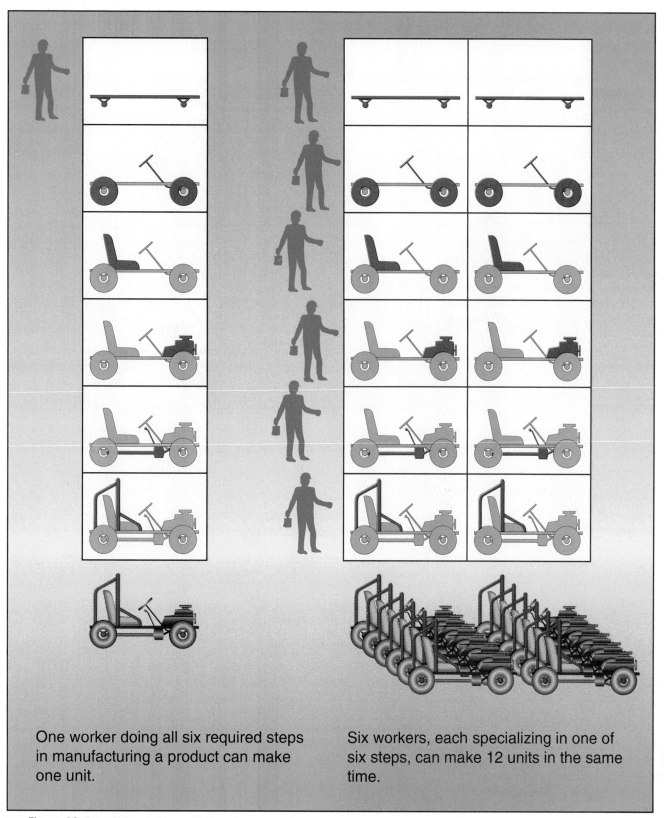

One worker doing all six required steps in manufacturing a product can make one unit.

Six workers, each specializing in one of six steps, can make 12 units in the same time.

Figure 10-4 With division of labor, fewer workers are necessary to manufacture high quality products.

Figure 10-5 Frederick Taylor used time studies in the 1880s to make factory work more efficient, which resulted in wage increases for the workers.

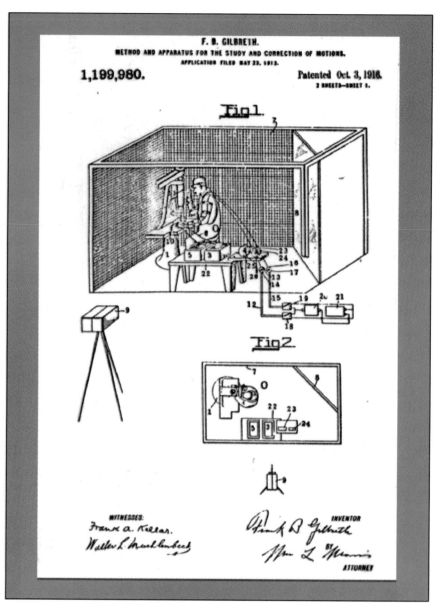

Figure 10-6 Frank Gilbreth made factory motion study measurements with his wife Lillian, and obtained this patent in 1911.

moving assembly line
a system that moves parts to the people who then complete their specific jobs with those parts

Henry Ford was an automobile manufacturer who made more cars in one year than all other U.S. manufacturers combined. Figure 10.7 shows one of his patent drawings. Ford was able to make so many cars because he invented the moving assembly line in 1913 for his factory in Dearborn, Michigan. In a **moving assembly line**, the worker stays in one place and the workpiece passes by. In Ford's moving assembly line, car frames were connected by one long chain. People attached parts as the frame slowly passed by. This improvement was so important that by 1924, half the world's automobiles were Fords. Practically all modern factories use moving assembly lines.

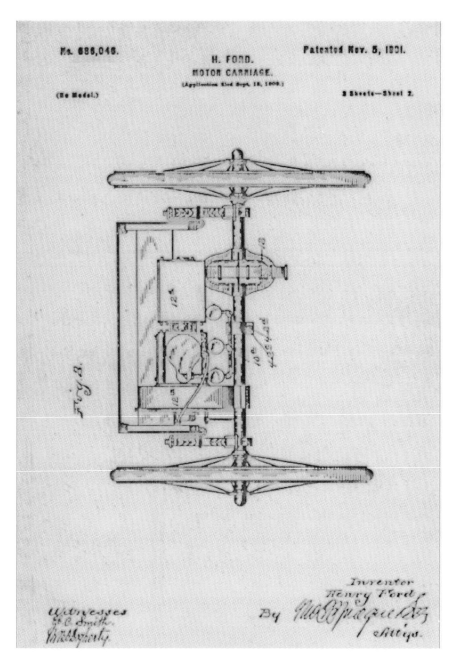

Figure 10-7 Henry Ford was an inventor and received many patents, such as this one in 1901 for a "Motor Carriage".

— Learning Time —

Recalling the Facts

1. What was the general name for blacksmiths, wagon makers, and shoemakers?
2. What was the Industrial Revolution?

Thinking for Yourself

1. List some things you use every day that were never inside a factory.
2. Was the Industrial Revolution really a revolution?

(Continued on next page)

3. What was the first product made in factories?
4. Eli Whitney divided factory work into many simple steps. What is this called?
5. Who helped develop factory charts and motion picture techniques?
6. Why could Henry Ford make more cars than anyone else?

3. List some of the problems an early factory might have had in changing from water power to steam engine power.
4. A product can be made more quickly when it moves on a moving assembly line, rather than the worker moving from product to product. Explain.

Applying What You've Learned

1. Write down, step by step, what you do to get ready for school in the morning. Compare your list with a classmate's list. Which one is better? Would a combination of the two lists be best? Explain. (This type of comparison is a simplified version of scientific management.)
2. Construct a project by the craft method. Then construct a similar project teamed with your classmates, using the division of labor approach. Compare the quality and speed of construction between the two projects.

Factories, Yesterday and Today

Who were some people who started well-known companies?

Some companies have colorful roots, or beginnings. Others have interesting stories about the people who started the company. The following list describes some of these beginnings. Many of these companies will be familiar to you.

- In 1839 Charles Goodyear found a way to make rubber useful as a waterproofing material. The company that took his name had nothing to do with Goodyear. The Goodyear Tire & Rubber Company of Akron, Ohio, was founded thirty-eight years after Charles Goodyear's death. See Figure 10.9.
- In 1844 Linus Yale invented a small tubular lock that made security available to everyone. His lock has been called the most important invention in lock-making history. The company he founded is Yale Security Incorporated of Charlotte, North Carolina. See Figure 10.10 on page 290.
- In 1853 Levi Stauss transported canvas for wagon covers to San Francisco from New York. Instead of making wagon covers, he sewed the canvas into durable trousers for gold miners. We now call those trousers Levi's. The company he founded is Levi

Figure 10-8 Jan Matzeliger invented a shoe sewing machine in the 1880s. His *lasting machine* was so important, it was the basic product made by the huge United Shoe Machinery Corporation.

Figure 10-9 Charles Goodyear often said that he discovered how to tame rubber's stickiness after accidentally dropping some on a hot stove.

Putting Knowledge To Work

Against All Odds

Jan Matzeliger was an African American immigrant who lived alone in Lynn, Massachusetts. See Figure 10.8. In the late 1800s, he invented a machine that helped revolutionize shoe manufacture. Matzeliger's invention was the basis of a great corporation (United Shoe Machinery Corporation). It produced several millionaires (himself not among them) and created work for thousands of Americans.

Matzeliger was born in South America and entered Philadelphia harbor as a seaman when he was twenty-one. Trained as a machinist, he learned to operate a machine for sewing shoes. He moved to Lynn because it was the center of the U.S. shoe industry.

The manufacture of shoes was slowed down by one complicated process: *lasting*, or shaping the top of a shoe and fastening it to the bottom of the shoe. Lasting was done by hand.

No one could build a machine to copy the way a person's hands and fingers handled the shoe leather. One factory owner spent $250,000 trying to make a lasting machine and failed.

Matzeliger took up the challenge. He had no friends to encourage him and no family to comfort him at night. He spent too much of his small factory earnings on his invention. He ate poorly and didn't spend much money to heat his small apartment. It took Matzeliger eight long years, but he finally invented a successful lasting machine. He received a patent in 1885 and signed over the rights to a manufacturer. Quietly working alone in his small apartment, Matzeliger succeeded where all the engineering talent that money could buy had failed.

Matzeliger died of tuberculosis four years later at the age of thirty-six.

Figure 10-10 Linus Yale invented a small lock in 1844 that made security available to all people.

Strauss and Company of San Francisco, California.

- In 1881 the first instamatic point-and-shoot camera, the Kodak, was offered to the public. The word *Kodak* has no special meaning at all. George Eastman made up a simple word that began and ended with the letter K. The company he founded is the Eastman Kodak Company of Rochester, New York.
- In 1887 Herman Hollerith invented a calculating system that used punched cards. He made the cards the size of $1 bills so that they could be stored in file cabinets made for banks. The company that Hollerith founded is now International Business Machines (IBM) of Armonk, New York. See Figure 10.11.

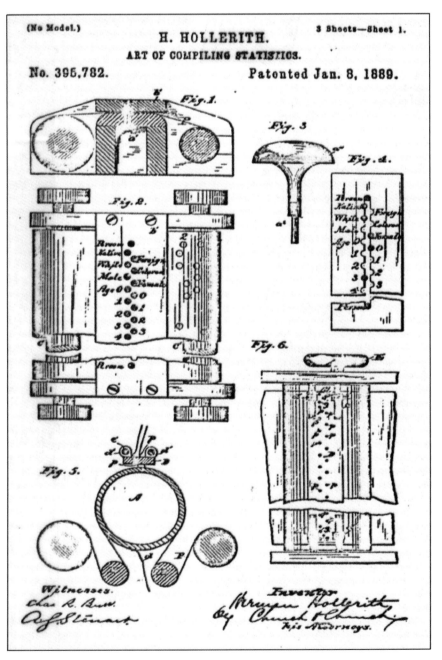

Figure 10-11 Machines for making punched cards were patented by Herman Hollerith in 1889. Punched cards for analyzing information were first used in the 1890 United States census.

290

- In 1907 Ole Evinrude invented an outboard motor for boats. He and his future wife were on a picnic. Evinrude became upset when a container of ice cream melted while he rowed a boat back to an island. The company he founded is now the Outboard Marine Corporation of Waukegan, Illinois.
- In 1913 A. C. Gilbert started the Erector Set Company. He had earned a medical degree from Yale University and won a 1908 Olympic gold medal in the pole vault (12 feet, 2 inches). He invented the erector set, the world's first construction toy. The company he founded is now Gabriel Industries of New York. See Figure 10.12.

Waukegan
(wȯ-'kē-gən)

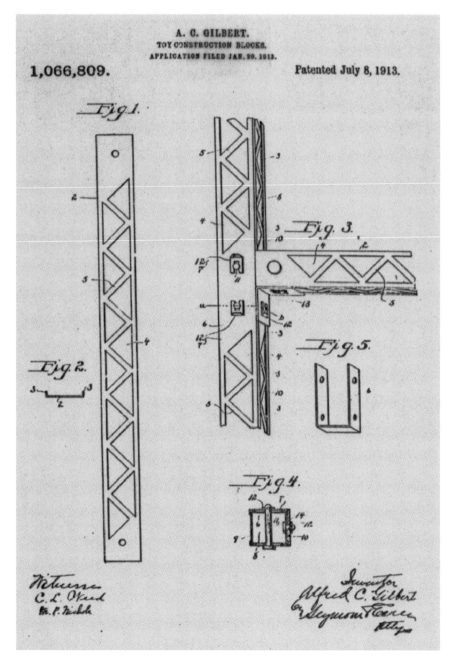

Figure 10-12 A.C. Gilbert received this patent in 1913 for "Toy Construction Blocks." At the time of his 1961 death, over ten million Erector Sets had been sold.

- In 1931 Clessie Cummins and one other person spent two weeks driving around the Indianapolis Speedway race track. To increase public knowledge of his diesel engine, they drove a truck nonstop for 14,600 miles. The company he founded is the Cummins Engine Company of Columbus, Indiana.
- Olive Beech and her husband, Walter, left their jobs in 1932 to build a small airplane that was almost as comfortable as a car. Their 1932 "Staggerwing" biplane met that goal. They cofounded the Beech Aircraft Company in Wichita, Kansas.
- Chester Carlson couldn't find anyone to manufacture the copier he invented in 1938. Companies didn't think anyone would want instant copies. The first easy-to-use dry copier was the Xerox 914 in 1959. The company that purchased Carlson's patent rights is now the Xerox Corporation of Rochester, New York.
- In 1977 Stephen Wozniak and Steven Jobs designed and built the Apple II computer in a garage. It was the first widely accepted personal computer. Jobs once picked fruit in Oregon, which is why they named their company the Apple Computer Company. It's located in Cupertino, California.

Practically all companies were started by one or two people who had a good idea and were willing to work hard. That is still true today. There may be small manufacturing companies or businesses in your town that are operated by hardworking individuals. Can you name some?

Tech Talk

The games of Monopoly and Scrabble were both developed in the 1930s. Monopoly uses street names from Atlantic City, NJ and Scrabble was originally named *Criss Cross*. These two games are the oldest and best selling board games of this century.

designing
the process of planning and drawing an idea

⸺ Teenage Factory ⸺

How could a teenager start a factory?

Someday you might come up with a good idea for a new product. What would you do about it? Maybe you would be like Elmer Sperry or George Westinghouse. Sperry invented a swiveling headlamp for locomotives in 1874 when he was only fourteen. He had many inventions and established the Sperry Gyroscope Company. Westinghouse patented the first of his 361 inventions, a steam engine, when he was only nineteen. He established the Westinghouse Corporation in 1869.

Suppose you have an idea for a new board game for two to four players. The board looks like the one in Monopoly. The game calls for a pair of dice and a colored button for each player. It has cards with questions, as in Trivial Pursuit. You think you might like to manufacture and sell it yourself.

1. You plan the game and then sketch out what you have in mind. Sketching your idea is **designing**—putting your idea on paper

so that others will know what you are thinking. The drawing is your *design*.

2. You want opinions from other people, so you talk to three of your teachers. They say that it seems like a good game. Each one tells you how he or she thinks you can improve it. See Figure 10.13A on the next page. This is market research. **Market research** is a way to find out what people will purchase. Your teachers think that the game is interesting enough for people to buy.

3. Since it's almost summer, you think you'll have enough time to make the game yourself. You plan to manufacture it in your room at home. You know you can't make it look as good as games made by large companies. However, you feel that the personal touch of a handmade game might help its popularity.

4. You can draw and color the playing surface on white paper, but you can't make the heavy cardboard back. You want the playing board to fold up like in other board games. You talk with the owner of a picture framing shop, and he says that he can make 100 cardboard backs for $1.20 each.

   A friend of yours has a computer with a printer that can print on lightweight cardboard. She agrees to print the game's question cards for $1.50 per set. She also says that she will print the instructions. See Figure 10.13B on the next page.

   Both the picture framing shop and your friend are your suppliers. A **supplier** is a person or company that makes something for your manufactured product.

   Of course you also want to *market*, or sell, the game. You check with four stores in your town. The owners all agree to offer your game for sale.

5. The framing shop requires an order of at least 100 cardboard game backs. So you make out a budget based on manufacturing 100 games.

100 cardboard game backs, $1.20 each	$120
100 packs of question cards, $1.50 per pack	150
100 pairs of dice, $1 per pair	100
400 colored buttons, 10 cents each	40
100 plastic bags to package the games, 25 cents each	25
Color pens, paper on which to draw the playing surface, rubber cement, and miscellaneous items	65
TOTAL	$500

   Your parents go with you to the bank and help you borrow $500 for the project. That $500 becomes your capital. **Capital** is money used in a business.

6. You contact your two suppliers and make arrangements to receive ten cardboard game backs and ten sets of question cards every week for ten weeks. You want to receive those items just in time to complete ten games a week. **Just in time** (JIT) is used by many manufacturers. With JIT, they don't have to store many items from their suppliers.

market research
the process of getting people's opinions about a product so that a company knows what changes to make or whether to sell the product

suppliers
people or companies that provide one or more of the parts for a manufactured product

capital
the money, goods, and possessions used in a business in order to make more money

just in time
a method of scheduling the arrival of materials at the time they are needed, so that it is not necessary to store those materials

Figure 10-13A Before making a new game, talk with teachers and other people to see what they think. This is market research.

Figure 10-13B A friend agrees to make some parts for your game. That friend becomes one of your suppliers.

production materials
everything necessary to make a product

added value
the increase in how much a piece of material is worth after it becomes part of a finished product

7. Before you draw the playing surface, you get your production materials together. **Production materials** are items used in making a product. This includes the white paper, color pens, ruler, and other items necessary to draw the playing surface.

Now you go into production and draw the playing surface. *Going into production* means you manufacture a product. When you are done, you will have **added value** to the white piece of paper. Because of your work, the blank paper will turn into a playing surface. Value added by the manufacturing process is one of the most important measurements made by the U.S. government.

294

Figure 10-13C Glue the parts together and check to see that everything looks good and is put together correctly. In a factory, such inspections are part of quality control.

Figure 10-13D Get all the game parts together, put them in one container, and your product is completed. The last step is to take it to a store to be sold.

Tech Talk

You may sometimes hear an adult talk about "working at the plant." *Plant* means the buildings at a factory. A factory can have several plants. The word probably came from *plantation*, a large land area where workers grow a single crop.

8. You draw the playing surface in your room at home. See Figure 10.13C. Your room is your factory. You carefully inspect the first playing surface you finish. You want to be sure that you did a good job. This inspection is called **quality control**. It means just what it says: You want to control the quality of your product.

   You make several more playing surfaces. You pick out the best ten and use rubber cement to glue them to the ten cardboard backs from the framing shop. You check the glued paper to be sure that you did it well. This is also quality control. Quality control is one of the most important parts of manufacturing.

quality control
the process of inspecting products to make sure they meet with all standards that have been set

295

9. It's time to assemble the parts of the game. You place four different-colored buttons in a small plastic bag. You add a pair of dice and a pack of question cards. You place the game board, instructions, and smaller plastic bag into a larger plastic bag. See Figure 10.13D on the previous page. This is like working on an assembly line in a factory. An **assembly line** is a line of people where each person completes a special task in assembling a product and then passes the product on to the next person.

10. You now have ten completed games. You drop off two or three at each of the four stores that agreed to market your game. The store owners will charge you $1 for each game they sell. This is their commission. A **commission** is payment based on sales.

You price the games at $9 each. How much profit will you make on each game? If you sell all of them, how much money will you make during the summer?

— Learning Time —

Recalling the Facts

1. Who was Levi Strauss, and what did he make that people still use today?
2. Who spent two weeks driving 14,600 miles around the Indianapolis Speedway race track?
3. How old was Elmer Sperry when he invented a locomotive headlamp?
4. What does it mean to design something?
5. What is market research?
6. What do the letters JIT mean?
7. How could you add value to a piece of paper?
8. What is an assembly line?

Thinking for Yourself

1. The Indianapolis Speedway race track is $2\frac{1}{2}$ miles long. How many times did Clessie Cummins go around the track in 1931, and what was his average speed?
2. In the teenage factory story, some teachers liked the game. Does this mean that it will sell well? Explain.
3. Name some items on a car that probably came from the car company's suppliers.
4. Name some ways that a sawmill adds value to trees and a furniture company adds value to lumber.

Applying What You've Learned

1. Name several manufacturing companies in your town or within a mile of your school. Use a telephone book for help, but limit your list to five.
2. Manufacture the game described in the preceding section. Work in groups to design the game boards, then pick the best design. Plan your manufacturing process and gather your materials. Then use division of labor and mass production to build the games. Sell the finished products.

Resources for Manufacturing

How is a factory organized?

Manufacturing companies use an organization of people, production materials, tools, and processes to make a product. It can be as simple as making a board game or as complicated as putting a car together. See Figure 10.14.

People

What are some jobs people do in factories?

Some companies, like General Motors (GM), the largest manufacturer in the United States, employ hundreds of thousands of people at many different plants. GM makes automobiles, automobile parts, trucks, buses, earth-moving equipment, and locomotives at over 100 factories.

Large companies have many engineers and technologists. Some design the products the companies make and are called design engineers or designer/drafters. Others decide the best way to manufacture the products and are called production engineers. Quality control engineers and technicians inspect the product to be sure that it's well made. Many other people operate manufacturing machinery, while others adjust, or set up, the machines. See Figure 10.15 on the next page.

Figure 10-14 People use processed wood and tools at this factory to manufacture window mouldings.

Figure 10-15 An industrial robot is an important piece of machinery in a modern factory. These people are being trained to operate the robot.

organization chart
a diagram of the jobs needed to make a product

All companies use organization charts to show how people fit into the manufacturing system. An **organization chart** is a block diagram of the jobs required to make the product. The chart shows everything at a quick glance. Figure 10.16 shows a simple one for the teenage factory. Here's what it means.

Block 1 - Your first step was to design the game. You decided...

Block 2 - What parts you would make and...

Block 3 - What parts you would purchase from suppliers.

Block 4 - After you made the playing surface, you checked it for quality.

Block 5 - You received the supplier's parts and glued down the playing surface.

Block 6 - You checked the glued surface for quality.

Block 5 - (Repeat) - You packaged the game in plastic bags.

Block 7 - You delivered the games to the stores for sale.

An organization chart is very important in manufacturing. It clearly shows each step. You can easily see how everything fits together.

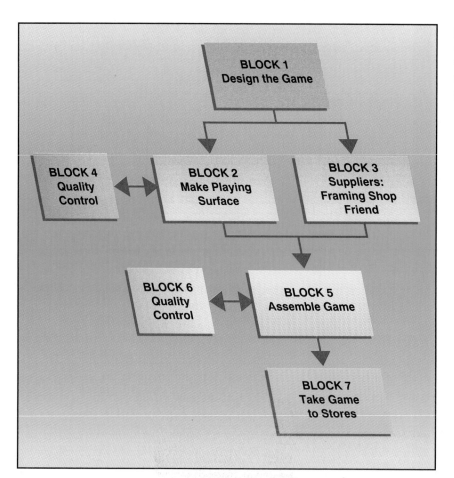

Production Materials

What are some of the different materials used to manufacture a product?

The materials a company uses to make its products are the production materials. They are different from *raw materials*. Raw materials are materials in the natural state. They include iron ore, trees, and cotton. See Figure 10.18 on page 302. Here are some of our important production and raw materials along with the major states of supply:

Production Material	Raw Material
Steel	Iron ore (Minnesota)
Aluminum	Bauxite ore (Arkansas)
Plastic	Petroleum (Texas)
Lumber	Trees (Oregon)
Textiles	Cotton (Texas)
Paper	Trees (Georgia)
Glass	Silica (Ohio)

bauxite
('bȯk-ˌsīt)

Car-Full Manufacturing

Automobile factories have hundreds of suppliers, thousands of employees, and cover 100 acres or more of land area. One factory can put together about 1,000 cars every day. The manufacture of a car starts with large rolls of sheet steel.

1. Metal Forming. After being cut to size, steel is bent to the proper shape. Large machines, called presses, form the roof, doors, fenders, and other body parts. See Figure 10.17a.
2. Body Assembly. Computer-controlled industrial robots automatically assemble the body parts together. People attach the doors and some other parts. See Figure 10.17b.
3. Painting. The completed car body is dipped into a tank filled with a special paint primer. The primer reduces rusting later in the car's life. Industrial robots paint the body, which is then dried in a large oven. See Figure 10.17c.
4. Trim. Seats, instrument panel, windshield, bumpers, grill, and other parts are attached on the trim line. At this stage, the car almost looks like a finished product. See Figure 10.17d.
5. Installing Engine. In practically all cars, the engine and transmission are one unit. The combination is called a *drivetrain*. The drivetrain is installed from underneath the car body. Tires and wheels are added. See Figure 10.17e.
6. Drive-away. All the parts are installed. The fuel tank is filled, and the engine is started for the first time. The car is driven off the assembly line. See Figure 10.17f.
7. Final Check. Everything on the completed car is given a final inspection. The car is then driven to the storage area before being delivered to the car dealer. See Figure 10.17g.

Companies stay in business by adding value to their production materials. Adding value can sometimes be a simple process. Take nail making, for example. A nail is a one-piece product made from a roll of strong steel wire. A nail-making machine puts a point at one end and flattens the other end. The machine can make about ten nails a second. Companies add value to steel wire by changing it into nails.

Other companies use several different materials and make a product with many different parts. An ordinary flashlight has ten parts: plastic base, top, lens, switch, lamp holder, spring, two flat pieces of metal, lamp reflector, and metal lamp conductor. See Figure 10.19 on page 303. The two batteries and lamp are purchased from suppliers.

Compared to making nails, it is harder to add value to the production materials purchased by a flashlight factory. The finished parts are more difficult to make, and the flashlight has to be put together. The assembled parts must work well together.

■ Figure 10-17 ■ a) The Ford Explorer utility vehicle uses many large parts that were shaped in presses at Ford's Louisville, Kentucky Assembly Plant. The parts are then welded together. b) Computer-controlled robots weld body parts together. c) One of the most common uses for an industrial robot is painting. d) Seats and interior trim parts are installed as the colorful vehicles travel slowly down a moving assembly line. e) This engine and transmission were installed from the top of the vehicle, before the body was added. In other vehicles, the engine and transmission are installed from the bottom. f) The Mazda Navajo, Ford Ranger, and Ford Explorer are all made in the same Louisville factory. g) After final checks, the cars are driven away. A modern automobile factory manufactures about 84 vehicles per hour.

Figure 10-18 (top) About seven *billion* pounds of cotton are grown in the United States every year. (bottom) The Bingham Canyon copper mine in Utah is one of our most important sources of copper.

Companies make products that have hundreds, or even thousands of parts. An airplane is such a product. However, you probably have some complicated products right in your own home. Your refrigerator is made from many different materials. A hair dryer has many parts, and so does a television. Can you think of other complicated products in your home?

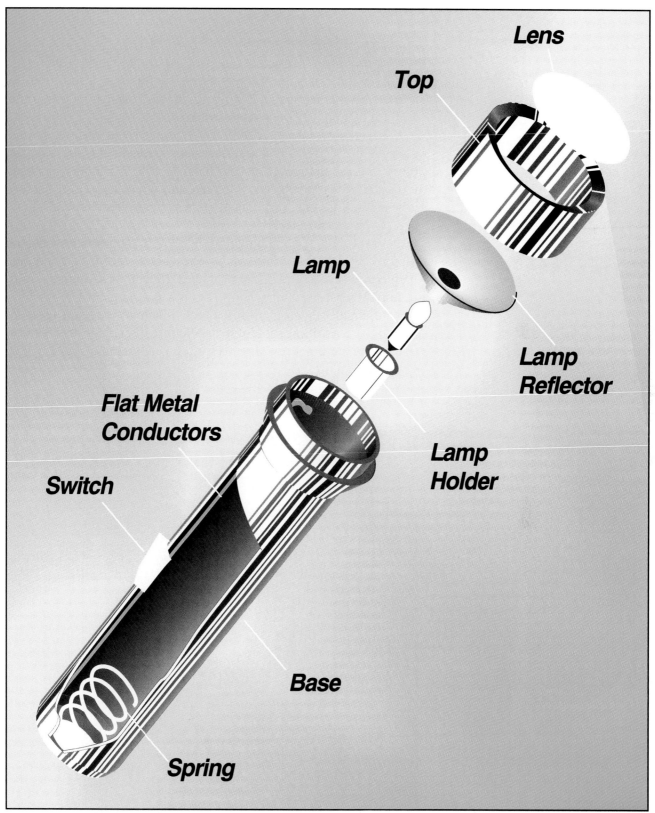

Figure 10-19 A flashlight has several parts made from different materials. All the parts have to fit well and work together.

Tools

What are tools, and how do people use them?

A *tool* is an instrument a person uses to do work. A computer can be a tool for some people. Proper clothing is a tool used by space-walking astronauts and fire fighters. Your pen is a tool that you use to complete your schoolwork.

In manufacturing, a tool usually means an instrument that changes the shape of material or fastens it together. *Hand tools* are those that use the power of your hand or arm. A hammer is a hand tool, and so are pliers, screwdrivers, and wrenches. Most machine adjustments are made with hand tools such as wrenches and screwdrivers. See Figure 10.20.

Large tools that are bolted to the floor and operated by electric motors are called *machine tools*. They bend, cut, drill, grind, and hammer materials into different shapes. Machine tools are the most important tools used in a factory. Some are as large as a room. Others are smaller, such as a drill press. Machine tools are used for such jobs as drilling holes in jet engines, bending steel for car doors, and cutting screw threads for the jack that may one day help you change a tire.

Figure 10-20 Hand tools are the ones people use most often. This man is using a hand-powered hammer to nail down a floor at a construction site.

A *power tool* uses a small motor for power. It can usually be held in your hands. Electric drills and electric screwdrivers are power tools. A table saw for cutting wood is an example of a power tool that can't be held in your hands. Some companies use power tools operated by compressed air. People use power tools for such jobs as grinding sharp edges from bicycle frames, assembling electronic parts, and sanding flat surfaces on wooden furniture. See Figure 10.21.

■ **Figure 10-21** ■ Some power tools operate from air pressure. This woman is using an air-powered hammer to nail down a floor at a construction site.

— Learning Time —

Recalling the Facts

1. What is the difference between a design engineer and a production engineer?
2. What is an organization chart?
3. How are production materials different from raw materials?
4. Is clothing a tool?
5. What kind of tools are used to make most machine adjustments?

Thinking for Yourself

1. Draw an organization chart that includes a company president and three vice presidents.
2. Is wood a production material or a raw material?
3. Name a machine tool that bends metal. Name machine tools that cut, drill, grind, and hammer metal.

(Continued on next page)

— Learning Time —

(Continued)

Applying What You've Learned

1. Pretend that you have invented a new machine. Go through all the steps necessary to obtain a patent. Ask some people to act as patent attorneys and patent office officials.
2. Design a time and motion study performance test. Time how long it takes to screw five different-sized nuts and bolts together. Do it several times for a good average. Draw some conclusions. Develop other performance tests using items in your shop.
3. Make a cardboard box to contain and protect a ceramic coffee cup, pair of sunglasses, or other small item. Start with a flat piece of cardboard and figure out how to cut and fold it. Test the package by dropping it to the ground.

Tech Talk

A *press* can form metal into unusual curved shapes. Some bathtubs are formed by presses. A *brake* bends the edges of sheet metal in a straight line. The outside shell of some refrigerators is formed by a brake.

Processes

What are some ways that people shape and fasten materials?

Companies manufacture products by using tools and production materials. The way a factory uses tools to add value to its production materials is called a process. A *process* is a system for doing something. Like all systems, it involves a number of steps or operations.

For example, suppose you want a personalized box to hold small items like keys, trading cards, patches, or jewelry. For your production materials, you might select a cigar box and a small sheet of self-stick plastic shelf liner. Scissors and a ruler would be your tools.

After measuring the box, you cut the plastic shelf liner and stick it to the outside of the box. The process could be done in different ways. You might decide to cover each side of the box separately, or cover two sides at once, or cover four sides at once. There are different ways to finish your project. Whichever way you do it, that will be your process.

Changing Size and Shape. As you just saw, processing often involves changing the shape of a material. The shape of wood, metal, and plastic is often changed by cutting, bending, or casting. See Figure 10.22.

Cutting means to use a sharp-edged tool. Hand cutting of production materials is done with a knife, saw, and other hand tools. It can also be done with a power tool such as an electric drill.

Bending means to shape a material, usually in a curve. It can be done by hand, as when people make woven wooden baskets.

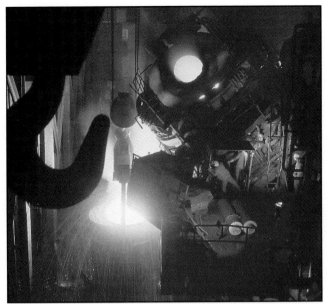

Figure 10-22 One of the most important processing operations is changing the size or shape of production materials.

Companies use large machines called presses and brakes to bend large pieces of metal.

Casting means to shape by pouring into a mold. You cast ice cubes by freezing water in an ice cube tray. Companies melt plastic and metal and pour them into molds. Many plastic toys are cast, and parts of some crank-type pencil sharpeners are made from cast metal.

Cutting, bending, and casting aren't the only ways to shape materials. For example, *forging* is hammering metal in shape. Blacksmiths made many parts this way. *Extruding* squeezes metal or plastic through an opening. It's much like squeezing toothpaste from a tube. Many tubes and bars are extruded. *Abrading* means to scrape or rub off. Filing, sanding, and grinding remove material by abrading.

Fastening. When a company manufactures a product, the different parts must be fastened together. Furniture, for example, is held together with nails, screws, and glue. Nails and screws are called *mechanical fasteners*. Glue is an *adhesive*. A hammer quickly drives a nail in place. A screw takes longer to use because a screwdriver has to twist it into the wood. Many companies use screws because they have more holding strength than nails. See Figure 10.23 on the next page.

Nails are the most common kind of wood fastener. They come in different sizes measured in *pennies*. An eightpenny nail, for example, is a medium-size nail often used in house construction. The name comes from its original cost. Many years ago in England, 100 eightpenny nails cost 8 pence (pennies). See Figure 10.24 on page 309.

Finishing nails have small heads. This makes them less notice-

Figure 10-23 A hammer is a striking tool that hits a nail into place. A screw has more holding strength and is twisted in place with a screwdriver.

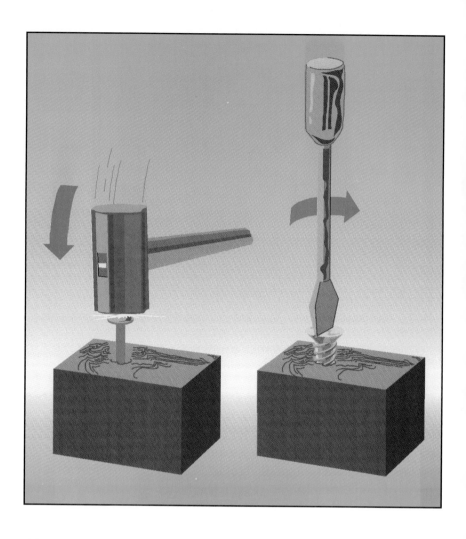

able. Look at the wooden frame around a door in a house. It is often held in place with finishing nails, and you have to look carefully to see them. They are sometimes covered up with a filler to hide them even more. Other nails have larger flat heads and are used where appearance is not important. They are called *box nails* or *common nails*. House construction requires several different sizes of common nails.

Screws are used to pull parts together. Wood screws have pointed tips so that they can cut their way through as they are turned. Machine screws are usually used with metal parts. They have flat tips. Nuts screw onto the end of the machine screw to pull the machine parts together. *Washers* fit between the nut and the material being fastened. Flat washers protect the material from being damaged as the nut turns on the machine screw. Lock washers keep vibration from loosening the nut. See Figure 10.25 on page 310.

Plastic and metal parts can also be fastened by melting. Some plastic audiotape cases are heated and melted together. No mechanical fasteners or adhesives are necessary. Many metal parts of a car body are welded (melted) together with high heat at one small spot.

It's called spot welding. Can you figure out why it's better to weld a car body together than to use machine screws?

Manufacturing processes are complicated and require many steps. But as you have just seen, most processes can be broken down into two major activities: shaping material and fastening it together.

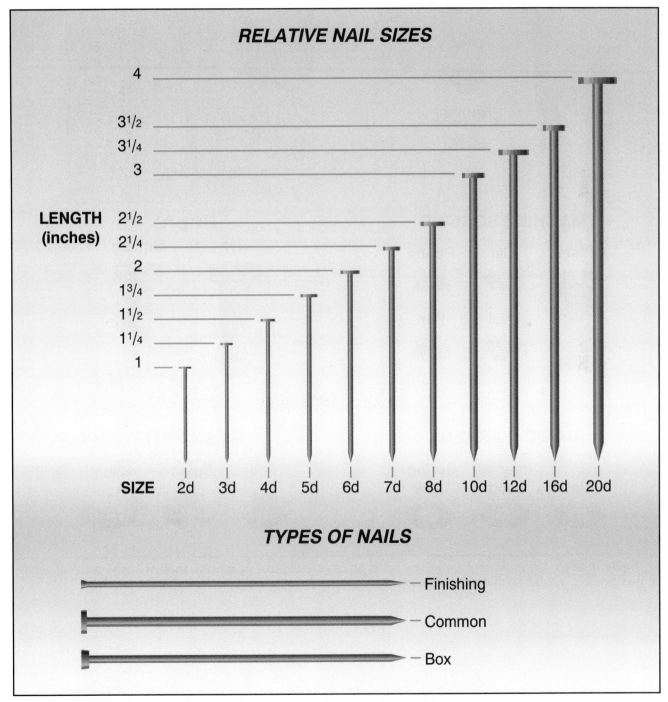

Figure 10-24 Nails come in many sizes and types. Numbers 4d and 6d are often used in woodworking classes.

309

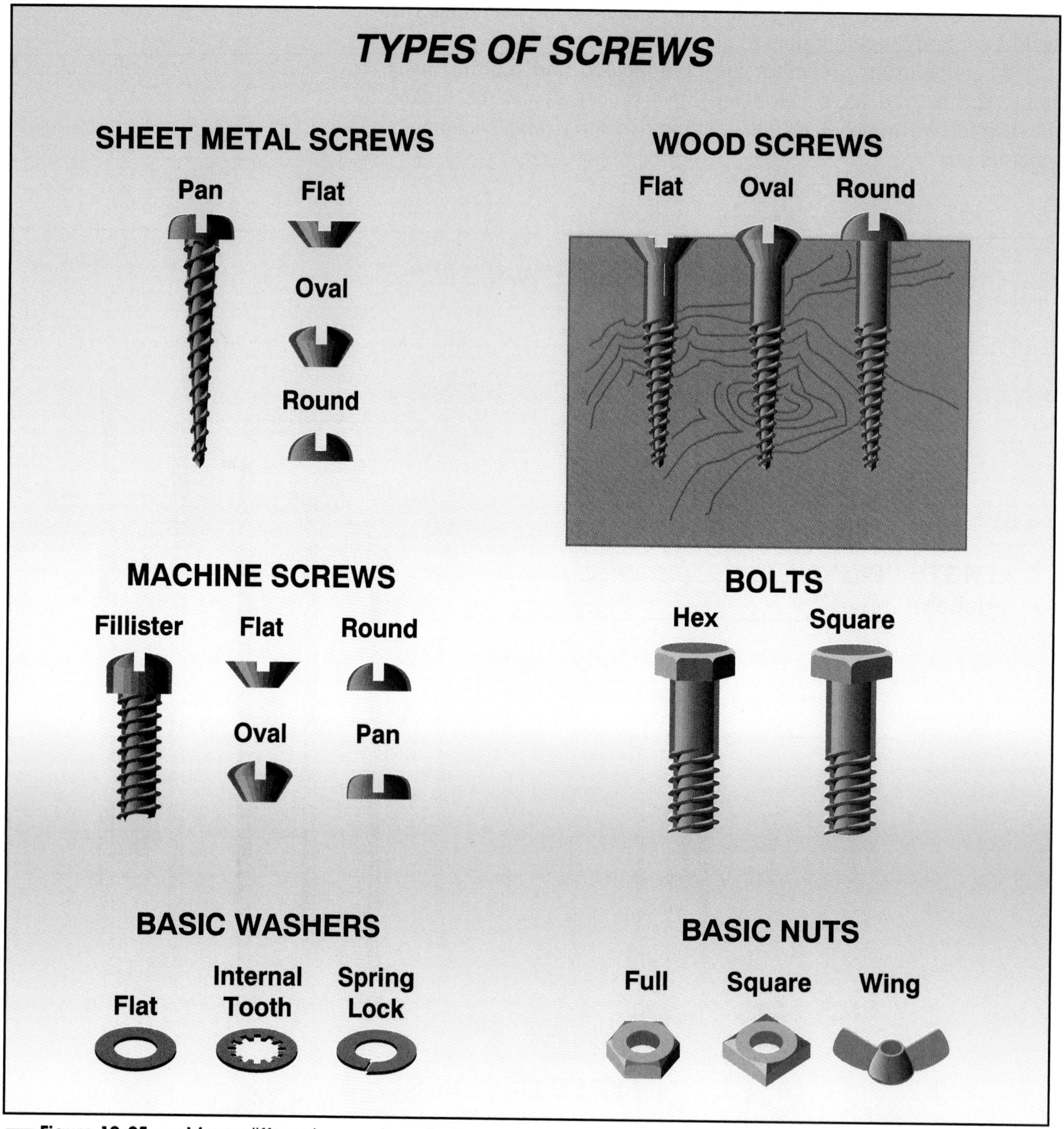

TYPES OF SCREWS

SHEET METAL SCREWS

Pan   Flat

Oval

Round

WOOD SCREWS

Flat   Oval   Round

MACHINE SCREWS

Fillister   Flat   Round

Oval   Pan

BOLTS

Hex   Square

BASIC WASHERS

Flat   Internal Tooth   Spring Lock

BASIC NUTS

Full   Square   Wing

Figure 10-25 Many different screw-type fasteners are used to assemble parts. Sheet metal screws, machine screws, bolts, and nuts hold metal and plastic parts together.

— Learning Time —

Recalling the Facts

1. What is a process?
2. What do sawing a wooden board and drilling a hole in plastic have in common?
3. Name two reasons for using washers when fastening materials together.

Thinking for Yourself

1. Name a method for shaping material that's not mentioned in this section.
2. Why does a screw have more holding strength than a nail?
3. Why is it better to weld a car body together instead of using mechanical fasteners?

Applying What You've Learned

1. Make a personalized box to hold small items. Use techniques outlined on page 306.
2. Make a casting of information chiseled into an old stone, such as a cornerstone of a post office or bank. Obtain permission from the owner or manager of the building. Then:
   a. Cover the stone face with one piece of heavy-duty aluminum foil. Carefully push the foil into all the names, dates, and other words on the stone. The shaped foil will be your mold.
   b. Completely cover the foil with foam-type shaving cream. Quickly place a large piece of heavy-duty cardboard onto the shaving cream. Slowly pull the foil away from the stone.
   c. Place the cardboard on the ground. Bend up the edges of the aluminum foil. Mix plaster of paris with water and pour it into the aluminum foil mold.
   d. Let the plaster harden, and peel away the foil. Paint or rub shoe polish on your casting.

Summary

Directions

The following sentences contain blanks. For each numbered blank, pick the answer that makes the most sense in the entire passage. Write your answer on a separate sheet of paper.

Manufacturing is carried out in large buildings called factories. All sorts of different products are made there and then they are sent to ___1___ , where people purchase them. People are consumers of manufactured products.

Before there were factories, people specialized in jobs known as ___2___ . Then came the Industrial Revolution and factories boomed. Eli Whitney used his idea of ___3___ in a rifle factory. He did not have one person making an entire rifle. Whitney had fifty workers making or ___4___ individual parts. Much

1.
a. playgrounds
b. courthouses
c. stores
d. churches

2.
a. crafts
b. hobbies
c. art
d. professions

3.
a. quality control
b. payment
c. perfection
d. division of labor

4.
a. selling
b. assembling
c. buying
d. getting

(Continued on next page)

(Continued)

later, Frederick Taylor used a ___5___ to measure factory operations. He developed standard ways of doing things.

Many companies with familiar names frequently have colorful stories about their roots, or beginnings. ___6___ Strauss, for example didn't make wagon covers from the canvas he brought to California. Instead, he made ___7___ and started a company with one of the most well-known names in the world. George Eastman was looking for a name to call his new product. He made up many different words until he finally decided to use ___8___. In modern-day America, Stephen Wozniak and Steven Jobs started the Apple Computer Company. They called it Apple because the word is easy to remember and Jobs once picked ___9___ in Oregon.

If someone is going to start a company or make a new product, they first need an idea. The first step is to put the idea on ___10___. Then everyone can see what the person has in mind. The final product will be made from production materials and parts made by ___11___. When everything is assembled, ___12___ will have been added by manufacturing.

Factories use tools and processes to manufacture products. A tool is used to change the ___13___ of material or hold it together. Large machine tools are important to manufacturing but there are other kinds of tools. A hammer, for example, is a ___14___ tool, and an electric drill is a ___15___ tool. A wrench is used to tighten scews that hold parts together. ___16___ are often used between a nut and the material being pulled together. When the product is completely assembled, it is put in a box or other package and sent to the consumer.

5.
a. scale
b. calculator
c. stopwatch
d. ruler

6.
a. Ole
b. George
c. Herman
d. Levi

7.
a. trousers
b. jackets
c. shirts
d. hats

8.
a. camera
b. Kodak
c. photograph
d. film

9.
a. fruit
b. onions
c. Brussels sprouts
d. tomatoes

10.
a. the floor
b. a computer screen
c. the table
d. paper

11.
a. machines
b. another department
c. suppliers
d. tomorrow

12.
a. weight
b. value
c. cost
d. time

13.
a. texture
b. cost
c. shape
d. color

14.
a. power
b. cheap
c. heavy
d. hand

15.
a. power
b. cheap
c. heavy
d. hand

16.
a. Adhesives
b. Screws
c. Washers
d. Nails

Key Word Definitions

Directions

The column on the left contains the key words from this chapter. The column on the right contains a scrambled list of phrases that describe what these words mean. Match the correct meaning with each word. Write your answer on a separate sheet of paper.

Key Word		*Description*	
1.	Manufactured product	a)	Finding out what people will purchase
2.	Factory	b)	Receiving items exactly when needed
3.	Consumer	c)	Finding the best way to do a job
4.	Craft	d)	Standard way of doing things
5.	Craftsperson	e)	Person who specializes in one job
6.	Industrial Revolution	f)	Worker completes small task and passes workpiece to next person
7.	Division of labor	g)	Company that makes something for another company
8.	Scientific management	h)	Person who uses items made in factories
9.	Time and motion study	i)	Changes caused by new manufacturing methods
10.	Moving assembly line	j)	Material used to make a product
11.	Designing	k)	How people fit into manufacturing system
12.	Market research	l)	Special type of job
13.	Supplier	m)	Product made in a factory
14.	Capital	n)	Dividing work into many simple steps
15.	Just in time	o)	Workpiece moves past the worker
16.	Production material	p)	Money used in a business
17.	Added value	q)	Putting an idea on paper
18.	Quality control	r)	Increase in value from manufacturing
19.	Assembly line	s)	Inspection during manufacture
20.	Commission	t)	Payment based on sales
21.	Organization chart	u)	Buildings in which products are manufactured

Going into Production

▬ Did You Know? ▬

You can manufacture a product all by yourself. One possibility is an adjustable bookshelf made from three small pieces of lumber and four short dowel rods. See Figure A. With such inexpensive pieces of wood, you can make three bookshelves and give one to a friend and one to a relative.

▬ Problems To Solve ▬

If you bought two identical items, you would probably find that they weren't *exactly* alike. There is usually a small difference in things like size, shape, or color. You will make three adjustable bookshelves and see those differences for yourself.

Resources You ▬ Will Need ▬

- Three pieces of wood, each about 1 inch thick, 6 inches wide, and 18 inches long; you can use pine or any other wood available in your area; the three pieces will be your bookshelf bottoms
- Six pieces of wood, each about 1 inch thick, 6 inches wide, and 8 inches long; again, you can use pine or any other wood; they will be your bookshelf ends
- Twelve ⅜-inch-diameter dowels, 1½ inches long
- Ruler
- Pencil
- Glue
- Hammer
- Hand drill with ⅜-inch drill bit
- Fine sandpaper
- Clear finish in an aerosol spray container

A.

A Procedure To Follow

1. Look at the assembly drawing in Figure B to get an idea of how to construct your bookshelves.
2. Pick up one of the 8-inch-long pieces of wood. Drill two ⅜-inch-diameter holes, 1 inch deep, in one end. The center of each hole should be 1 inch from the edge. Repeat for the other five bookshelf ends.
3. Place a small amount of glue on the end of a 1½-inch-long dowel rod. Using the hammer, tap the glued end all the way into one of the holes. Repeat for the other eleven dowel rods and holes.
4. Pick up one of the 18-inch-long pieces of wood. You will drill twelve ⅜-inch-diameter holes in the bottom. See the drawing for their location. Drill each hole just a little deeper than ½ inch—say ⅝ inch. Repeat for the other two bookshelf bottoms.
5. Wait for the glue to dry on the ends of the dowels. Then see how well the dowel rods in the bookshelf ends fit the holes in the bookshelf bottom. Find the best combination of bookshelf ends and bottoms.
6. Use sandpaper to smooth all the surfaces and edges. Using proper ventilation, spray one or two coats of clear finish from the aerosol spray container.

What Did You Learn?

1. The book shelf ends probably fit into some holes better than others. That's normal. No matter how hard we try, it's quite difficult to make two parts exactly the same. What could you have done to improve the quality of your adjustable bookshelf?
2. Suppose that you wanted to organize your own teenage factory to make and sell thirty adjustable bookshelves. In making that many, what would you do differently?

B.

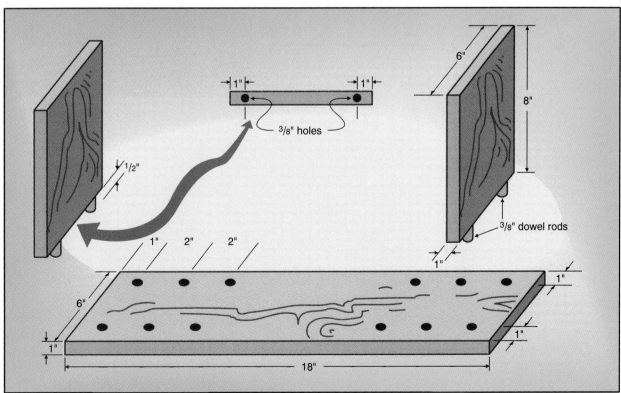

315

A Closer Look at Manufacturing

Did you know that more manufactured products can be found in the United States than in most other countries? The reason is simple. America is the largest producer of manufactured items in the entire world. See Figure 11.1 on the next page. Manufacturing is the most important economic activity in the United States.

America's 360,000 factories make about one-fourth of the world's manufactured items. Some products made in the United States each year are:

- About 12 million cars and light trucks
- Enough textile fabric to cover the state of New Jersey
- Computers worth $86 billion

Countries with few manufacturing companies are called **developing nations**. Their workers make products using old-fashioned tools and techniques. Rug making, for example, is usually done by hand and takes a long time. Using old systems of manufacture, workers in developing nations can't make very many products. See Figure 11.2 on the next page.

In a **developed nation** like the United States, machines rapidly manufacture high-quality products. Rug making in developed nations is usually done with machines. Well-trained workers using high-speed machines make enough rugs each year to cover the entire floor in about 6 million homes.

— Key Words —

In this chapter you will learn and study the meanings of the following important technology terms.

DEVELOPING NATION
DEVELOPED NATION
PROTOTYPE
SCHEDULE
ROBOTICS
CNC
CAD
CAM
CIM
STANDARD
METRIC MEASUREMENT SYSTEM
U.S. CUSTOMARY MEASUREMENT SYSTEM
DUAL DIMENSIONING
OSHA
NIOSH
WHOLESALER
RETAILER

■ **Figure 11-1** ■ Automobile tires are among the thousands of different products manufactured every day in the United States.

■ **Figure 11-2** ■ Many undeveloped and underdeveloped nations still make products by hand.

Research and Development Department

☰ *How does a company research and develop a product?*

American companies make everything from aluminum foil to zippers. To help you learn more about how manufacturing works, let's create an imaginary manufacturing company and observe it in action. Suppose that the company is named the Z. Z. Zipper Company. Its top executives are thinking of manufacturing a folding boat covered with a zippered nylon fabric. To the best of their knowledge, no one has ever made such a product. Let's see what happens.

How can the company decide if they want to manufacture a zippered boat? That's the job of their research and development (R & D) department. Research is an activity that looks at brand new ideas. Development uses research to create new products. Research is the looking, and development is the doing. See Figure 11.3.

The research group at Z. Z. Zipper does market research to see if current boat owners would be interested in buying a zippered folding boat. They also look for a new soft plastic that could be made into a leakproof zipper. The development group makes sure that the new plastic is strong enough. They also choose a specific nylon fabric for the boat. The research group and the development group work together closely. That's why they are in the same department.

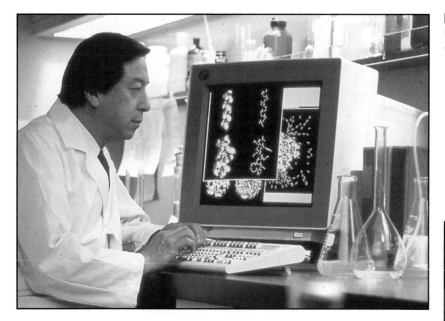

■ **Figure 11-3** ■ Research chemists in laboratories look for new products or new ideas for old products.

■ **Figure 11-4** ■ Design engineers often use computer equipment to visualize new products before the item is manufactured.

—— Design Department ——

≡ *What are some of the jobs of the design department?*

The design department takes over after the R & D department has completed its work. Design engineers decide how big something should be. They decide the color, the material, the shape, and everything else that is part of a product's design. See Figure 11.4.

The design department at Z. Z. Zipper obtains the R & D department's information. The design department has to answer an assortment of questions:

- What should be the length of the boat?
- The nylon fabric will wrap around a frame, much like an umbrella. Should the frame be wood or aluminum?
- Should the boat frame fold up or come completely apart for storage?
- Should the nylon fabric be in two pieces or three pieces?

Some design engineers think the boat should be 12 feet long, and others think 15 feet is better. Some prefer an aluminum frame instead of a wooden frame. Most think the frame should come apart for storage, but a few think it should fold up. They are almost evenly split between using two or three pieces of nylon fabric. How can they resolve these differences of opinion?

One way is to make an experimental prototype. A **prototype** is a handmade test model of the product. The design department asks their technicians to make two prototype boats. One boat will be 12 feet long with an aluminum frame that comes apart. It will be cov-

prototype
a handmade test model of a product by which later stages or designs are judged

We use the word *prototype* to mean a test model. The unusual word comes from the Greek words *protos*, which means "first," and *typos*, which means "a model." A prototype is the first model.

ered with three pieces of nylon fabric. The other boat will be 15 feet long with a fold-up wooden frame. It will be covered with two pieces of nylon fabric. By making two prototypes, the design engineers can find answers to all their important questions. Can you see why only two prototypes are necessary?

Once the experimental prototype boats are completed, technicians test them. They see how easily each boat can be assembled and taken apart. They check for water leaks and frame strength. The technicians take the boats to a nearby lake and use them as an average consumer would. They ask opinions from experienced boat owners.

After the tests are over, the design engineers look at all the information they have obtained. They discuss the test results among themselves and come up with a final design. They decide that the boat will be 15 feet long and will have an aluminum frame that comes apart. The two-piece nylon fabric is chosen because it is easier to use than the three-piece fabric.

The engineers draw their plans with computer-aided design (CAD) equipment. See Figure 11.5. The plans are stored in the computer's memory and are later used by the production department.

Putting Knowledge To Work

A Useful Material

Plastic is the most surprising manufacturing material in the world. It can replace wood in furniture, and yet it can also replace glass in windows and eyeglasses. Plastic textiles replace wool and cotton in clothing. Some automobile body parts are made of plastic.

One use would never have been predicted a few years ago. Special high-strength plastic replaces some metal parts inside experimental automobile engines. One of those engines is made by the Polimotor Research Corporation of Franklin Lakes, New Jersey. The engine is powerful enough to operate a compact car and weighs only half as much as an all-metal engine. It also runs more smoothly and makes less noise than a regular engine. It is not an all-plastic engine, as some newspapers have reported. Many of the parts are still made of metal.

The special plastic is strengthened with fibers of glass or carbon. The Trek Bicycle Company of Waterloo, Wisconsin, makes a bicycle with a one-piece plastic frame. Professional cyclists like such bicycles because they weigh less and are stiffer than bicycles with steel frames. The cyclists say it takes less energy to pedal a bike with a lightweight stiff frame. Bicycles ridden by the U.S. Olympic cycling team have frames made of fiber-strengthened plastic.

Modern industrial plastics are not the kind used for milk containers, toys, or pens. They are strong and well-manufactured materials that are finding new uses every day.

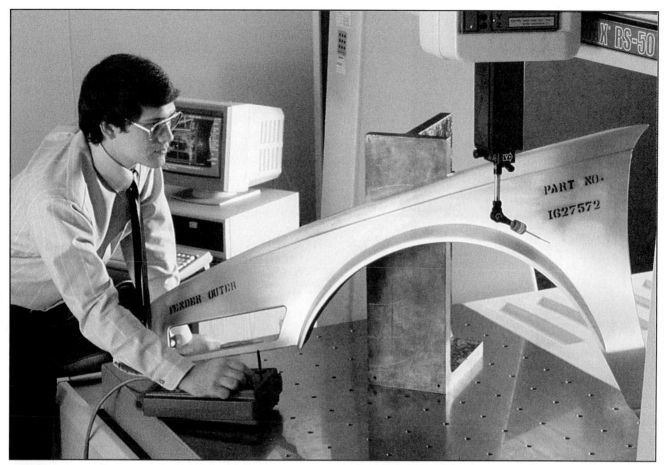

Figure 11-5 ■ Technologists and engineers produce designs on computer screens and then store the finished drawing in the computer's memory.

— Learning Time —

Recalling the Facts

1. What is the most important economic activity in the United States?
2. Is America a *developing* nation or a *developed* nation?
3. What is the difference between research and development?
4. Name four things in a product's design that are decided by the design department.
5. What is a prototype?

Thinking for Yourself

1. Name three developing nations.
2. Name three developed nations.
3. Is research related to development? Explain.
4. Design engineers are employed by a company that makes outdoor furniture. Name four specific things they might have to decide.

(Continued on next page)

Applying What You've Learned

1. Before constructing a full-size prototype, companies sometimes build small models (10 or 12 inches long) of a new product. Use plastic straws to construct a model of the frame for a folding boat. See if you can figure out some problems to avoid when making the prototype.
2. Plastic milk containers are made of low-strength plastic, but you might be able to recycle them into some useful items. From used milk containers, make prototypes of your own design for the following items:
   a. A large funnel
   b. A comfortable handle attached to a scrub brush
   c. A colorful light, by attaching colorful items to the outside and putting a small lamp inside
   d. A game, using a ball and the cut-out bottom of one or more milk containers

   Share some of your ideas with your classmates.

Production Department

schedule
a timed plan that includes the work to be done and when it should be done

Figure 11-6 This award-winning Texas Instrument plant in Sherman, Texas has a clean, pleasant, and well-lit production department. The design department is located on the second level.

How is the production department different from the design department?

The production department is responsible for actually making the company's product. See Figure 11.6. It must be a quality product and manufactured to a specific schedule. A **schedule** is a timed plan, such as twenty boats a day.

The production engineers decide that one supplier will provide nylon fabric cut to the correct size, and another supplier will manufacture the aluminum frame. The Z. Z. Zipper Company will make the zipper, attach it to the nylon fabric, assemble all the parts, and put them into a box for marketing.

The production department tells the purchasing department the names of the two suppliers. The purchasing department contacts the suppliers and arranges to have nylon fabric and aluminum frames delivered according to a just-in-time (JIT) schedule. Z. Z. Zipper will make twenty boats a day. They want their suppliers to make daily deliveries of twenty sets of nylon fabric and twenty boat frames.

The fabric will come from the N. N. Nylon Company. It's a company that Z. Z. Zipper has dealt with for eight years. They have been pleased with N. N. Nylon's fabric.

To find a frame supplier, three production engineers visit several companies. The engineers choose the A. A. Aluminum Company because that company makes high-quality products and uses modern manufacturing methods.

Robotics

Are industrial robots like the robots used in movies and television shows?

Robotics is a part of technology that deals with the use of industrial robots. An industrial robot has one mechanical arm and is controlled by a computer. See Figure 11.7. These robots are not the walking and talking type you may have seen in motion pictures. Industrial robots are machine tools. The end of the robot's arm might have a *gripper* to grip items and move them from place to place. The end also can have a paint sprayer or other devices for different jobs.

At the A. A. Aluminum Company, a robot bends aluminum for the boat frame. Another robot welds two pieces together. Electricity heats the metal until it is soft enough to weld, or stick together. Some robots move parts from one machine to another. They are *pick-and-place robots* because they pick up an item and place it somewhere else. Finally, another robot paints the boat frame. Manufacturers use more industrial robots for welding and painting than for any other purpose. See Figure 11.7 again. There are about 40,000 industrial robots at work in U.S. factories.

Robots do jobs that are hazardous, boring, or otherwise unpleasant for people. They can be easily reprogrammed to do other tasks. Suppose that A. A. Aluminum needs to make 20 boat frames for Z. Z. Zipper and 100 aluminum tent poles for T. T. Tent Company. After completing one order, A. A. Aluminum's robots could be reprogrammed in minutes—and ready to complete the next order.

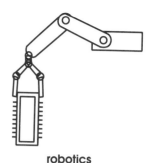

robotics
technology that deals with the design, manufacture, and use of robots in industrial and automated situations

Tech Talk

The scientist Isaac Asimov has written about 200 books for both young people and adults. In the March 1942 issue of *Astounding Science Fiction*, he had one of his characters say, "Now let's start with the three fundamental rules of robotics." That was the first time the word *robotics* ever appeared in print.

■ **Figure 11-7** ■ This computer-controlled industrial robot works under the supervision of a qualified worker.

CNC, CAD, CAM, and CIM

≡ *How are computers used in modern factories?*

CNC (Computerized Numerical Control)
machine tool operation controlled by numerical commands from a computer

CAD (Computer-Aided Design)
a method of planning or creating a product using a computer

CAM (Computer-Aided Manufacture)
a system that uses computers to operate the machinery in a factory

CIM (Computer-Integrated Manufacture)
the use of one computer system to control the design, manufacturing, and business functions of a company

CNC, CAD, CAM, and CIM are groups of letters that refer to industrial activities that use computers. The letter C stands for *computer*.

With **CNC**, or computerized numerical control, machine tools operate by commands from a computer. With **CAD**, or computer-aided design, designers and engineers use computers to draw plans for the parts. With **CAM**, or computer-aided manufacture, machine tool operators program computers to operate all the machinery. And with **CIM**, or computer-integrated manufacture, all the computers in the company are linked together, or integrated.

A. A. Aluminum draws up their plans with CAD equipment in the engineering offices. The company first used computers on the factory floor many years ago with CNC machine tools. With CNC, operators type instructions into a computer. The machine tool follows the step-by-step instructions. See Figure 11.8.

The company ties their CNC equipment together with a system known as CAM. With CAM equipment, operators write software programs. The parts being manufactured automatically go from one machine to another. It's all controlled by computers. See Figure 11.9.

A. A. Aluminum is just starting to use CIM. With CIM, their design and production departments can communicate instantly. The purchasing department can tell their just-in-time (JIT) suppliers when to deliver production materials. The marketing department can plan when to start selling products. Management can direct the entire company from one location. CIM is a company-wide process and is more difficult to start up than CAD or CAM.

Figure 11-8 ■ Computer-controlled equipment can assemble even delicate electronic devices.

Figure 11-9 ■ Computer-guided carts move in-process automobiles from place to place.

Quality Control

How do people control the quality of products made in a factory?

A. A. Aluminum, like Z. Z. Zipper and all other companies, is concerned about quality control. Companies build quality into their products during manufacture. They don't just inspect the final product and throw away all the bad ones. That would be expensive and wasteful. See Figure 11.10.

The people at A. A. Aluminum inspect the boat frames many times during the manufacturing process. If a problem is found, the operators can quickly adjust the robots or the other machine tools. Good quality control means that there are fewer parts that have to be scrapped, or thrown away. It also means that the product is the best the company can make.

Many companies check the quality control used by their suppliers. A supplier must pass a test before a company will buy their parts. After passing the test and being approved, a supplier receives a Supplier Certification Award. All U.S. automobile companies, for example, require their suppliers to have such approval. See Figure 11.11 on the next page.

Figure 11-10 ■ Inspection is an important part of quality control. These optical parts must pass several industrial standards before they leave the factory.

■ Figure 11-11 ■ Associated Spring is an approved supplier for many companies. Companies present certificates and plaques like these, to all their certified suppliers.

■ Putting Knowledge To Work ■

Lasers on the Line

Modern manufacturing is so different from what it once was that it is difficult to keep up with the changes. Worker involvement, computers, and industrial robots are all examples of new ways of doing things. Now, lasers are a part of the ongoing Industrial Revolution.

Manufacturers use lasers in two ways. One way is to cut metal or drill holes. Another is to weld parts together. Automobile manufacturers are using lasers more and more. Chrysler uses lasers to weld transmission parts together at their Kokomo, Indiana, plant. Cadillac uses them to trim body metal in Flint, Michigan. Volvo welds body parts together with lasers in Gothenberg, Sweden.

Laser welding is faster than conventional welding and results in a stronger joint, some-times over twice as strong. When coupled with an industrial robot, laser welders easily work with complex shapes like automobile roofs, dashboard panels, and floor pans. Two pieces of metal with a wavy line can easily be welded together. Another advantage of using lasers is dependability. Laser equipment operates for a long time before it needs adjustment.

For more information about lasers in manufacturing, you can visit two Internet Web sites. Try "www.convergent-energy.com" for Convergent Energy in Sturbridge, Massachusetts. Rofin-Sinar of Hamburg, Germany, and Detroit, Michigan, is at "www.rofin-sinar.com". Both companies make laser welders or cutting tools. You can also try the Society of Manufacturing Engineers at "www.sme.org".

— Learning Time —

Recalling the Facts

1. What department is responsible for actually making the product?
2. What department contacts the suppliers and arranges to have production materials delivered?
3. How is an industrial robot different from the walking and talking type seen in motion pictures?
4. How did the pick-and-place robot get its name?
5. How does a supplier earn a Supplier Certification Award?

Thinking for Yourself

1. Why is CIM so difficult to start up?
2. Production engineers are employed by a company that makes outdoor furniture. Name four specific things these engineers have to decide.
3. Name as many departments as you can that would be linked by a company-wide CIM system. For each department, why would they want to know what the other departments were doing?
4. Why would a company not care if their suppliers had a Supplier Certification Award? Give some examples.

Applying What You've Learned

1. Design and build your own protective container for an egg drop. Drop a container holding a fresh egg from 6 to 10 feet onto the floor. Some recommended rules to consider: (1) Make size restrictions for the container. (2) Everyone cleans up their own mess. (3) The egg will be put into the container at school, so the container cannot be built around the egg. (4) Every container must have solid sides, no fair just wrapping it with foam rubber. (5) To promote creative packaging techniques, try to exclude the use of padding.

Industrial Standards. A. A. Aluminum can pick from many different kinds of aluminum from their supplier. They can choose an aluminum that is easily bent, as are steel paper clips. They can choose an aluminum that is more rigid, like steel sewing needles. The company has been in business a long time and knows their production materials very well. They choose a No. 6063 aluminum tube for Z. Z. Zipper's boat frame.

When you use a pencil, you can pick one with soft lead, medium lead, or hard lead. Wooden pencils are almost always stamped with a *No. 1* or *No. 2* or other number. No. 1 is a soft lead and makes a dark line. No. 2 is the most common type of lead and makes an average line. Higher numbers make lighter lines. The numbers on the pencils are a standard. A **standard** is a system used

standard
a guideline set up as a rule for comparison of such things as quantity, value, weight, or quality

N	Smooth-surface designed for a natural finish. Free of open defects. Allows not more than six well-made repairs per 4 x 8 panel.
A	Smooth, paintable. Not more than eighteen neatly made repairs. may be used for natural finish in less demanding applications.
B	Solid surface. Shims, circular repair plugs and tight knots to 1" permitted. Wood or synthetic patching material may be used. Some minor splits permitted.
C *(Plugged)*	Improved C veneer with splits limited to 1/8" width and knotholes and borer holes limited to 1/4" x 1/2". Admits some broken grain, Synthetic repairs permitted.
C	Tight knots to 1 1/2". Knotholes to 1". Synthetic or wood repairs. Discoloration and minor sanding defects permitted. Limited splits allowed.
D	Knots and knotholes to 2 1/2". Limited splits are permitted.

Figure 11-12 Sheets of plywood are graded according to a standard that rates quality. The letters are stamped on each sheet.

for comparison. See Figure 11.12. A No. 2 pencil made by one pencil company will make a line that looks just like a line made by a No. 2 pencil from another company. A. A. Aluminum picks No. 6063 aluminum tubing because it is a strong and lightweight metal.

Several national groups have established industrial standards for aluminum, steel, plastic, gasoline, and many other materials. One group is the *American National Standards Institute (ANSI)*, and another is the *American Society for Testing and Materials (ASTM)*. The *Society of Automobile Engineers (SAE)* set up a standard way for labeling automatic transmissions in cars. Look at the letters behind the steering wheel on a car with an automatic transmission. They are always in the order *P*(ark), *R*(everse), *N*(eutral), *D*(rive), *L*(ow). See Figure 11.13. Some cars have minor changes like a D1 and D2 or an L1 and L2, but the pattern is still the same.

Measurement Units. Most American companies manufacture products using the **metric measurement system.** Metric measurements are those based on meters and centimeters. The United States is the only major developed nation that also uses measurements in feet and inches. Such measurements are part of the **U.S. Customary measurement system.**

American companies want to sell products to European countries and other nations that use the metric system. They also want to sell their products in the United States, where people are more familiar with the U.S. system of measurement. Companies do this by **dual dimensioning**. Their products are made using both the metric and U.S. Customary measurement systems. See Figure 11.14.

The Z. Z. Zipper boat is 15 feet long, or about 4.6 meters long (1 meter = 3.28 feet). The marketing department can advertise its length as 15 feet for U.S. consumers or as 4.6 meters in other countries.

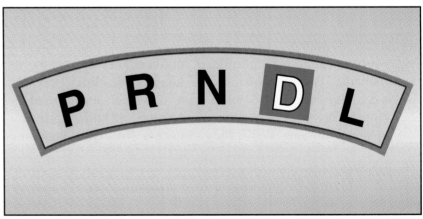

Figure 11-13 All automatic transmissions have the same gear shift pattern. It is a standard developed many years ago by the SAE.

Safety

≡ *How is the government involved with factory safety?*

Nothing is more important in a factory than the *safety* of the people who work there. Safety means freedom from injury or any danger of injury. To keep everyone aware of safety, many colorful and eye-catching signs are posted around the A. A. Aluminum factory. The posters have clever statements, like "A thick skin is good, but protective clothing is better" or "Safety is no accident."

Eye protection must be worn when a person uses any tool, even a hand tool. A hammer hitting metal might chip a piece from the hammer or metal workpiece. A face shield or safety glasses protect a person's sight. Since 1947, the Wise Owl Club has recognized employees and students who saved their sight by wearing eye protection. The club is part of the National Society to Prevent Blindness.

Other types of protective equipment include hard hats, ear plugs, gloves, and safety shoes with steel tops under the leather. For air that has dust or vapors, people wear special filters over their nose and mouth to protect their lungs.

Many factory safety rules are required by federal or state laws. The federal government's *Occupational Safety and Health Administration* (**OSHA**) establishes safety rules and checks up on companies. The *National Institute of Occupational Safety and Health* (**NIOSH**) approves protection equipment such as hard hats and safety glasses. See Figure 11.15 on the next page.

You can make sure that your shop safety sense is as good as that used at A. A. Aluminum and other companies. Just follow five simple steps.

1. Make sure you follow safe procedures.
2. Wear the proper safety equipment.
3. Learn how to use a tool before you start. Check with your teacher for instructions.
4. Inspect tools regularly. If the tools are worn or damaged, they should be immediately repaired or replaced.
5. Work slowly and always think about safety.

■ **Figure 11-14** ■ Standard sizes for modern ten-speed bicycles are often measured from the center of the crank to the horizontal tube. A typical dual-dimensioned measurement is 56 centimeters or 22 inches (1 cm = 0.393 inches).

OSHA (Occupational Safety and Health Administration)
the government agency that sets safety rules and checks to make sure the rules are being followed by companies

NIOSH (National Institute of Occupational Safety and Health)
the agency that approves for use protective equipment, such as safety glasses, hard hats, and steel-toed shoes

—— Marketing Department ——

≡ *How does a company sell the products it makes?*

The final step in the manufacturing system is handled by the marketing department. They find buyers for the company's product.

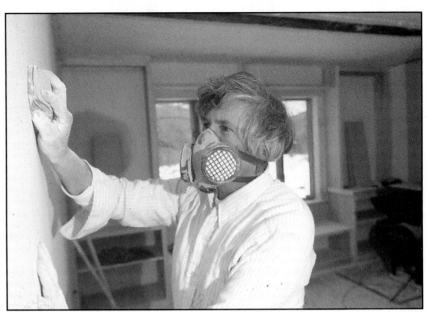

■ **Figure 11-15** ■ All industrial safety equipment, like the respirator worn by a drywall installer, must be approved by NIOSH.

The production department at Z. Z. Zipper assembled all the boat parts and put them into a strong cardboard box. Twenty packaged boats went to the company's warehouse every day. See Figure 11.16. To sell the boats, the marketing department contacted the B. B. Boat Sales Company.

B. B. Boat is a wholesaler. A **wholesaler** is a company that purchases large amounts from a manufacturer and then sells smaller amounts to retailers. A **retailer** is a company, or store, that sells products to consumers. Examples of retailers are Sears, Wal-Mart, and the stores in your town and the rest of America. See Figure 11.16.

Trucks from B. B. Boat arrive at Z. Z. Zipper's warehouse every Wednesday afternoon. The boxed boats are loaded, and the truck goes on to another manufacturer to pick up other items. The wholesaler has agreed to buy 2,000 boats. Once that number is reached, B. B. Boat will decide if it wants to buy more.

wholesaler
the merchant who buys large quantities from a manufacturer and sells smaller quantities to retailers

retailer
a merchant who buys products from a wholesaler and sells them to final consumers

——— Planning for the Future ———

How do different departments in a company prepare for manufacturing new or improved products?

Although the boats are in production, there are still many important jobs to be done by all departments.

• The R & D department is looking for new products to manufacture. If B. B. Boat only buys 2,000 boats, Z. Z. Zipper will have

Figure 11-16 (left) A wholesaler purchased many windows from the factory and stored them in this warehouse. (right) A retailer purchases smaller quantities from a wholesaler and sells them to consumers like you, your relatives, and your friends.

to make some other product. The R & D department is investigating reversible zippered seat covers for cars. One side would be in colors and patterns for teenagers. The other side would be in colors and patterns for adults.

- The design department is working on making the boat more convenient to use. A small wrench is now used to put the boat frame together. The design department is looking for a way to use metal pins that don't require tools.
- The production department wants a better way to attach the zipper to the nylon fabric. They glue and sew them together. The glue takes too long to dry. The production department is thinking about using an adhesive tape instead of glue.
- The marketing department always looks for new places to sell the company's products. They write advertisements for boating magazines and visit wholesalers. Two of the employees recently flew to Glasgow, Scotland, to try and sell their company's 4.6-meter boat.

—— Careers in Manufacturing ——

What could you expect to do if you went to work in a factory?

Mary Lou Schoenfeld wasn't exactly thrilled when her father decided to take her to his company's first open house. Mary Lou had never been inside a factory. The plain-looking buildings with fences didn't look too inviting. See Figure 11.17 on the next page.

When they entered her father's building, Mary Lou was surprised to see how bright it was. Colorful safety posters were everywhere. The clean, painted concrete floor had no litter of any kind. Many of the machines were orange or yellow.

There were many other people there and Mr. Schoenfeld seemed to know almost all of them. Mary Lou and her father stopped near his machine tool, and Mr. Schoenfeld tried to explain how it worked. He saw that his daughter didn't understand what he was saying, so he explained, "It's really not as difficult as it looks, Mary Lou. Learning to operate a machine tool is just like learning to play a new game. It seems difficult at first. But once you know the rules, it's easy."

They went to the engineering offices, where Mary Lou talked with Delores Royse. Ms. Royse operated CAD equipment that looked like a personal computer. She had a college degree in engineering technology and had been with the company only eight months. The company sent her for a six-week training course at the company that manufactured the CAD equipment. Her company paid for the 2,000-mile airplane trip, hotel room, and all meals. Ms. Royce told Mary Lou that many companies send their employees for special training.

After lunch in a modern cafeteria that looked like a fast-food restaurant, the two Schoenfelds stopped at the personnel department. That's where people fill out job application forms. The com-

■ **Figure 11-17** ■ Modern factories seem like mysterious places until you go inside and see how everything works together.

pany had a three-ring binder with pages that described their current job openings. Here are some of the manufacturing jobs that were available:

CNC Machine Tool Operator - Must be familiar with CNC turret lathes and milling machines.

Reliability Engineer - Experience with quality control plans, testing parts, and writing reports required.

Product Support Technician - Investigates production problems and maintains robotic equipment as necessary.

Software Programmer - Must be familiar with programming systems used in a CAM factory.

When the day was over, Mary Lou had met many new people and had a much better idea of what her father did at the factory. She had learned that factories aren't mysterious and confusing places. She even had a good cheeseburger in the cafeteria.

— Learning Time —

Recalling the Facts

1. What is a standard?
2. What is dual dimensioning?
3. Why must eye protection be worn even when using hand tools?
4. Name two federal government organizations involved with factory safety.
5. What is the job of the marketing department?

Thinking for Yourself

1. Why is an industrial robot classified as a machine tool?
2. Companies sometimes say, "Quality is built into our product." What does that mean?
3. Some pencils are labeled as $1\frac{1}{2}$ or $3\frac{1}{2}$. Why?
4. Is the metric measurement system or the U.S. Customary measurement system better? Explain.
5. Some companies can be both a wholesaler and a retailer. Explain.

Applying What You've Learned

1. Establish R & D, design, production, quality control, and marketing departments. Build several prototype products for evaluation. Go into production with one of them.
2. Conduct a *Consumer Reports* type of analysis on a simple consumer product. Compare different brands. Some possible products are small batteries, clear tape, and paper towels.

Summary

Directions

The following sentences contain blanks. For each numbered blank, pick the answer that makes the most sense in the entire passage. Write your answer on a separate sheet of paper.

The United States is the largest producer of manufactured items in the world. America's factories make about ___1___ of the world's manufactured products. This is possible because America is a developed nation. ___2___ rapidly manufacture many high-quality products.

Companies are organized into many departments. Each department has its own job to do. The ___3___ and development department, for example, looks at new ideas and creates new products. After the company decides to make a new product, ___4___ engineers decide how it will look. They are usually responsible for building a prototype, a handmade ___5___. Once all the testing is completed, the ___6___ department decides how to manufacture the product. It must be a quality product made to specific standards.

Companies use many modern methods to make their products. One of the newest techniques is called ___7___. It uses an industrial robot that has one ___8___ and is controlled by a computer. Robots do jobs that are hazardous or boring for people. An industrial robot can be part of CIM, computer-___9___ manufacture. CIM uses computer technology to link all the departments within one company.

Companies use industrial standards to manufacture the best products they can. The measurement standards used by most U. S. companies include the U. S. Customary and ___10___ measurement systems. An organization called ASTM establishes standards for production materials. ASTM stands for American Society for ___11___ and Materials.

Nothing is more important in a factory than worker safety. For example, ___12___ protection must be worn when a person uses any tool, even a hand tool. Many federal organizations check up on companies to be sure they are following the rules.

1.
a. one-half
b. one-third
c. one-fourth
d. one-fifth

2.
a. Machines
b. Robots
c. Unskilled workers
d. Retired workers

3.
a. quality
b. research
c. organization
d. production

4.
a. production
b. quality
c. manufacturing
d. design

5.
a. product
b. mock-up
c. test model
d. clay model

6.
a. production
b. development
c. quality
d. personnel

7.
a. quality control
b. electronics
c. drafting
d. robotics

8.
a. motor
b. outlet
c. mechanical arm
d. machine tool

9.
a. involved
b. integrated
c. implemented
d. instituted

10.
a. technical
b. modern
c. metric
d. British

11.
a. Tooling
b. Technical
c. Trying
d. Testing

12.
a. eye
b. hearing
c. hand
d. foot

Key Word Definitions

Directions

The column on the left contains the key words from this chapter. The column on the right contains a scrambled list of phrases that describe what these words mean. Match the correct meaning with each word. Write your answer on a separate sheet of paper.

Key Word	*Description*
1. **Developing nation**	**a)** Technology of industrial robots
2. **Developed nation**	**b)** Computer-aided design
3. **Prototype**	**c)** Country with many manufacturing companies
4. **Schedule**	**d)** System used for comparison
5. **Robotics**	**e)** National Institute of Occupational Safety and Health
6. **CNC**	**f)** Measurements based on meters
7. **CAD**	**g)** Sells products to consumers
8. **CAM**	**h)** Occupational Safety and Health Administration
9. **CIM**	**i)** Computerized numerical control
10. **Standard**	**j)** Country with few manufacturing companies
11. **Metric measurement system**	**k)** Measurements based on feet and inches
12. **U.S. Customary measurement system**	**l)** Handmade test model
13. **Dual dimensioning**	**m)** Purchases large amounts from a manufacturer
14. **OSHA**	**n)** Computer-aided manufacture
15. **NIOSH**	**o)** Timed plan
16. **Wholesaler**	**p)** Using both metric and American measurements
17. **Retailer**	**q)** Computer-integrated manufacture

PROBLEM-SOLVING ACTIVITY

Testing Products

Resources You Will Need

- Flashlight
- Six batteries that fit your flashlight: two of brand A, two of brand B, and two of brand C; make sure they all have the same ratings printed on the outside (for example, "Size D Alkaline" or "Size AA Heavy Duty")
- One or two plastic grocery bags
- Clock

Did You Know?

A product manufactured by one company is usually a little different from the same product manufactured by another company. Even though they all try to meet the same standards, sometimes one company makes a better product than the others. Consumers Union checks different products like cars, door locks, stereo equipment, and toothpaste. They publish their findings in the monthly magazine *Consumer Reports*.

Problems To Solve

Are the plastic trash bags made by one company better than those made by another? Is there a difference in the sticking ability of clear tapes? Are paper towels that are made from recycled paper just as good as paper towels that aren't? Should you purchase Brand A's hand soap or Brand B's? Make up your own tests and find out.

Check some past issues of *Consumer Reports* to see what they tested. You may get some good ideas. You can probably find past issues in your school or town library.

Suppose that you wanted to find out which brand of locally available flashlight batteries is the best. You decide to test three different brands.

A.

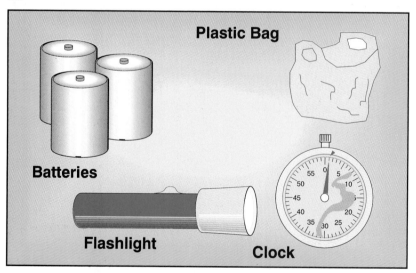

Plastic Bag

Batteries

Flashlight

Clock

A Procedure To Follow

1. Write down which brand was cheapest. An expensive set of batteries might last only a little longer than a less expensive set. Use your own judgment to determine the importance of cost. Neatly write down all your information. See Figure A.

2. Your first test might be to see which set of batteries produces the strongest light. One way to do this is with white or light brown plastic bags from your local grocery store. See Figure B. See how many layers of plastic it takes to completely block the light from a flashlight that has brand A batteries inside. Keep folding the plastic over and over. It's not unusual to use twenty or more layers. Repeat for brands B and C. Neatly write down all your information.

3. Perhaps another test would be to see how long it takes to wear out the batteries. See Figure C. Put two new brand A batteries in the flashlight and turn it on. Check the time. Keep the flashlight on until it no longer puts out much useful light. "Useful light" is something you will have to decide on your own. Just do your best. Discard the batteries and repeat for brands B and C. Neatly write down all your information.

What Did You Learn?

1. Based on your testing, which brand provided the most value? Why do you think so?
2. What other tests could you have conducted on the batteries? For example, how would you have decided which brand is best in very cold weather?
3. Did you have any trouble reading the information your wrote down? How important was it to clearly write down all the information from the test?

B.

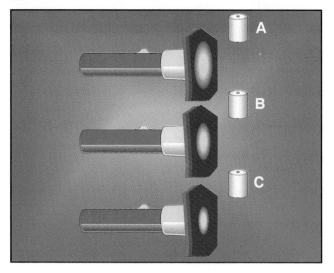

C.

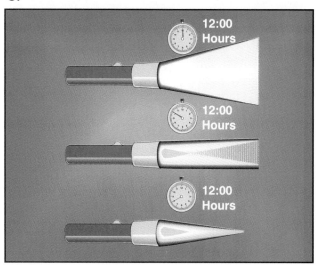

Construction

— **Key Words** —

In this chapter you will learn and study the meanings of the following important technology terms.

SITE
SKYSCRAPER
GEODESIC DOME
FLOOR PLAN
CONCRETE
PRECAST
FOUNDATION
FOOTING
MORTAR
JOIST
SUBFLOOR
STUD
GABLE ROOF
RAFTER
FIBERGLASS
CAULKING
DRYWALL
PREFABRICATED
SUBGRADE
ASPHALT
SUSPENSION BRIDGE
TRUSS BRIDGE
CANTILEVER BRIDGE
GIRDER BRIDGE

Suppose that it's raining outside. The rain is falling really hard. The wind is howling. It's pitch dark. You're not worried because you know you'll stay warm and dry inside. Perhaps you're inside a house, apartment, or mobile home. Or, you might be inside a school, restaurant, bus terminal, or hospital. See Figure 12.1 on the next page. All of the buildings that protect us from bad weather were assembled by people who work in the construction industry.

Another important part of construction is building roads and bridges. Cars and trucks travel swiftly from city to city because people have constructed smooth, level roads. Strong bridges allow trains, trucks, and cars to cross gorges, rivers, and large bodies of water. Other large construction projects include building dams, airport runways, and rocket launchpads.

There is one major difference between construction and manufacturing. Manufacturing is carried out in a factory. Production materials are all brought to the factory. Each construction project takes place at a specific location, or **site**. Materials are brought to the project.

To many people, the word *construction* means building a house or commercial building. You might be more familiar with houses. After all, single-family houses make up about two-thirds of the households in the United States. For that reason, much of this chapter will be about house construction. See Figure 12.2 on the next page.

■ **Figure 12-1** ■ (left) Some people live in tall apartment buildings or condominiums.   (top right) Control towers are built to hold both people and machines.   (lower right) Schools are built to be comfortable and meet your educational needs with classrooms, cafeterias, laboratories, auditoriums, and other important rooms.

—— The History of Construction ——

What types of construction have all civilizations wanted?

Ancient civilizations were interested in the same types of construction projects as we are today. They wanted comfortable buildings, good roads, and safe bridges.

Early Buildings

How would you describe some of the first buildings ever built?

What comes to mind when you hear the word *caveman?* People hunting dinosaurs? People wearing animal skins? Dinosaurs were gone from the earth long before prehistoric people arrived. However, prehistoric people probably did wear animal skins. They also lived for a time in dark, dusty, smoky caves. It was the only shelter they knew about. "Cavemen" were just beginning to use their brains to figure out how to construct shelters.

Two million years ago, people were hunters and not farmers. They soon learned to build simple shelters from woven sticks and animal skins. The lightweight shelters could travel with them as they searched for new sources of food.

About 4,500 years ago, Egyptians built pyramids that are one of the wonders of the ancient world. The ruins of 35 pyramids still

■ **Figure 12-2** ■ More people live in single-family houses like this one, than in any other kind of a housing unit.

stand near the river Nile. Each was built to protect the body of a king after he died. Some pyramids are so large that their square base covers the area of ten football fields.

Around 900 A.D., Toltec natives of central Mexico also built pyramids. See Figure 12.3. The one at Cholula is among the largest structures in the world. It has large steps and a temple on its flat top. Ordinary people back then didn't live in such grand buildings. They lived in houses that looked much like today's. Natural materials such as wood and stone were used in their construction.

During the American colonial period, many trees grew in the eastern half of the United States. That's why most early houses were made from wood. Log houses were quite popular because they could be easily put together. Log houses, however, were wasteful because so much wood was necessary for one house. Their construction also required a great deal of strength to position the heavy logs.

A new type of house construction started to appear in the 1840s. Instead of being made from logs or large wooden beams, these new houses were made of lightweight sticks. It was soon called stick construction. The frames went up quickly and provided both safe and strong dwellings. Practically all modern houses are built this way. See Figure 12.4 on the next page. Even **skyscrapers** use a type of stick construction.

The first skyscraper was the 1885 Home Insurance Building in Chicago. At ten stories high, it was not as tall as some other buildings. However, the Home Insurance Building was the first to use a metal frame as a basic part of its design. The outside walls were connected to the metal frame. The walls did not support the building as in log houses. The Home Insurance Building was torn down in 1931.

Toltec
(tăl′tĕk)

Cholula
(chə-′lü-lə)

■ **Figure 12-3** ■ The pyramid at Cholula, Mexico built by the Toltec natives, is one of the largest structures in the world.

Tech Talk

The word **skyscraper** was first used in 1884. It has come to mean a tall building with an inside frame, usually of strong steel. This definition disqualifies the pyramids. At 450 feet in height, pyramids may scrape the sky, but they don't have internal frames.

Figure 12-4 The structure of a house looks as if it's made from many different sticks. This method of construction is sometimes called stick construction.

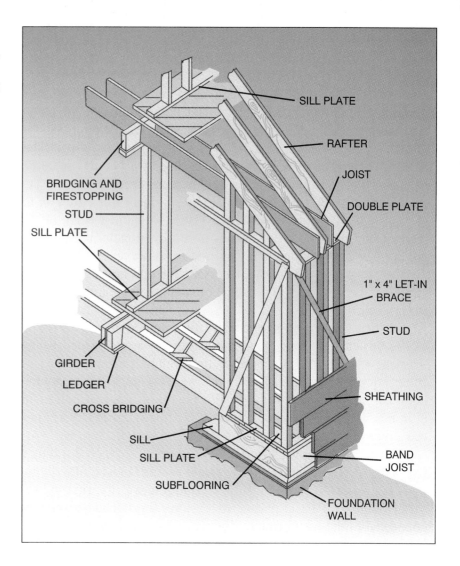

SILL PLATE

RAFTER

JOIST

DOUBLE PLATE

BRIDGING AND FIRESTOPPING

STUD

SILL PLATE

1" x 4" LET-IN BRACE

STUD

GIRDER

LEDGER

CROSS BRIDGING

SHEATHING

SILL

SILL PLATE

SUBFLOORING

BAND JOIST

FOUNDATION WALL

Early Roads

How were early roads constructed?

Most ancient roads were little more than dirt paths. The first really great road builders were the Romans. They started building roads about 300 B.C. The Romans dug a wide, road-sized trench a foot or more deep. They filled it with rocks and topped it off with a pavement of flat stones. The roads were higher in the center so that rainwater would drain off and be carried away by ditches along the sides. The Romans built over 50,000 miles of roads, and some are still used today. See Figure 12.5.

The ancient Romans discovered that roads could more easily support heavy wagons if the roads were covered with a durable material. This is called *surfacing* a road. The Romans used flat stones, but some early American roads were surfaced with logs or planks laid crosswise. Both types of road were bumpy and had to be traveled very slowly. Wagons could be damaged by too much shaking. Horses' hooves sometimes became stuck between the logs or

Figure 12-5 This ancient Roman road near Rome, Italy is still used today, about 2,000 years after it was constructed.

planks. The first section of the National Road in the eastern United States, now U.S. Highway 40 or Interstate Highway 70, was made from logs.

Around 1800, the British developed a method for easily making a smooth road surface. They used tar, which came from crude oil. George McAdam was the first to heat the mixture and spread it over a thick layer of crushed rocks. It is still a common way of surfacing roads, driveways, and parking lots. Americans commonly call the material asphalt or blacktop. British call it macadam, or use the brand name Tarmac.

Early Bridges

Where were the first metal bridges made?

The Romans not only gave us long-lasting roads, but also gave us strong, well-designed bridges. They developed the arch—wedge-shaped stones locked in a curve. The arch distributed weight sideways as well as down. See Figure 12.6 on the next page. Like their roads, some Roman bridges have lasted for centuries.

During the late 1700s, improvements in metal manufacturing greatly reduced the price of iron. That's when people started to make bridges out of iron. One hundred years later, bridge builders started to use steel, a much stronger material. The first major all-steel bridge was built across the Mississippi River at St. Louis in 1874. The bridge appeared so delicate that twenty-seven of the country's leading engineers said it would quickly collapse or be washed away by a flood. Neither happened. The Eads Bridge has safely spanned the most powerful and frequently flooded river in America for more than 100 years. It was named for its designer and chief engineer, James Buchanan Eads. See Figure 12.7 on the next page.

Tech Talk

Asphalt is an unusual word with an unusual history. The ancient Greeks found the material near the Dead Sea and used it to cover the stone walls of their buildings. The Greek word *asphaltus* means "to secure." The Dead Sea was once known as Lake Asphaltites. Modern highway engineers sometimes refer to *asphaltic concrete*. It's just another term for asphalt.

Figure 12-7 The beautiful Eads Bridge, the world's first major all-steel bridge, is a short distance up the Mississippi River from the St. Louis Arch.

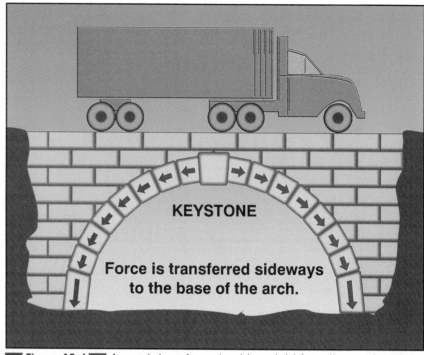

KEYSTONE

Force is transferred sideways to the base of the arch.

Figure 12-6 An arch transfers a truck's weight from the center of a bridge to the two sides of the arch.

— Learning Time —

Recalling the Facts

1. What is the major difference between construction and manufacturing?
2. Why were the Egyptian pyramids built?
3. What new type of house construction began in the 1840s?
4. What was the first skyscraper?
5. What is macadam?
6. Give the name and location of the first major all-steel bridge.

Thinking for Yourself

1. What kind of materials might be brought to a construction site for a discount store?
2. What is the world's tallest concrete and/or stone structure not using an internal frame?
3. Who was president when the interstate highway system was started?
4. Why are rocks added to tar to make macadam? In other words, what purpose do the rocks serve?

Applying What You've Learned

1. Make a teepee or other early native American dwelling. Discuss construction problems that the early people may have encountered.
2. Measure your walking pace so that you can estimate distances for construction purposes. Determine the size (floor area) of various rooms or buildings.

(Continued on next page)

— Learning Time —

(Continued)

3. Make a type of road surface material from chopped tires and plaster of paris. Evaluate the product.
4. Build a full-size arch from wooden blocks or foam plastic.

Putting Knowledge To Work

The Statue of Liberty

After 100 years of greeting every ship entering Upper New York Bay, the beautiful lady was showing her age. Her iron frame was badly rusted and weak. Her streaked green copper skin had many missing fasteners. The 151-foot-tall Statue of Liberty needed repair. The work began in 1983.

Lady Liberty was a gift from the French government in 1885. Gustave Eiffel (g$\overline{oo}$s′täf ī′fəl), designer of the Eiffel Tower in Paris, designed her supporting frame. She arrived in 214 crates and was named Liberty Enlightening the World.

The first step in the modern restoration project was to check her outside skin. The original deep brown color had aged to a light green. Using sound waves to measure the copper's thickness, the workers found it was in good condition.

The most challenging task was to repair the badly rusted and weakened internal iron frame. A large outer framework was built to hold Lady Liberty together while workers removed damaged iron ribs. They found asbestos material between the skin and frame. Because airborne asbestos fibers are linked to health problems, the workers wore face masks and carried their own air supply. That greatly hampered their efficiency.

Much of the iron frame was replaced with stainless steel. To keep the copper skin from buckling while the frame was being repaired, only a few sections of the frame were replaced at one time.

The worst rusting was found in Lady Liberty's upraised arm. It may have been caused by incorrect assembly of the statue after she was shipped from France. The torch was also incorrectly modified in 1916. The problems caused large water leaks, and the torch was so badly damaged that it couldn't be repaired. Its shape was computer analyzed, and a new torch was built.

Lady Liberty was given a stronger frame, a new and brighter torch, new elevators, new stairs, and windows. Wearing her familiar green dress, she was officially reopened to the public on Independence Day in 1986.

345

Building a House

How difficult is it to build a house?

Constructing a house can't be too difficult. Until fairly recently, many people built their own. During the 1800s, there were no construction companies in many regions of the United States. Pioneer farmers had to construct shelters for themselves and their animals. Neighbors helped, and everyone worked until the walls were up and the roof nailed down.

Those early amateur house builders had no power tools, few metal nails, and no printed plans. In spite of those difficulties, many houses they built are still standing. The basic construction techniques they used are still used today.

In this section, you will learn how a modern house is put together. Since there are so many different kinds of houses, it is easier to explain just a single type. One of the more popular styles in America is a single-level house with a *crawl space* underneath. A crawl space is the area under the floor, and it has just enough space for a person to crawl. See Figure 12.8. Heating ducts, water pipes, sewer pipes, and electrical wires are located in the crawl space.

Whether constructing houses, roads, or bridges, planning is the first and most important step. A well-designed house is like a good pair of shoes. Both should fit into the surroundings. (You wouldn't play baseball wearing heavy hiking boots.) Both should be exactly what the owner wants. (You may want blue running shoes.) And both should be the right price. (You might not be able to afford very expensive shoes.) In house construction, these conditions can all be met by careful planning. Planning means choosing a good place to

Figure 12-8 Water pipes, electrical wires, and other items are under the floor of many houses. It's commonly called the crawl space.

Putting Knowledge To Work

The Geodesic Dome

Most houses are rectangular in shape—but not all. One dramatically different house shape came from the fertile mind of Richard Buckminster Fuller (1895-1983). He was always interested in affordable housing. During the 1920s, he built an all-metal house that could be shipped in a tube 16 feet long and 5 feet in diameter. His dymaxion (dī-măx′ē-ən) house ("dynamic" and "maximum") wasn't a commercial success, but his geodesic (jē-ə-′dĕs-ĭk) dome was.

Fuller's patented **geodesic dome** was a sphere made from many small pyramid-shaped frames. Combining the two shapes resulted in a lightweight structure of enormous strength. The word *geodesic* means a surface made from short, straight bars forming a geometric shape. Geodesic domes use only a tiny fraction of the amount of material used in ordinary structures.

Perhaps the largest example of a geodesic dome was the twenty-story, 250-foot-diameter structure used for the U.S. Pavilion at Expo 67, the 1967 World's Fair in Montreal, Canada. See Figure 12.9. It was the last large dome that Fuller personally worked on. But geodesic domes aren't used just for large structures. Some houses use the geodesic dome too.

Houses can take advantage of the unusual dome shape in several ways. Much less construction material is used. No complicated structural construction techniques are needed. Interior columns are not necessary, so large rooms can be easily made. It's a good design for seaside houses because geodesic domes can safely withstand winds as high as 150 mph. Geodesic domes are now recognized as a significant American contribution to architecture.

build the house, selecting a house design, and figuring out how to pay for the house. Planning also means deciding on which materials to use, how the house will be heated, and obtaining any necessary building permits.

Choosing a Good Place

Can you build a house anywhere?

A house is built on a *construction site*, or *building site*. All sorts of locations are possible. A house can be built in a busy city neighborhood or along a quiet country road. It can be built on the side of a hill, on a sandy beach, or just about anywhere. See Figure 12.10 on the next page.

Some building sites are better than others. Before choosing one, there are some questions that need to be answered.

- Are electricity lines close by? How about city water, natural gas for heating, and telephone lines? If city sewers are not available, the house must have its own treatment system.

Figure 12-9 The huge 20-story geodesic dome at the U.S. Pavilion of the 1967 Montreal World's Fair may have been the largest geodesic dome ever built.

Figure 12-10 Construction in some locations requires unusual techniques, such as building this structure in a sandy area.

floor plan
sketches that show the size and location of rooms in a building

- Will the house face south? Houses facing south are sometimes less expensive to heat than houses facing north. Can you figure out why?
- How much will property tax cost?
- Most construction sites require moving dirt. How much dirt will have to be moved? Houses on hillsides are more expensive to build than those on level land.

Selecting a House Design

Why do we have different house designs?

The basic design of a house should fit in well with its surroundings. A log house might look out of place in a city neighborhood of brick houses. However, it might look fine along a tree-lined country road with fewer houses. Sometimes local laws restrict the type of houses that can be constructed in certain areas. These laws protect the home owners already there. These home owners can be sure that the neighborhood will always look about the same.

After choosing the general design, the next step is to determine the size of the house. *House size* means its floor area. The smallest houses have about 1,000 square feet of living area. An average-size house might have 1,200 to 1,500 square feet. See Figure 12.11. However, houses and apartments can be many different sizes.

Rough sketches are made of the floor plan. A **floor plan** shows the location and sizes of all the rooms in the house. Careful

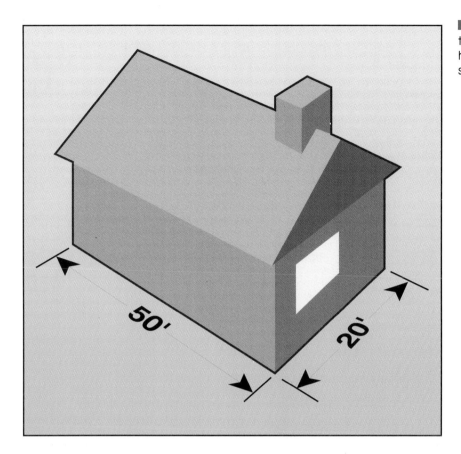

Figure 12-11 House size means the floor area. The floor area of this house is 50 feet X 20 feet = 1,000 square feet.

drawings are then made from the sketches. See Figure 12.12. The drawings are commonly called *blueprints*. When the drawings are reproduced, they turn out as white lines on blue paper. Architects can draw the blueprints, but many people save money by using standard floor plans. Housing magazines and building supply stores are sources for the plans.

Paying for the House

How do average people pay for expensive houses?

A house is the most expensive purchase made by most families. Houses usually cost too much for people to pay cash. Banks, savings and loan associations, and other financial institutions lend money to build houses. Home buyers make a down payment on the purchase price and borrow the rest. The cost of borrowing money is called *interest*. Banks and others advertise it as a percentage called *interest rate*. It is not unusual for interest rates on houses to be around 10 percent. The loan is paid back with monthly payments for twenty to thirty years.

Monthly payments depend on interest rate, amount of money borrowed (called the *principal*), and how long the loan will last. Precise monthly payments are figured out by the bank. Sometimes the payments include house tax and fire insurance. The chart below

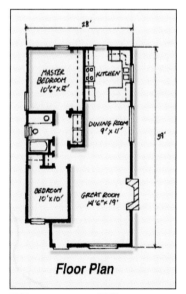

Floor Plan

Figure 12-12 Floor plans such as these can be purchased from housing magazines, newspapers, or building supply stores.

will give you an idea of monthly payments over a twenty-year period. It is only for principal and interest. It does not include taxes and fire insurance.

Look for the amount borrowed (the principal) along the top. Next look for the interest rate along the left. The two meet at the monthly payment. For example, suppose you borrowed $75,000 at 8 percent. Your monthly payment would be $627 for twenty years. What would be the monthly payment for a $50,000 loan at 7 percent interest?

Interest		Principal		
	$50,000	**$75,000**	$100,000	$125,000
6%	358	537	716	896
7%	388	581	775	969
8%	418	**627**	836	1,046
9%	450	675	900	1,125

Materials

What are houses made out of?

Practically any material can be used for a shelter. Native Americans used animal skins for their tentlike teepees. Sod houses in the northern plains were made from dirt. Some Eskimos still make temporary hunting shelters from snow or ice. Some retired persons live in travel trailers made from plastic and aluminum. See Figure 12.13. People have always made shelters from whatever was available.

Modern construction techniques use three materials more often than any others. They are wood, concrete, and steel.

Wood. Wood has always been used for house construction in the United States. It is the basic building material in about 70 percent of the houses. America is the world's second largest user of wood after Russia. One of wood's advantages is energy savings. It takes less energy to change trees into wood for construction than to make concrete or steel.

Wood is light enough to be easily carried, and it can be quickly fastened together with nails. Wood buildings can be more easily altered than concrete or steel buildings. Wood is also a renewable resource. New trees can be planted as more wood structures are built. Concrete and steel are not renewable resources.

Concrete. A wet **concrete** mixture of portland cement, gravel, sand, and water looks like a thick gray paste. It's poured into molds to make sidewalks, stairs, outside windowsills, bird baths, and many other items. Depending on the size of the molded item, concrete hardens in one or more days. Concrete is used to make commercial buildings because it is strong and can be precast. **Precast**

■ **Figure 12-13** ■ Many retired persons live in travel trailers like this 30-foot long polished aluminum Airstream.

concrete building parts are made in a factory. They are brought to the construction site for assembly. See Figure 12.14.

Only a small amount of concrete is used in house construction. It is used in the foundation. The **foundation** is the very bottom of a house, and it supports the entire weight of the house.

Steel. Bridges and skyscrapers use large amounts of structural steel. It's easy to see the structure in bridges. The steel beams and other pieces are completely open to view. In skyscrapers, outside walls hide the strong steel beams inside.

Some commercial buildings have thin steel for the outside walls and roof. They are called sheet metal buildings. Large ones can be built in just a few days. They have strong steel frames and a concrete floor. A backyard steel storage shed is an example of a small sheet metal building.

Like concrete, only a small amount of steel is used in house construction. Of course, nails are made of steel, as are some outside doors. However, as part of the average house structure, steel is used only for some floor supports.

foundation
the bottom support of a house, or other building, that rests on the ground

Tech Talk

Several different wood-related words are used by people in the construction industry. *Trees* grow in the forest. *Wood* is the material under the bark of trees. *Lumber* is wood that has been cut into useful shapes. *Boards* are thin pieces of lumber, usually less than 2 inches thick.

Figure 12-14 Buildings like this Ohio State University Office Tower use granite faced precast concrete walls that attach to a steel structure. The walls were made by Bluegrass Art Cast of Winchester, Kentucky.

— Learning Time —

1. Why did pioneers have to construct their own homes?
2. Name four items that pass through the crawl space under a house.
3. What is a building site?
4. Why do local laws restrict the type of house that can be built in certain neighborhoods?
5. Suppose you borrowed $100,000 at 9 percent interest for twenty years. What would be the monthly payment?
6. Name four advantages of wood as a building material for houses.
7. What four materials are mixed together to make concrete?

Thinking for Yourself

1. Would it be less expensive to heat a house located on top of a hill or in a valley? Explain.
2. What is the floor area of the living space in your home? Don't include garages or storage areas like unused basements.
3. Why don't people pay back house loans in five or ten years instead of twenty or thirty?
4. Why do lending institutions require house builders to carry fire insurance?
5. Suppose that you lived in a geodesic dome house. From your standpoint, list some advantages and disadvantages of living in a dome.

Applying What You've Learned

1. Make a small geodesic dome. Later, build a large one. Test both by hanging weights from the center.
2. Make a realistic floor plan for a realistic geodesic dome house.
3. Evaluate floor plans in a housing magazine.

Lumber Sizes

What are the different lumber sizes used in building a house?

Many years ago, people weren't concerned about the size of lumber for house construction. They used axes and hand saws to cut logs for wall supports, roof supports, doors, and other purposes. There were no standards, so an exact size wasn't important.

Times have changed. Modern lumber is cut to careful sizes and sold in thicknesses and widths called *nominal dimensions*. The word *nominal* comes from the Latin word for "name." So a nominal dimension is just a name for a lumber size.

For example, one common nominal board size is 1 × 6, pronounced "one-by-six." This means that the board is close to being 1 inch thick and 6 inches wide. It's actual size is $\frac{3}{4}$ inch thick and $5\frac{1}{2}$

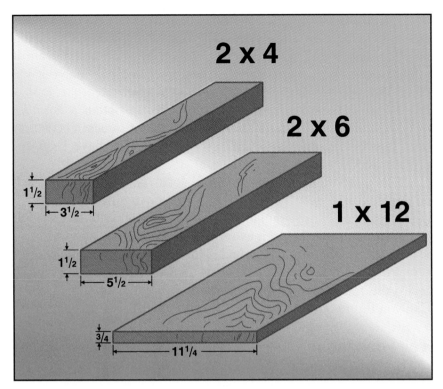

2 x 4

2 x 6

1 x 12

$1\frac{1}{2}$

$3\frac{1}{2}$

$1\frac{1}{2}$

$5\frac{1}{2}$

$\frac{3}{4}$

$11\frac{1}{4}$

Figure 12-15 The nominal size of lumber is only close to the actual size. The actual size of a "2 X 4" is smaller than two inches by four inches.

Tech Talk

Most countries use metric measurements. Their equivalent to 2 × 4 is 50 × 100 mm, and their equivalent to 2 × 6 is 50 × 150 mm. American sheet stock, like plywood and wall paneling, is 4 × 8 feet. The metric equivalent is 1200 × 2400 mm.

inches wide. Can you see why it's easier to call it a "one-by-six"? See Figure 12.15.

The following list gives the lumber sizes used in most wood houses. Floors, walls, and roofs use nominal 1-inch lumber. Two-inch lumber is used in the frame, or structure, of the house. Lumber can be purchased in many different lengths, but the most common are 8 feet, 12 feet, and 16 feet.

Nominal Size	Actual Size
1 × 6	$\frac{3}{4} \times 5\frac{1}{2}$ inches
1 × 12	$\frac{3}{4} \times 11\frac{1}{4}$ inches
2 × 4	$1\frac{1}{2} \times 3\frac{1}{2}$ inches
2 × 6	$1\frac{1}{2} \times 5\frac{1}{2}$ inches
2 × 8	$1\frac{1}{2} \times 7\frac{1}{4}$ inches
2 × 10	$1\frac{1}{2} \times 9\frac{1}{4}$ inches

Think of the nominal size of lumber as only a name, not an actual size. Saying "I want to buy three eight-foot-long two-by-fours" is the same type of sentence as "I want to buy three eight-foot-long push brooms."

Building Permits

Why do local governments require special permits to construct a building?

Figure 12-16 Local governments must approve building plans to be sure the finished structure will be safe and strong.

footing
the bottom part of a foundation, made of hardened concrete, under the foundation wall

Figure 12-17 The foundation of a house is made from concrete poured into a form made of lumber. The lumber is removed after the concrete hardens.

Cities and other local governments want to be sure that houses are built to be safe and strong. They make regulations affecting building design and construction. Can you imagine how unsafe it might be to live in a house with no windows and only one outside door? What problems can you see with such a house?

Local governments require all builders to obtain a *building permit*. A building permit authorizes the builder to begin construction. It is issued only after the construction plans have been approved. See Figure 12.16.

The plans must meet or exceed local *building codes*. Each community has its own set of construction regulations called a building code. These regulations may differ from city to city. For example, some northern cities require a house foundation to be at least 4 feet down into the ground. This distance keeps the foundation from being moved by freezing in winter. In some southern cities, the foundation has to be only 2 feet down into the ground. Can you figure out why?

The Foundation

How is a foundation constructed?

In house construction, the foundation starts with a trench dug around where the outside walls will be located. Its depth depends on the local building code, and its width is usually 16 inches. See Figure 12.17. Boards are placed along the inside of the trench, and concrete is poured to a thickness of about 8 inches. The hardened concrete is called the **footing**.

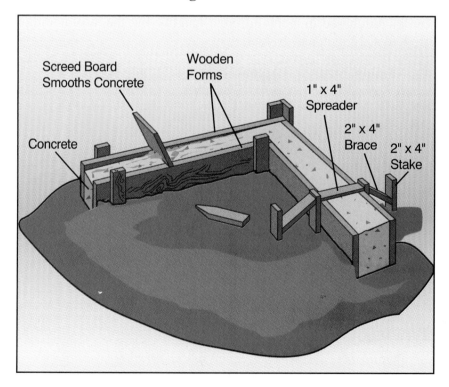

The foundation also includes a low wall called the *foundation wall*. The wall is made from poured concrete or precast concrete blocks that have large holes inside to reduce their weight. Depending on the specific house and its location, the foundation wall has about three rows of blocks. See Figure 12.18. The blocks are carefully positioned on the footing and fastened together with mortar. **Mortar** is similar to concrete, but it doesn't have gravel in it.

The Floor

How is a floor attached to a foundation?

Once the foundation is completed, the floor is the next step. The holes in the concrete blocks are filled with mortar, and J-shaped or L-shaped bolts are placed into the wet mortar. After the mortar hardens, a nominal 2-inch-thick piece of lumber is bolted to the concrete blocks. Floor supports will be nailed to this wooden *sill plate*. See Figure 12.19 on the next page.

mortar
a mixture similar to concrete, made of sand and cement, used to glue or fasten concrete blocks together and sometimes used in plastering

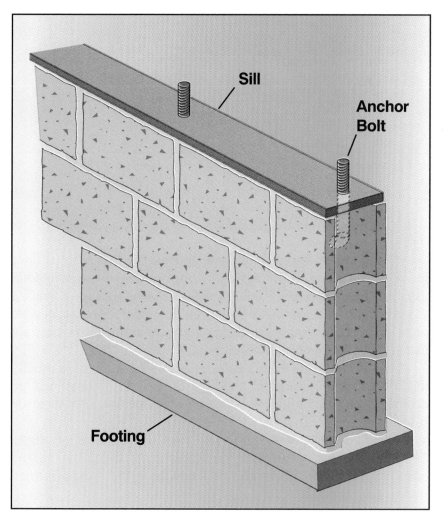

Figure 12-18 Three rows of concrete blocks are fastened together with mortar to make the foundation wall. The floor is built on top of the wooden sill plate bolted to the wall.

Sill

Anchor Bolt

Footing

Tech Talk

Joist is an unusual word that's based on an old French word for "bed" or "beam." **Joists** are parallel beams that hold up a floor or ceiling.

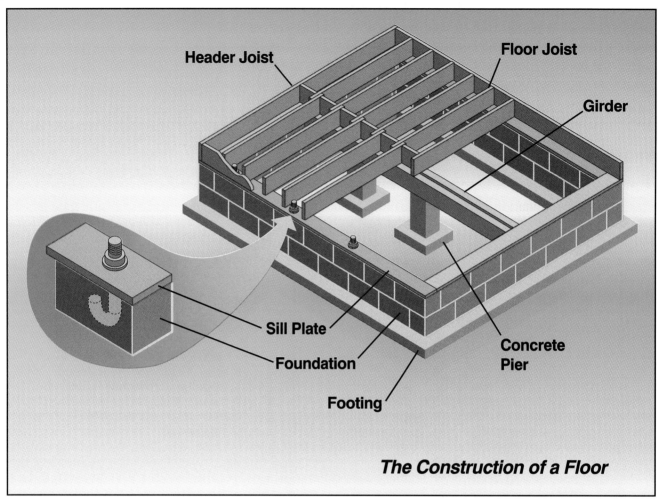

Header Joist

Floor Joist

Girder

Sill Plate

Foundation

Concrete Pier

Footing

The Construction of a Floor

■ **Figure 12-19** ■■ Floor joists are nailed to the sill plate and center girder. This method makes a rigid floor.

The floor is supported underneath by 2 × 8 or 2 × 10 floor joists. They extend from the front of the house to the rear and are usually 16 inches apart. When the foundation walls are far apart, a wooden or steel center beam is built at the middle. The beam rests on small concrete supports called *piers* or *pads*. The floor joists are nailed to the sill plate and center beam.

A **subfloor** is nailed to the floor joists. The subfloor is usually made of plywood sheets 4 × 8 feet in size. Plywood sheets go down quickly. See Figure 12.20. The subfloor can also be made from 1 × 12 boards.

During construction, the floor takes a great deal of abuse. Tools are dropped on it, nails are accidentally driven into it, and paint is spilled. The subfloor absorbs that wear, and later, a finished floor is attached to it.

subfloor
the first layer of flooring, usually made of plywood, that takes much abuse during the construction of the building

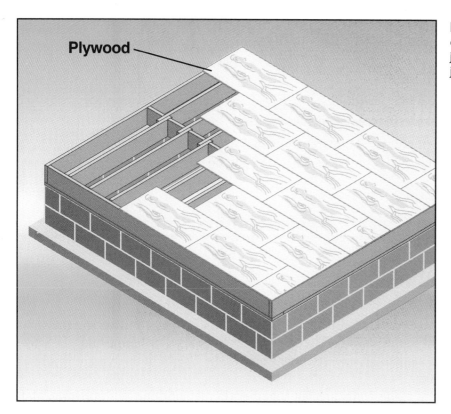

Plywood

Figure 12-20 Staggered sheets of plywood are nailed to the floor joists to make a subfloor. Staggered joints make a strong floor.

— Learning Time —

Recalling the Facts

1. A 2 × 6 is really $1\frac{1}{2}$ by $5\frac{1}{2}$ inches in size. Why do we call it a 2 × 6?
2. What is nominal 1-inch lumber usually used for? What is nominal 2-inch lumber usually used for?
3. Why do local governments require special permits to construct a building?
4. What's an average foundation depth for a house in America?
5. How many rows of concrete blocks are in an average foundation?
6. When would a floor have a wooden or steel center beam?

Thinking for Yourself

1. Can you make two 1 × 6 boards by cutting a 2 × 6 in half? Explain. Can you make one 2 × 6 by nailing two 1 × 6 boards together? Explain.
2. Would a 16-foot 2 × 8 cost twice as much as two 8-foot lengths? Explain.
3. Some local governments have a building requirement that door and window area total no less than 10 percent of the total floor area. What's the reason for such a requirement?
4. Floor joists have to be carefully positioned if the subfloor will be 4 × 8 plywood sheets. Explain.

(Continued on next page)

Tech Talk

A hunting knife fits into a leather cover called a *sheath*. The word means "protective covering." Wall *sheathing* is a protective covering for a house.

stud
one of the upright, or vertical, wall supports of a building. Usually placed 16 inches apart.

Figure 12-21 Exterior walls are nailed together on the subfloor. Several people raise the wall and nail it in place.

Walls

Why are walls put together on the floor?

With a solid subfloor on which to work, constructing the walls is the next step. They are made from 2 × 4 lumber that is 8 feet long and spaced 16 inches apart. The height of an average wall is 8 feet.

The walls are put together on the subfloor, with openings left for windows and doors. It is much easier to construct a wall on a flat floor, where it can be easily supported. When the wall is finished, it's raised into position by two or more people and nailed to the floor. See Figure 12.21. The vertical 2 × 4 wall supports are called **studs**.

As each wall is built, it's nailed to those already put up. The walls are strengthened by ceiling joists connecting the top of the front wall to the top of the back wall. Ceiling joists resemble floor joists, except they're 8 feet higher up.

Once the walls are up, a sheet of $\frac{1}{2}$-inch-thick 4 × 8-foot plywood is nailed at each outside corner. The plywood braces the corners and makes the entire structure more rigid.

The rest of the outside walls can be covered with the same type of plywood. However, it is also common to use stiff $\frac{1}{2}$-inch-thick plastic insulation. It looks like the styrofoam plastic used in some beverage cups. The insulation keeps the house warmer in the winter and cooler in the summer. No matter what material is nailed to the walls, it is called *wall sheathing*.

The Roof

What are the two ways that roofs are made?

The roof must keep water out when it rains. It also must be strong enough for people to walk on, and it must be able to support heavy snow loads. Most roofs have two sloping sides that meet at the ridge. This style is called a **gable roof**. The shape makes a strong and secure cover for the house.

One common way to make a roof is by building a sloped frame. **Rafters** made from 2 × 6 or larger lumber are nailed to the top of the walls and meet at a peak. They make up the basic roof structure. See Figure 12.22. A ridge board at the peak acts like a

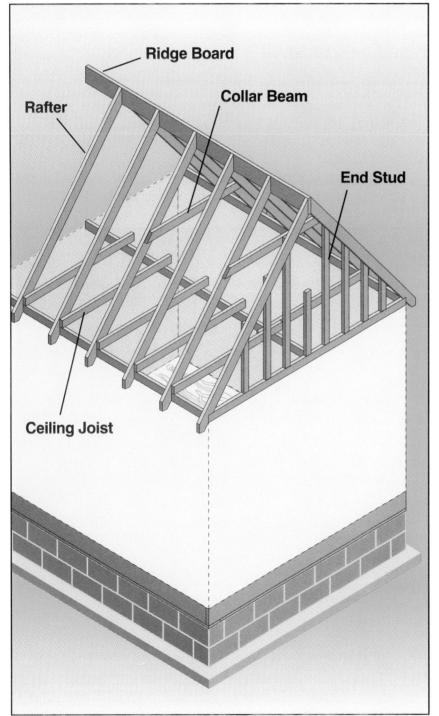

Figure 12-22 Rafters make up one standard type of roof structure. Rafters nailed to the top of the walls and a ridge board, provide the sloping structure for the roof.

Ridge Board

Rafter

Collar Beam

End Stud

Ceiling Joist

central support for the rafters. The rafters are covered with 4 × 8-foot sheets of $\frac{1}{2}$-inch plywood or 1 × 12 lumber. This covering is called the *roof decking*.

Many roofs are made from prefabricated wooden parts called trusses. Trucks deliver the trusses, which are placed on top of the walls by a small crane. See Figure 12.23. After being nailed in place, trusses are covered with roof decking. The major advantage of trusses is that the roof can be made much more quickly, sometimes in about half a day.

Siding and Roofing

What are some siding and roofing materials?

The external appearance of a house is what gives it character. Some houses just seem to look better than others. It might be because the builders or owners carefully picked siding material and roofing material.

Many different kinds of siding material can be fastened to the outside walls of a house. Wood, plastic, metal, and brick are only a

■ **Figure 12-23** ■ Construction workers frequently use special steel fasteners to connect structural members. This worker is preparing the house for roof trusses.

few examples. Wood siding is put up one board at a time. The special boards are usually 5 or more inches wide and thicker on one edge. When they're nailed in place, one board overlaps the next.

Wider sections of plastic or aluminum siding are manufactured to resemble wood siding. Like wood siding, they are nailed in place, but they go up more quickly. Also, they don't need to be painted.

Bricks can also cover the outside walls. The first row rests on part of the foundation. The brick veneer, as it is called, is fastened to the wall with special strips of metal. One end of a short metal strip is nailed to the wall, and the other end is placed in the mortar between rows of bricks. When the mortar dries, the metal strips hold the bricks firmly against the wall sheathing. See Figure 12.24.

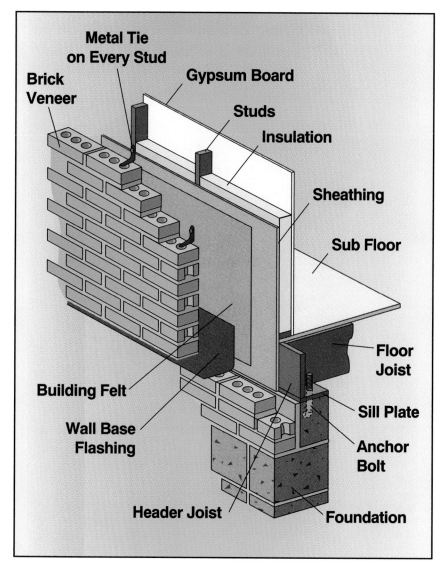

Figure 12-24 Brick veneered houses have the bricks attached with metal ties in the mortar between the rows. The interior walls are covered with gypsum, another name for drywall.

Metal Tie on Every Stud
Brick Veneer
Gypsum Board
Studs
Insulation
Sheathing
Sub Floor
Floor Joist
Building Felt
Sill Plate
Wall Base Flashing
Anchor Bolt
Header Joist
Foundation

Asphalt shingles are nailed to the roof decking over a layer of heavy asphalt paper called felt underlayment. The two layers provide good sealing against rain and snow. The first row of shingles is applied near the edge of the roof, and the following rows overlap each other. This method makes sure that water runs off the roof instead of under the shingles. See Figure 12.25. Shingles are made with many small stones on their surface. The stones strengthen the shingles and also reflect some of the sun's heat.

Once the windows and doors are installed, the outside is pretty much completed. A person walking by would think the house could be occupied. However, a bit of interior work needs to be completed before anyone moves in.

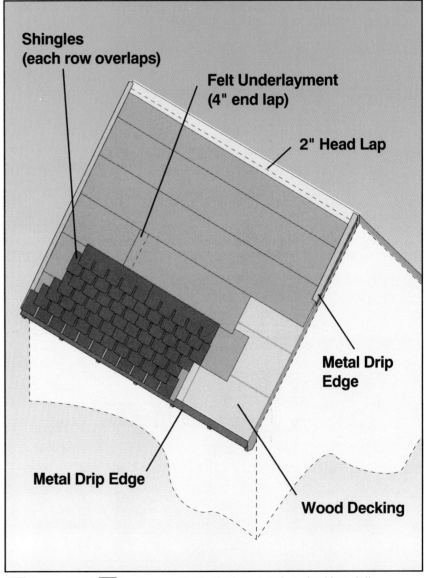

Shingles
(each row overlaps)

Felt Underlayment
(4" end lap)

2" Head Lap

Metal Drip
Edge

Metal Drip Edge

Wood Decking

Figure 12-25 Roof covering includes wooden decking, felt underlayment, and overlapped shingles. Metal drip edges keep rain water from damaging the ends of the wooden decking.

Insulation

What does insulation look like, and how does it work?

caulking
a toothpaste-like material used to seal cracks in a house

Insulation is like a blanket. The more you have and the better it's tucked in, the warmer you will be in the winter. The most popular type of insulation is a fluffy type of fiberglass. It's made in long rolls $3\frac{1}{2}$ inches thick and just the right width to fit the 16-inch spacing between floor joists, ceiling joists, and wall studs. Some wall sheathing is also insulation. Used together, the two different insulations provide a thicker "blanket" for the house.

The fluffy fiberglass roll has waterproof paper attached to one side. The paper faces the inside of the house and is a moisture barrier. It keeps moisture away from the wooden framing materials. See Figure 12.26. Moisture would cause the wood to become wet and perhaps rot.

Cracks always appear in a wood house during its construction. Those cracks can let in cold air. **Caulking** is used to seal them up. Caulking looks a little like tooth paste and is applied with a special tool called a caulking gun. Sealing the cracks is like tucking in the house blanket.

Figure 12-26 House insulation is often fluffy. Twelve inches or more in the ceiling provides a thick blanket against the cold.

— Learning Time —

Recalling the Facts

1. You want to add a wall inside a house. Would you build the wall structure from 1 × 6 lumber or 2 × 4 lumber?
2. What's the purpose of nailing plywood sheets to exterior walls at the corners?
3. What are the two ways of building a roof?
4. Name four types of siding.
5. Why do roof shingles overlap each other?
6. What is the purpose of a moisture barrier on fiberglass insulation?

Thinking for Yourself

1. Why do some floors require 2 × 10 floor joists, but a wall requires only 2 × 4 studs?
2. What are some advantages of wood siding over plastic or metal siding?
3. When nailing shingles to a roof, why don't people start at the top of the roof?
4. Name some types of insulation not mentioned in this section.

Applying What You've Learned

1. Design a storage shed or doghouse.

(Continued on next page)

2. Write a computer program to determine the energy loss through a wall. You need to know:
   a) The number of degree days in your area; call your weather service to find out (it's about 2300 in Dallas, 1500 in Los Angeles, and 5000 in New York)
   b) The "R" value of the wall insulation
   c) The area of the wall
   Here's the approximate formula:
   Energy loss = [Degree Days × 24] × [Wall Area/(3.4 + R value)]
   The units on Energy Loss will be BTUs, British Thermal Units. Compare energy loss through walls with different insulation R values, and through walls of different areas.

Inside Walls

How are inside walls made?

Building contractors try to *close up* the house as quickly as possible. They want to complete the outside so their people can work on the inside without worrying about the weather. There are many jobs to finish. One of the biggest jobs is to build and cover the inside walls.

Inside walls are made just like the outside walls. The frames are assembled on the floor, raised, and nailed in place. Workers use 4 × 8-foot sheets of $\frac{1}{2}$-inch-thick drywall to cover both the walls and ceiling. **Drywall** is made from plaster sandwiched between two layers of sturdy paper. See Figure 12.27. Even if a wall will be cov-

drywall
the inside covering of walls and ceilings, made from plaster and sturdy paper

■ **Figure 12-27** ■ Each 4 X 8 foot drywall sheet is heavy and brittle. The sheets are carefully installed by trained and experienced workers.

ered with paneling, drywall is frequently used under it. The heavy drywall sheets are nailed to the walls and ceiling. A plasterlike material called *spackling* fills in the cracks between the sheets and covers the nail heads. However, the inside walls and ceiling can't be covered until the utilities are installed.

Utilities

What type of utilities are used in most houses?

When construction workers refer to *utilities*, they mean electricity, natural gas, water, and sewage disposal. These services are provided by businesses called public utilities. See Figure 12.28.

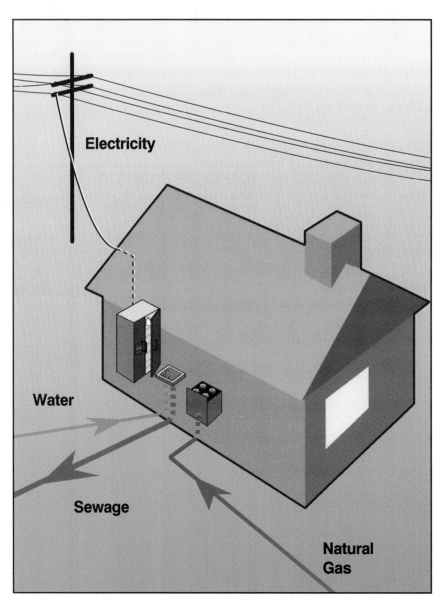

Figure 12-28 Public utilities include electricity, natural gas, water, and sewage disposal.

Electricity

Water

Sewage

Natural Gas

Electricity is used for heating, lighting, cooking, air conditioning, and other purposes. Natural gas is used for heating and cooking. Water is used for drinking and cleaning. Sewage disposal removes wastes from the house.

Electricity. Electricity comes from the power company and enters the house through a heavy insulated wire called a service entrance cable. A watt-hour meter located outside measures how much electrical power is used. About once a month, a meter reader checks the reading on the watt-hour meter, and the house owner receives a bill from the power company. See Figure 12.29.

Electricity branches to different parts of the house from the electrical service panel. One branch, or circuit, might be the kitchen. Another circuit might include two bedrooms. And so on. A large appliance that uses a lot of electrical power, like an electric stove or an electric dryer, would have its own circuit. Wires extend from the service panel to each room of the house. They pass through holes drilled in the wall studs.

Each circuit can carry only a certain amount of electrical power. If the power gets too high, a circuit breaker trips and stops the electricity. Appliances that create heat use more electrical power than motors, lights, or electronic equipment. Toasters, for example, use more electrical power than electric fans, fluorescent lights, or television sets. If you were to turn on three toasters on the same circuit, the circuit breaker would trip. The circuit breaker is a safety device that keeps the wires inside the wall from overheating and possibly burning down the house. There is one circuit breaker for every electrical branch.

Many homes still use *fuses* for the same purpose. Fuses have short pieces of lead wire inside. The lead melts when electrical flow reaches the rated value, 15 or 20 amperes. With a melted conductor, the electricity stops.

There is one main circuit breaker inside the electrical service panel. It shuts off all the electricity in the house. You should know how to trip it. Suppose that a serious electrical problem developed in your home. If you didn't know exactly what to do, flipping the main breaker would stop all electricity flowing into the house. See Figure 12.30.

Natural Gas. Natural gas is used in home heating, hot water heating, and cooking. More U.S. homes are heated with natural gas than by any other method, including electricity and fuel oil.

Many heating systems use a system called forced air. Natural gas, for example, is burned inside a furnace. A fan blows air over hot parts inside the furnace. Metal ducts in the crawl space connect each room to the furnace. The heated air is forced through the ducts and warms the house.

Another method heats water in a boiler. The water doesn't actually boil. It's heated to about 180°. The hot water is sent to each

■ **Figure 12-29** ■ Meter readers determine how much electricity a house uses in one month. The home owner usually receives an electricity bill every month.

room through pipes, and radiators transfer heat from the water to the room. There are no metal ducts. The water is then sent back to the boiler for reheating.

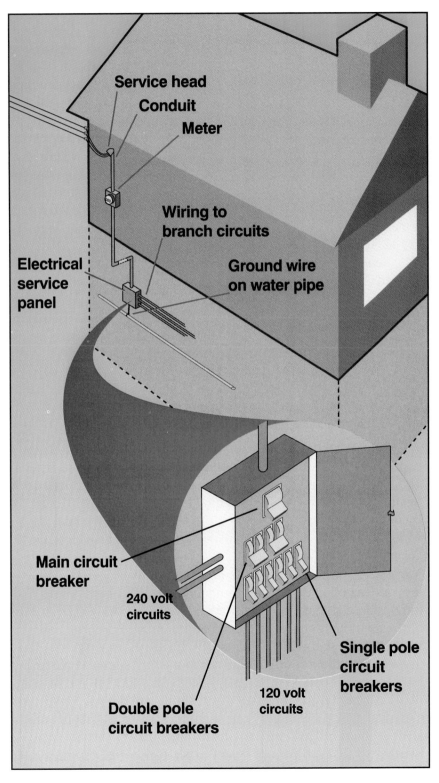

Figure 12-30 The main circuit breaker stops all electricity in a house. Trip the switch if there is a serious electrical problem and you don't know what else to do.

Figure 12-31 The shutoff valve for this natural gas meter is in the pipe on the right. Turn it a quarter turn to shut off the flow of natural gas.

Figure 12-32 Each sink has its own shutoff valve coming out of the wall. The one on the left is for hot water and the other for cold water.

All heating systems have a main shutoff valve. Natural gas systems have valves near the regulator, outside the house. Fuel oil systems have valves in the main fuel line, inside the house. You should know how to turn off the fuel to the heating system. A serious house fire could be prevented if you know how to turn off the source of fuel. See Figure 12.31.

Water and Sewage. Water and sewage services are usually provided by the same utility. Pressurized water is delivered to the house through underground pipes. It enters the house through pipes in the crawl space and then branches out to the bathrooms, kitchen, and other places where water is used. The water pipes are typically $\frac{3}{4}$ or $\frac{1}{2}$ inch in diameter. They can be made from copper or plastic.

The main water shutoff valve is in the water line where it enters the house. You should know where it's located. Suppose you accidentally kick and break the water line under a sink. If you quickly find and turn off the main water valve, no harm will be done. Otherwise, the water could do quite a bit of damage. See Figure 12.32.

Sewer pipes are much larger than water lines and are not pressurized. The lines are all sloped so that natural gravitational flow removes the waste water from the house. The plastic or cast-iron pipes are 3 inches in diameter and larger. There is usually no sewage shutoff valve. Since sewage is not under pressure, a leak would not cause much damage.

Manufactured Housing

Where are manufactured houses made?

Most of the residential construction in the United States is done on site. Only a small portion uses parts manufactured in factories. Houses, or parts of houses, made in factories are said to be **prefabricated**. The Italian inventor Leonardo da Vinci referred to such a construction technique before 1500. He wrote about the advantages of house parts being made at one location and put together at another.

Factory production of residential dwellings uses less energy, less raw material, and less labor than on-site stick-built houses. Factories also can make more residential dwellings in less time. Most of the apartment houses in Russia and in many European countries were made of prefabricated parts.

Some ordinary-looking houses in the United States are assembled from parts made in a factory. Walls are framed and insulated on an assembly line. Windows and siding are added. Water lines, electrical wires, heating systems, and complete bathrooms are installed. Interior walls are completed. The house is usually made in two halves. This allows the parts to be transported on trucks. See Figure 12.33.

Figure 12-33 ■ Manufactured houses are heavy and must be moved by trucks to the set up location.

Two house halves are delivered to a building site where the foundation has been completed. Large cranes position the parts on the foundation, and everything is bolted together. Connections are made to the utilities, and the house is ready to be moved into. The time from delivery to move-in is just a day or two.

Mobile homes are the most popular type of manufactured house in the United States. There are about 4 million of them. See Figure 12.34.

Figure 12-34 ■ Mobile homes are the most popular type of manufactured house in America.

—Learning Time—

Recalling the Facts

1. Why do building contractors try to close up a new house as quickly as possible?
2. Name four public utilities.
3. Why is an electric stove on a circuit by itself?
4. Many furnaces are called forced-air furnaces. Why?
5. What force removes the waste water from a house?
6. What is the most popular type of manufactured house?

Thinking for Yourself

1. What are some differences between constructing inside and outside walls?
2. Why is it good to have many electric circuits in a house?
3. It's usually easier to heat a two-story house with a boiler than with forced air. Why?
4. All waste water lines inside a house's living area have J-shaped pipes called water traps. There's one under every sink, for example. What's the purpose of the water trap?

(Continued on next page)

Applying What You've Learned

1. Prefabricate different-sized birdhouses. Distribute them in kit form with assembly information to interested persons.
2. Locate the main electrical circuit breaker in your home. Also, find out how to turn off the water supply and fuel supply for your heating system. Make neat signs and post them for all your family members to see.

Major Construction Projects

≡ *Are houses the only type of construction project?*

People who build houses make up only part of the construction industry. Many other people are employed building larger structures. Many major construction projects are one of a kind. The designer can't simply purchase the plans from a building supply magazine. Careful designing by qualified civil engineers and architects ensures the safety of the structures. Government-funded buildings can be constructed only after all the blueprints are approved by professional engineers.

Skyscrapers

≡ *Are all tall buildings constructed the same way?*

The word *skyscraper* might cause you to automatically think about New York City. New York has many tall buildings, but so do other cities. Chicago, for example, has three of the world's five tallest skyscrapers; New York has the other two. Toronto has the sixth tallest skyscraper.

The supporting structure of a skyscraper is usually structural steel. The entire weight of the building rests on the internal frame, not on the walls. In fact, skyscraper walls are sometimes called *curtain walls,* meaning that they merely hang on the structure. That's why skyscrapers can have outside walls of glass and other unusual materials. See Figure 12.35.

Not all skyscrapers are constructed alike. The 1,454-foot Sears Tower in Chicago uses over 100 concrete piles for support. It's the tallest building in the world. The twin towers of the World Trade

Figure 12-35 Taking only 14 months to construct, the Empire State Building in New York City was opened to the public in 1931. These workers hoisted an American flag to signify completion of their part of the structure.

Center in New York have a metal mesh skin that supports much of the building's weight. At 1,350 feet, it's the second tallest building in the world. Architects and engineers are always looking for ways to make tall buildings both strong and beautiful.

Roads and Highways

What is underneath the surface of a highway?

Incredible as it may sound, there are almost 4 million miles of roads in the United States. That includes everything from rural roads to city streets to interstate highways. The government spends billions of dollars every year to maintain existing roads and construct new ones.

All paved roads are made in three layers. See Figure 12.36 on the next page. The **subgrade** is the natural soil along the road. If it's not level or firm enough, heavy machines scrape and pack the soil. Next comes the *base*. A common base is sand or gravel. It provides support for the surface and keeps water from collecting underneath. Water could freeze and break the pavement. The final layer is the *surface*. The surface is a smooth top, higher at the middle to drain off water.

The surface material depends on the type of traffic that will use the road. Many modern highways are made from concrete—a mixture of portland cement, gravel, sand, and water. It is easy to mix, doesn't have to be heated like asphalt, and dries to a hard, durable road surface. Concrete can be strengthened with steel bars or mesh placed in the wet concrete. That is why it's used where heavy traffic is expected. America's interstate highway system was started in 1952 and was designed to be all concrete. The 42,500

subgrade
the soil or rock that forms the first layer of a road

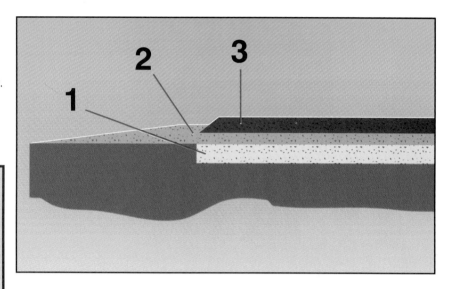

■ Figure 12-36 ■ Paved roads are made in three layers. Layer 1 is the subgrade, or packed soil. Layer 2 is the base of sand or gravel, and layer 3 is the surface of concrete or asphalt.

asphalt
a brownish-black, flexible material made from tar or pitch found in nature and used to make and repair some road surfaces

suspension bridge
a roadway that is hung from large cables and goes across a wide span, or space

truss bridge
a bridge that is held together by steel beams that are fastened together in triangular shapes

Edinburgh
(ĕd′n-bûr′ə)

cantilever bridge
(kăn′tə-lē′vər)
a bridge made of two self-supporting beams, each of which is fastened to the ground at one end. The beams meet in the middle of the bridge.

girder bridge
a bridge made of beams that rest on the ground on either side of the span

miles of interstate highways follow routes that connect 90 percent of the major cities. Airport runways and interstate highways are usually surfaced with concrete about 10 inches thick.

Asphalt is used for other roads as well as to repair worn concrete roads. It is difficult to get new concrete to stick to old concrete. New concrete sometimes breaks away after a short time. However, asphalt is more flexible and sticks better to concrete. Many interstate highways have been resurfaced with asphalt.

Bridges

What are the different kinds of bridges?

Bridges are among the most beautiful human-made structures. The Golden Gate Bridge in San Francisco is an example of a beautiful suspension bridge. **Suspension bridges** hang from large cables and are used to bridge over wide spans. When it opened in 1937, the Golden Gate Bridge had the world's longest span (4,200 feet) and the highest supporting towers (746 feet). See Figure 12.37.

A **truss bridge** is one held together with steel beams. The beams are fastened together in the shape of many triangles. The Firth of Forth Bridge near Edinburgh, Scotland, is a **cantilever bridge** strengthened with trusses. A cantilever is a self-supporting beam that is securely fastened to the ground at one end. Two cantilevers meet in the middle to make a bridge. The mile-long Firth of Forth Bridge, opened in 1890, carries railroad traffic 158 feet above the water. Its strong design withstands high winds in the area. The Eads Bridge, mentioned earlier in this chapter, is also a cantilever bridge with trusses.

A **girder bridge** has no obvious external structure. A beam, called a girder, connects one side to the other. The bridge ends rest on the ground. They are not securely fastened, as are cantilever bridges. Girder bridges are frequently supported by piers partway

Figure 12-37 ■ Few people will disagree that San Francisco's Golden Gate Bridge is one of the most beautiful suspension bridges in the world.

along the span. Many interstate highway bridges use this design. See Figure 12.38 on the next page.

Tunnels

How are tunnels made through mountains and under bodies of water?

Bridges go over natural barriers, but tunnels go under them. Tunnels are constructed through mountains and under water. Modern tunneling methods use blasting with explosives and drilling with huge machines. A large metal tube called a *shield* fits inside the tunnel as it is drilled.

Some tunnels are quite long. The Saint Gotthard Tunnel for automobiles goes 10.1 miles through the Alps in southern Switzerland. The Seikan Tunnel for trains connects the Japanese islands of Honshu and Hokkaido. Opened in 1988, it is 33 miles long, with 14.6 miles underwater.

One tunnel project stands above all others as being the most expensive private construction project in history. It's the 32-mile tunnel under the English Channel between Folkestone, England, and Calais, France. It was a joint venture among British and French companies. The project started in 1986, and scheduled train service began in November 1994. At 23.6 miles underwater, it's the longest undersea tunnel ever built. See Figure 12.39 on page 375.

Tech Talk

The Seikan Tunnel took twenty-four years to construct. The name means "to have command of the sea."

Saint Gotthard
('gät(h)-ərd)

Seikan
(sī-kän)

Honshu
('hän-(ˌ)shü)

Hokkaido
(hä-'kīd-(ˌ)ō)

Calais
(kă-'lā)

373

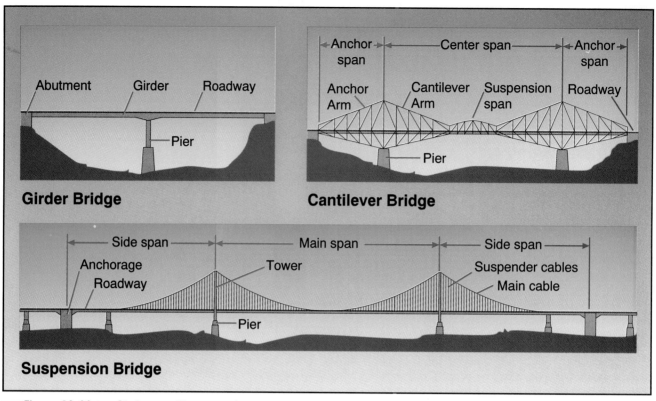

Figure 12-38 Girder, cantilever, and suspension bridges are the most commonly seen in the United States.

The project involves three tunnels: two for high-speed electric trains and one between them for maintenance. Massive 27-foot-diameter boring machines turned cutting bits at a slow 3 revolutions per minute to remove material. Drilling was carried out from both countries at the same time until British and French workers broke through and met under the English Channel in 1990. The tunnel varies in depth from 90 to 480 feet below the bottom of the seabed.

The London-to-Paris route is one of the busiest in the world. Sixteen high-capacity passenger trains leave London's Waterloo Railway Station every day for the 3-hour-and-15-minute trip. With interiors similar to those in modern airplanes, each train carries up to 794 passengers at approximately 200 mph.

Construction Careers

What kinds of jobs are available in the construction industry?

There are indoor jobs and outdoor jobs. There are jobs using construction tools, and there are jobs using pencil and paper. There

Figure 12-39 For the first time in history, England is connected to Europe with a tunnel. The 32-mile English Channel tunnel is the most expensive private construction project ever. This photo shows the tracks being laid.

are jobs requiring a college education, and there are jobs you learn "on the job." In short, there are all kinds of jobs.

Carpenters, cement masons, brick layers, electricians, plumbers, roofers, and heating system specialists all work on houses. Specialized jobs include heavy equipment operators, who move dirt and other items for major construction projects. Iron workers put skyscrapers and bridges together. Architects and engineers design the projects. These are just some examples of the many different careers.

You may have seen people building houses and think you'd like to be a carpenter. It's interesting work, but carpenters do more than just build houses.

• They build wooden forms in which to pour concrete for high-rise buildings.
• They erect scaffolding at large projects.
• They build low-rise apartment houses.
• They install finish trim inside houses and apartments.
• They construct prefabricated building sections in factories.

One way to become a competent and well-respected professional carpenter is to go through an apprenticeship program. An apprentice is a person learning a new trade. The U.S. Department of Labor recommends a four-year training period, or about 8,000 working hours. The terms of the apprenticeship are included in a written agreement with a local Carpentry Joint Apprentice Committee. Training experience consists of learning how to do the following jobs:

Layout	500 hours
Form building	1,200
Rough framing	1,200
Outside finishing	800
Inside finishing	1,500
Care and use of tools	500
Welding	500
Plastics	300
Acoustics	1,000
Miscellaneous	500
TOTAL	8,000 hours

Carpenters will always be needed to build housing units as the United States continues to grow. Also, carpenters will be needed in the maintenance departments of factories, businesses, and large residential projects, as well as for remodeling commercial and residential buildings. On the whole, the employment outlook for carpenters and all construction employees is excellent.

— Learning Time —

Recalling the Facts

1. What city has three of the world's five tallest buildings?
2. Why can skyscrapers have walls of glass?
3. How many miles of roads are in the United States?
4. Why is concrete used for roads that get heavy traffic?
5. Why aren't concrete highways usually repaired with new concrete?
6. What kind of bridge is the Golden Gate Bridge?
7. Once a tunnel is drilled with a boring machine, why don't rocks continue to fall from the top of the tunnel?

Thinking for Yourself

1. Name the five tallest buildings in the world and the cities where they're located.
2. What skyscraper was featured in the first successful monster movie, the 1933 *King Kong*? What skyscraper was featured in the 1976 *King Kong* remake?
3. Concrete highways are poured in sections with tar strips between the sections. What is the purpose of the strips?
4. The English Channel tunnel is 32 miles long, but only 23.6 miles are under the sea. Why the difference?

Applying What You've Learned

1. Build a model of a well-known skyscraper.
2. Use plaster of paris as if it were portland cement. (a) Mix with fine gravel, and check for

(Continued on next page)

strength improvements. (b) Imbed wide mesh screening in the wet plaster, and check for strength improvements. (c) Try other possibilities.

3. Design and construct a small precast girder bridge. Use strengthened plaster of paris.

— Summary —

Directions

The following sentences contain blanks. For each numbered blank, pick the answer that makes the most sense in the entire passage. Write your answer on a separate sheet of paper.

Construction is carried out at a building site that changes as the project is completed. An important part of construction is building ___1___ . The first was the Home Insurance Building in Chicago. It was the first to use ___2___ as part of its basic design. Roads are also included in construction projects. The first really great road builders were the ___3___ . They were the first to make roads higher at the center to drain off water. The British developed a material for surfacing roads that we still use today. In America, we call that material ___4___ . One other major construction item includes bridges. The first major all-steel bridge was built across the Mississippi River at the city of ___5___ . It was named for the builder, James Buchanan Eads.

Since you may be more familiar with houses, this chapter was mostly about house construction. Before a house can be built, rough sketches are made of the ___6___ . This helps show the shape and size of all rooms. Afterwards, those rough sketches are turned into ___7___ . They can also be purchased from housing magazines and building supply stores.

Wood is the basic building material in most houses. It is strong, fairly light weight, and can be quickly fastened with ___8___ . Concrete is also a strong material but is usually only used in the ___9___ . Steel is the only metal used in most houses. It is sometimes used for ___10___ supports.

1.
a. toys
b. monuments
c. cars
d. skyscrapers

2.
a. a metal frame
b. concrete
c. wood
d. blocks

3.
a. British
b. Egyptians
c. Romans
d. Americans

4.
a. concrete
b. asphalt
c. macadam
d. decking

5.
a. St. Louis
b. Chicago
c. Des Moines
d. Houston

6.
a. payment plan
b. outside
c. floor plan
d. windows

7.
a. blueprints
b. whiteprints
c. models
d. supervisors

8.
a. glue
b. rope
c. mortar
d. nails

9.
a. walls
b. ceiling
c. foundation
d. kitchen

10.
a. wall
b. floor
c. roof
d. ceiling

(Continued on next page)

Summary

A floor is supported by 2 × 8 or 2 × 10 joists. A _____11_____ is nailed to the floor joists. Some floors have a center _____12_____ that rests on concrete supports. Once the floor is finished, work can begin on the walls.

Walls are nailed together on the floor and raised into position. The vertical 2 × 4 supports are called _____13_____. The walls are held in position with ceiling joists. The outside of the walls are covered with plywood or stiff insulation board. The covering is called _____14_____. Roof supports are nailed to the top of the walls. The roof can be made with rafters or trusses and covered with a plywood or board decking. The house's exterior is finished off with shingles and siding.

All paved roads are made in three layers. The top layer is called the _____15_____. Concrete is frequently used as the top layer. The highway can be strengthened with steel _____16_____ or mesh placed in the wet concrete. Many highways are resurfaced with asphalt. It sticks better and is _____17_____ than concrete.

One of the most beautiful bridges is the Golden Gate Bridge in San Francisco. Unlike the Golden Gate Bridge, the Firth of Forth Bridge in Scotland is known as a _____18_____ bridge. Many bridges are further strengthened with trusses. A truss is made of steel beams fastened together in the shape of many _____19_____. Many interstate highway bridges have no obvious external structure. They are known as _____20_____ or beam bridges. Each major bridge is designed for a specific place; there are no standard plans.

Modern tunneling methods use explosives and drilling with huge machines. A metal tube called a _____21_____ fits inside the tunnel as it is constructed. The English Channel Tunnel is sometimes called the Chunnel. It has a total of three tunnels and passengers ride _____22_____. It connects England with France.

11.
a. drywall
b. door
c. foundation
d. subfloor

12.
a. strip
b. beam
c. location
d. 2 × 6

13.
a. joists
b. studs
c. rafters
d. insulators

14.
a. drywall
b. cardboard
c. windows
d. sheathing

15.
a. covering
b. surface
c. grade
d. base

16.
a. bars
b. beams
c. clips
d. chips

17.
a. cheaper
b. prettier
c. more flexible
d. stronger

18.
a. cantilever
b. beam
c. suspension
d. girder

19.
a. squares
b. circles
c. lines
d. triangles

20.
a. cantilever
b. beam
c. suspension
d. girder

21.
a. pipe
b. tube
c. shield
d. tunnel log

22.
a. cars
b. trains
c. trains and cars
d. maglevs

Key Word Definitions

Directions

The column on the left contains the key words from this chapter. The column on the right contains a scrambled list of phrases that describe what these words mean. Match the correct meaning with each word. Write your answer on a separate sheet of paper.

Key Word	Description
1. Site	a) Vertical wall support
2. Skyscraper	b) Natural soil along a road
3. Geodesic dome	c) Fastens concrete blocks together
4. Floor plan	d) Bridge held together with steel beams
5. Concrete	e) Specific construction location
6. Precast	f) Fluffy type of insulation
7. Foundation	g) Mixture of cement, gravel, sand and water
8. Footing	h) Bridge with ends that rest on the ground
9. Mortar	i) Plaster sandwiched between sturdy paper
10. Joist	j) Concrete building parts made in a factory
11. Subfloor	k) Roof with a triangular-shaped end
12. Stud	l) Self-supporting beam bridge
13. Gable roof	m) Shows location and sizes of rooms in a house
14. Rafter	n) Shelter made of many small, pyramid-shaped frames
15. Fiberglass	o) Material for sealing cracks
16. Caulking	p) Flexible road surface material
17. Drywall	q) Hardened concrete under a foundation wall
18. Prefabricated	r) Supports under a floor
19. Subgrade	s) Tall building with internal frame
20. Asphalt	t) Bridge that hangs from large cables
21. Suspension bridge	u) Support under the roof decking
22. Truss bridge	v) Floor that absorbs wear during construction
23. Cantilever bridge	w) House parts made in a factory
24. Girder bridge	x) House support in contact with the ground

PROBLEM-SOLVING ACTIVITY

Building a Truss Bridge

▬ Did You Know? ▬

Truss bridges were developed during the 1500s. They use a framework called trusses to support themselves and the people using the bridge. The parts of a truss are arranged in the form of many triangles.

▬ Problems To Solve ▬

Much of the strength of a bridge comes from its design. A bridge made of lightweight materials can be very strong. Can you make an 18-inch-long bridge from balsa wood strips that will support a partly full bucket of sand?

▬ A Procedure To Follow ▬

1. Look up truss bridge designs in an encyclopedia or a technical book. Make a full-scale drawing of a bridge that is exactly 18 inches long, approximately 4 to 6 inches tall, and approximately 4 to 6 inches wide.

Resources You ▬ Will Need ▬

- Six 36-inch-long pieces of ³⁄₁₆-inch square balsa strips
- Quick-drying glue or cement
- Pencil
- Paper
- Ruler
- Razor blade
- 2- to 3-gallon plastic bucket with handle
- Sand
- Nylon cord
- Small hardwood plate to rest on bottom of bridge for attaching bucket
- Scale to weigh the bridge and bucket
- Safety goggles and long-sleeved shirt

A.

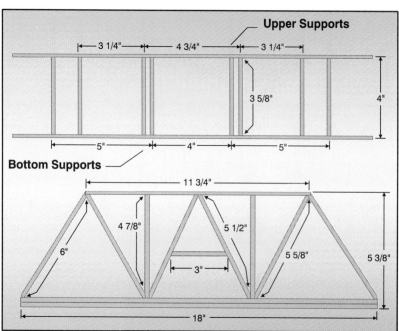

2. Construct the bridge using the ³⁄₁₆-inch square balsa strips. To improve stiffness, glue two strips together for the main horizontal stringers. See Figure A. All joints must be flush, as in full-size bridges. No overlapping joints are allowed. Do not use metal items of any kind.

3. Your design must be able to accommodate the wooden plate used for loading the bridge. At the middle of the sides, leave an opening at least 1½ x ½ inches.

4. After the glue has completely dried, weigh your bridge. Then see how much load it can carry before it fails. Place two tables 16 inches apart, and have the bridge span the gap. Put the wooden plate on the bridge and tie the bucket to it. Have the bottom of the bucket about 4 to 6 inches above the floor. See Figure B.

5. Put on your safety glasses and long-sleeved shirt. Slowly pour sand into the bucket until the bridge breaks. It will probably fail quickly and might scatter broken balsa wood. Make sure the bucket of sand doesn't tip over when the bridge fails and the bucket drops to the floor. Weigh the bucket to see how much load your bridge carried.

What Did You Learn?

1. How much weight did your bridge carry in comparison to its own weight? Was it five times more? Ten times more? More than that? Can you see why the design of a bridge can help it carry very heavy loads?

2. Where did your bridge begin to fail? Was it at the center, the edges, or the top? How could you change your design so that your bridge would carry a heavier load?

B.

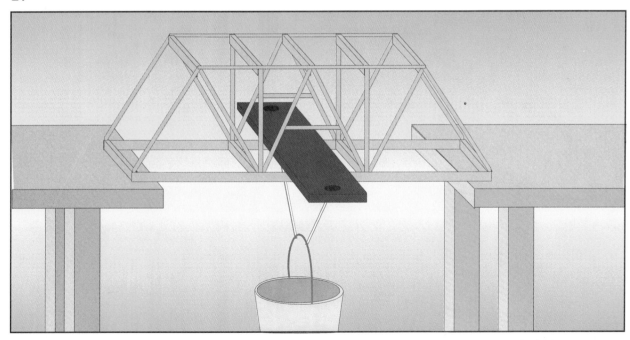

Far to Go

Joshua Slocum was the first person to sail around the world alone. It took him three years, from 1895 to 1898. In 1969, Neil Armstrong, Edwin Aldrin, Jr., and Michael Collins traveled to the moon and back. The journey took eight days. As you learn about transportation technology, think about the journeys you want to make. How will you get there?

Chapter 13

Generating Power for Transportation

You probably recognize the movie hero shown in Figure 13.1 on the next page. Some people might call this man powerful. What do you think the word *powerful* means? Does it mean muscular? Athletic? Mighty? Probably any of those words could be used to describe a person's strength.

Now suppose that you want to describe a powerful engine. Would you use the same words that you would use to describe a powerful person? An engine is not usually muscular or athletic or mighty. You would probably use different words. For example, you might describe an engine as being strong or tough.

Try to count all the engines that you and your family operate. You might first think of the engine in a lawn mower or car. Some people also use snowblowers, chain saws, outboard motors, and motorcycles. We have many engines at our disposal. See Figure 13.2 on the next page.

Our modern society has become dependent on engines. Engines produce the power we want. Life would be difficult if we couldn't generate power. Without power for transportation, you might have trouble getting to school. There would be no trucks or trains to deliver food to grocery stores. You couldn't visit relatives or friends who lived far away. Maybe you can think of other problems that would occur in a world without engines.

Key Words

In this chapter you will learn and study the meanings of the following important technology terms.

MOTOR
ENGINE
BOILER
EXTERNAL COMBUSTION
 ENGINE
COMBUSTION
VACUUM
CONDENSER
TURBINE
INTERNAL COMBUSTION
 ENGINE
PISTON
CYLINDER
CYCLE
STROKE
RECIPROCATING MOTION
ROTARY MOTION
CRANKSHAFT
JET ENGINE
THRUST
PROPELLANT

■ Figure 13-1 ■ Superman is a timeless superhero who came from Krypton in 1938 with "powers and abilities far beyond those of mortal man."

■ Figure 13-2 ■ One engine-powered product that people use is a motorcycle such as this Kawasaki Zephyr 750.

The History of Transportation Power

How did your ancestors get from place to place?

Our earliest ancestors used only animals, flowing water, and wind for their transportation power. Horses were used for the Pony Express in 1860. Letters traveled by horseback from St. Joseph, Missouri, to Sacramento, California. Sending a letter cost $1 per half ounce, and it took ten days to complete the 1,966-mile trip.

Flatboats floated with the current on the Ohio and Mississippi rivers. A flatboat trip from Pittsburgh, Pennsylvania, to New Orleans, Louisiana, covered 1,950 miles. It was a five-to-six-week journey.

Scheduled ship travel between New York City and Liverpool, England, began in 1818. Sailing east to Liverpool took three to four weeks. Traveling took a great deal of time. People looked for faster ways to get from place to place.

Major advances in the development of power for transportation came about because British coal mines in the early 1700s were flooding with groundwater. People were looking for a way to pump the water out of the mines. As the miners went deeper and deeper, the groundwater seeped in and made it more difficult to remove coal. The miners scooped water into buckets and carried it out with pack mules.

A blacksmith named Thomas Newcomen heard of the problem and invented an engine to operate a pump. The engine didn't look like modern ones. It was the size of a small room and operated very

A Human-Powered Airplane

Airplanes are usually powered by engines, not by human beings. Yet designer Paul MacReady built an airplane in 1977 that used a person's legs to turn the propeller. It was the first human-powered airplane to fly a course set by the British Royal Aeronautical Society. The flight path was a figure eight around two pylons a half mile apart while clearing a 10-foot-high T-shaped pole at the start and finish. For his accomplishment, MacReady was awarded an $87,000 prize.

Professional cyclist Bryan Allen piloted MacReady's plane, called the *Gossamer Condor* at an airport outside Shafter, California. The pusher-type propeller was in the rear, behind the cockpit, just like in the Wright Broth-er's 1903 first successful airplane. Allen pedaled a bicycle chain that slowly turned the large propeller. He developed 0.43 horsepower during the $6\frac{1}{2}$-minute record-making flight and was totally exhausted at the end. Most people develop far less power.

The plane was 30 feet long and had an enormous 96-foot wingspan, 3 feet longer than a DC-9 jet airplane. Made of lightweight metal and plastic, it weighed only 70 pounds. Its size and light weight made it quite sensitive to wind gusts. If the wind had been blowing more than 2 mph, it would have been impossible to keep the *Gossamer Condor* on course.

---

slowly. See Figure 13.3 on the next page. Newcomen's engine was the first engine in the world. It was a steam engine.

Water is a liquid, and steam is a gas. Newcomen knew that steam took up a lot more space than the water from which it came. In fact, 1 gallon of water makes 1,670 gallons of steam. He put water in a large metal container that looked like a covered pot. It was called a **boiler** because water boiled inside and was converted to steam.

Newcomen's engine, like all early steam engines, used a solid fuel, such as coal or wood. A coal fire underneath the boiler heated the water until it turned to steam. Steam pressure made the engine operate. The engine operators preferred to use coal rather than wood because a bushel of coal has over six times the energy of a bushel of wood. It took much less coal to keep an engine going.

The Newcomen engine was not significantly improved until James Watt made some changes in 1764. He made it small enough to fit inside a building and powerful enough to operate factory machinery. Watt-type steam engines were used in factories before they were used in boats, locomotives, or automobiles.

Solid fuel, like coal and wood, takes up quite a bit of space. It is also difficult to transport from place to place. Technologists

boiler
an enclosed vessel, such as a pot or kettle, in which water is boiled and converted to steam

began to experiment with ways to use a liquid fuel. Liquids fit into any container and can be easily moved through pipes. The only liquid fuel available came from the small amount of crude oil that would seep to the surface of the ground. Native Americans used it for cooking and lighting hundreds of years before European settlers arrived.

Because the amount of crude oil found on the ground was so small, it cost about $20 a barrel. No one knew how to get more. In 1859, a retired railroad conductor showed people how to do it. Edwin Drake used a steam engine to power the drill for the first successful oil well, shown in Figure 13.4. It was in the northwestern part of Pennsylvania, near Titusville. So much oil was produced that its price dropped to just 10 cents a barrel in 1862.

The wells provided a regular supply of crude oil that was refined into gasoline and kerosene. Both of these are liquid fuels that can be used to operate heat engines. At that time, kerosene was more important than gasoline because kerosene was widely used in lamps. Today, kerosene is used in portable heaters during the winter. It is also the fuel used by jet engines.

Only a few liquid-fueled engines were made in the middle 1800s, and most were one of a kind. It took special tools and lots of

Figure 13-4 Edwin Drake (right) drilled the world's first oil well in Pennsylvania. He discovered oil 69 feet below the surface.

time to make an engine. A person usually built one or two for personal use. Most of the engines were crude and did not run very well.

Nikolaus Otto, a German technician made the first successful gasoline engine in 1876. See Figure 13.5. It was used in a car manufactured by Karl Benz. The Benz car of 1886 was the first automobile to be patented.

The diesel engine was named for the German inventor Rudolf Diesel. His first successful engine ran in 1893. Diesel engines operated differently from gasoline engines and were much larger. They used kerosene for fuel. Early diesel engines powered ships and electrical generators.

Robert Goddard operated the first successful liquid-fueled rocket engine in 1926. See Figure 13.6 on the next page. It burned

Diesel
('dē-zəl)

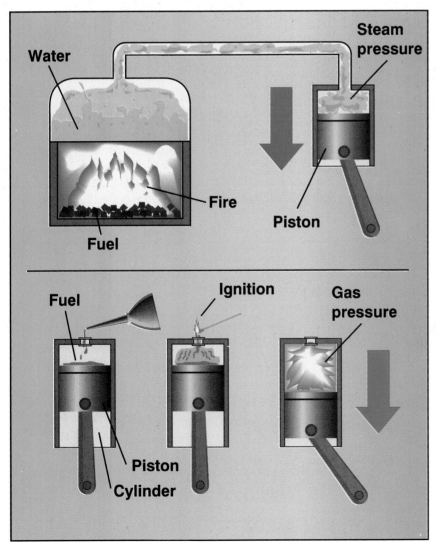

■ **Figure 13-5** ■ Steam engines (top) produced power when steam pressure pushed on a piston. Gasoline engines (bottom), like the one made by Nikolaus Otto, used an internal fire to develop gas pressure that pushed on a piston.

Tech Talk

Do you know what kerosene is? It's a fuel oil made from crude oil. Kerosene was originally a trademark. The name came from the Greek word *keros*, which means "wax." Most wax comes from oil. The British word for kerosene is *paraffin*, a word we use to mean wax.

Figure 13-6 Robert Goddard made the first liquid-fueled rocket. He is the only American who invented an internal combustion engine.

alcohol and liquid oxygen to produce a thrust force. Goddard's rocket shot 41 feet into the air after he fired it near his home in Worcester, Massachusetts.

Many people in England, Germany, and America worked on early jet engines. The first successful jet engine was used in the German Heinkel He-178 experimental warplane shown in Figure 13.7. Using kerosene for fuel, it flew for the first time in 1939.

Figure 13-7 The 1939 Heinkel He-178 was the first jet-powered airplane to fly.

— Learning Time —

Recalling the Facts

1. Name two common types of solid fuel. Name two common types of liquid fuel.
2. What are two advantages of liquid fuel over solid fuel?
3. What was the world's first engine?
4. How many gallons of steam could be made from 5 gallons of water?
5. Name two German engine inventors. Name one American engine inventor.

Thinking for Yourself

1. What was the average speed, in miles per hour, of each of the following?
   a. A Pony Express rider
   b. A river flatboat
   c. A ship sailing from New York to Liverpool
2. Why were steam boilers only partly filled with water?

(Continued on next page)

Applying What You've Learned

1. List all the engines or motors you have operated in your life. For example, suppose that your family has owned three lawn mowers. If you have operated all of them, that counts as three engines. Keep separate lists for engines and motors.
2. Make a small lift pump for water, like that operated by Newcomen engines. Materials would include a 1- to 2-inch-diameter hard plastic tube and a wooden piston with a leather flapper valve.
3. With a map of the United States, determine how long it would take to travel between certain cities in the 1800s. Use Pony Express and river flatboat times.

External Combustion Engines

What are some types of external combustion engines?

Any power plant with a fire outside the engine is an **external combustion engine**. Steam engines use heat from burning coal or wood to change water into steam. This burning is also called **combustion**. Since the fire is under the boiler, which is outside of the engine, steam engines are external combustion engines.

Steam Engines

Where does the steam come from?

The Newcomen engine was the first engine in the world to work without animal, water, or wind power. See Figure 13.8 on the next page. A worker opened the steam valve. Steam entered the cylinder and forced the piston up. The worker then closed the steam valve and opened the water valve. Water sprayed into the cylinder, cooled the steam, and changed it back to water. Remember that water takes up less space than steam, so a weak **vacuum** was created in the cylinder. This means that the pressure inside the cylinder was less than atmospheric pressure. Atmospheric pressure then pushed the piston back to its original location. The worker closed the water valve and opened a drain valve. The worker then repeated the operation over and over.

The piston went up and down only about five times per minute. In spite of its massive size, the engine developed about as much power as a lawn mower engine. It was used only for pumping water from mines.

external combustion engine
(kəm-ˈbəs-chən)
an engine that converts the fuel to energy outside the engine itself

combustion
burning

vacuum
a space that contains less pressure than is in the atmosphere

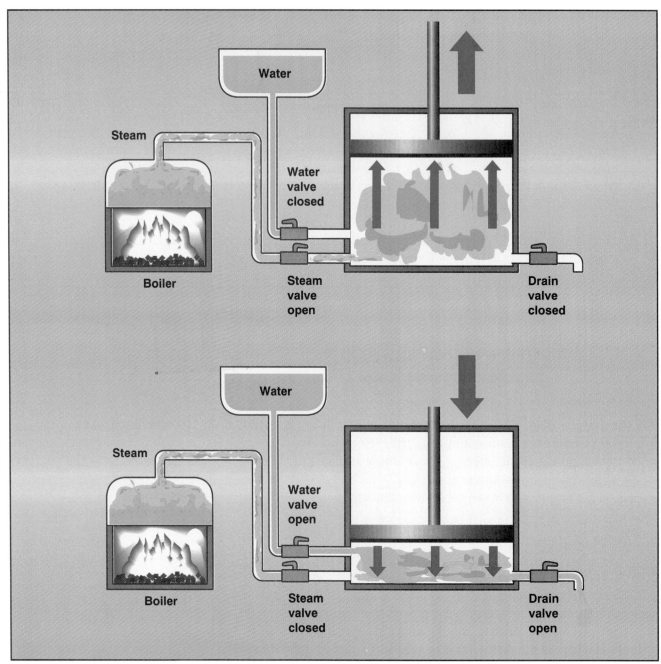

■ Figure 13-8 ■ A worker opened and closed valves to operate a Newcomen steam engine of about 1712. The piston moved up (top) when the steam valve was opened, and the water and drain valves were closed. The piston moved down (bottom) when the steam valve was closed, and the water and drain valves opened. The steam inside the cylinder condensed and lowered the pressure slightly. Atmospheric pressure pushed the piston down.

condenser
an instrument that changes steam back into water

James Watt's improvements to Newcomen's steam engine made it practical for transportation. His first major change was the addition of a **condenser**. His condenser changed steam back to water outside the engine. Watt's method allowed the engine to stay hot all the time. In Newcomen's engine, the entire cylinder was cooled. Lots of steam energy was wasted in reheating the cylinder after it had been cooled with water.

Watt also invented the *double-acting piston*. He used a special valve to direct steam pressure to one side of the piston and then to the other. See Figure 13.9. The piston did double duty, and Watt obtained twice as much power from the engine. These changes so improved the engine that many people think that Watt invented the steam engine. He didn't. He improved it.

There are no significant uses for steam engines in modern-day America. Their efficiency is half that of gasoline and diesel engines. They are large and don't produce much power. However, many people enjoy seeing old steam engines because they provide a link with our technological past. You can sometimes see antique steam-powered locomotives or tractors operating at museums or outdoor shows.

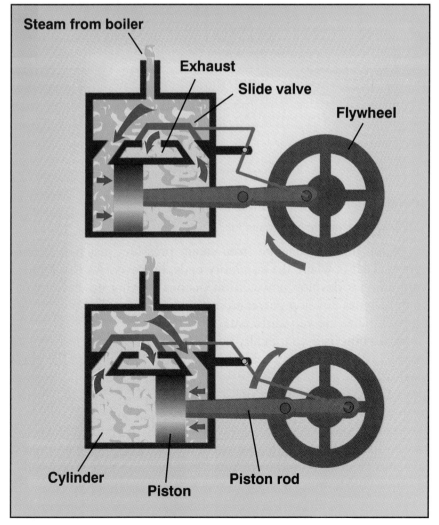

Steam from boiler

Exhaust

Slide valve

Flywheel

Cylinder

Piston

Piston rod

■ **Figure 13-9** ■ James Watt's major improvement to Newcomen's steam engine was adding a double-acting piston in about 1765. As the slide valve moved back and forth, steam pressure went to one side of the piston and was removed from the other side.

Tech Talk

The International Electrical Congress met in Chicago in 1893. They established the *watt* as a unit of power. Named for James Watt, 746 watts equal 1 horsepower.

James Watt

James Watt was a 28-year-old laboratory technician at Glasgow University in Scotland. The physics department owned an operating model of a Newcomen steam engine. One day in 1764, it stopped running, and Watt was asked to repair it. Even though Newcomen engines had been pumping water for over fifty years, it was the first steam engine that Watt ever saw.

Watt quickly fixed the engine and began to think of ways to improve it. His first small experimental engine was made from brass and worked very well. Watt was a cautious worker, and it took him ten years to build a full-size and satisfactory engine.

He formed a partnership with manufacturer and businessman Matthew Boulton. It was an ideal relationship because Boulton encouraged Watt to build more engines and to design engines of higher power. The company they formed was located in Birmingham, England. By 1880, when Watt's patents had run out, he had about 500 engines operating in England.

Watt's engines almost completely replaced Newcomen's. His engines were so much better that the Newcomen engine was all but forgotten. Watt became financially successful and was soon incorrectly looked upon as the inventor of the steam engine. The engine that Watt repaired is on public display at Glasgow University.

Steam Turbines

What are steam turbines used for?

Steam turbines operate from steam pressure, just like steam engines. That is where the similarity ends, however. As you know, steam engines develop power from pistons moving up and down. Steam turbines develop power from spinning disks. The two kinds of power plants are very different.

A **turbine** is a continually spinning disk. It works just like a pinwheel. Blow on a pinwheel and it spins. You could call the pinwheel a breath turbine because your breath makes it spin. Steam from a boiler spins steam turbines, as shown in Figure 13.10. There are also water turbines and gas turbines.

Steam turbines are very large and are used in electrical plants to produce electricity. About 85 percent of the electricity made in America comes from steam turbines. Steam turbines also power oceangoing ships. Ship turbines develop as much as 150,000 horsepower.

turbine
('tər-,bīn)
a disk or wheel that is made to turn continuously by air, water, or steam currents

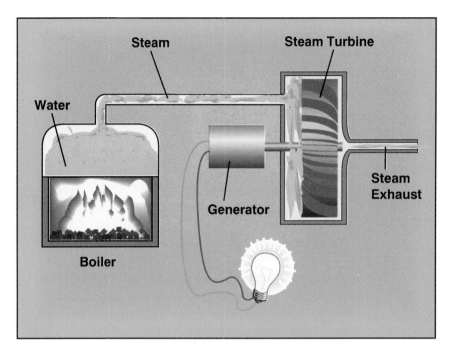

Water

Steam

Steam Turbine

Generator

Steam Exhaust

Boiler

— Learning Time —

Recalling the Facts

1. What made the piston in Newcomen's engine move back down to where it started?
2. How powerful was a Newcomen engine?
3. What was Watt's first major change to the steam engine?
4. A steam turbine operates differently than a steam engine. Explain.

Thinking for Yourself

1. How does a vacuum make a vacuum cleaner work?
2. Newcomen's engine used a vacuum. Watt's did not. Explain.
3. Do nuclear power plants use steam turbines to generate electricity? Explain.

Applying What You've Learned

1. Make a scale drawing of an entire Newcomen engine. Assume that the piston is 2 feet in diameter and 6 feet long.
2. Either design your own paddle wheel/turbine or make one of styrofoam and plastic spoons. Turn it with water from a faucet. If you design your own paddle wheel, compare its effectiveness with designs done by other students in your class.

———— Internal Combustion Engines ————

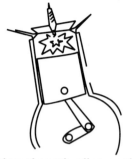

internal combustion engine
an engine in which the fuel is burned inside the engine

piston
a cylinder-shaped object that slides inside a cylinder of an engine. A downward stroke creates power.

cylinder
in an engine, the tube in which a piston slides

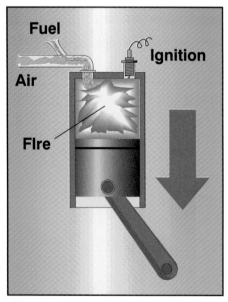

■ **Figure 13-11** ■ All internal combustion engines, like this gasoline engine, produce power from an internal fire.

Why is this engine called an internal combustion engine?

Did you know that there's a fire inside the engine of your family car? It's what makes the engine operate. You can't see the flames because they're deep inside the engine. This internal (inside) fire means that your family's car is powered by an **internal combustion engine.** Gasoline, diesel, gas turbine, and rocket engines are all internal combustion engines. See Figure 13.11. Although you can see the fire in some jet engines and all rocket engines, the combustion is still inside the engine.

Each engine type is designed for a specific type of transportation. See Figure 13.14 on page 399. Gasoline engines are primarily used in automobiles. However, they are also used in boats and small airplanes. Diesel engines are used mainly to power trucks and locomotives, but they are also used in boats and some cars. Gas turbine engines are used in all modern, large airplanes and some trucks. Rocket engines are used only for space travel. Electric motors turn the wheels in diesel-electric locomotives and are used in experimental automobiles.

Most engines that we use in our daily lives create power from a *piston* sliding inside a *cylinder*. A **piston** is a plug that just barely fits inside a **cylinder**, which is a tube closed at one end. Pressure at the closed end of the cylinder forces the piston down a few inches. This very small downward motion is enough to cause automobiles to travel on land, boats to move through water, and small airplanes to fly in the air.

The cylinder pressure in an automobile engine is the same type of pressure you develop when you blow up a balloon or pump up a bicycle tire. However, your lungs or a tire pump don't develop enough pressure to operate an engine.

In a steam engine, the space above the piston is filled with pressurized steam that comes from a boiler. The steam is directed to the cylinder and pushes the piston down. Another way to move the piston is with a controlled fire inside the engine.

Fuel and air are placed inside the cylinder. Look at Figure 13.11 again. Burning the mixture builds up a high pressure in a very short time. The burning time is about 0.003 second. The gas pressure pushes the piston down with great force. The piston is connected to a disk that's forced to rotate. The motion of the rotating disk turns the wheels of a car or truck. It can also turn the propeller of a boat or airplane.

Engine Cycles

Are engine cycles like other types of cycles?

When an event happens over and over again, we say that it

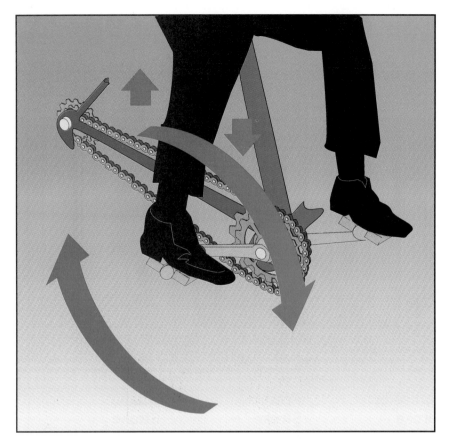

goes in **cycles**. The seasons of the year cycle. There are also life cycles, food cycles, and business cycles, just to name a few.

It may sound strange, but a bicycle also cycles. In pedaling a bicycle, your legs go up and down, repeating the motion over and over again. See Figure 13.12. A downward motion of your leg makes one stroke. A **stroke** is movement in one direction. Think of a canoe paddle stroke or the stroke of a pen.

An upward motion of the same leg is next. This motion also makes one stroke. Your legs make two strokes before repeating their motion. With a little imagination, we could say that your bicycle is operated by a two-stroke human power plant. Your legs deliver power to the rear wheel, which is called the *driving wheel*. Wheels under power are called driving wheels. The other wheels are trailing wheels.

Four-Stroke Cycles. Much like your legs pedaling a bicycle, the pistons inside an engine move down and up. The most popular type of engine is the *four-stroke cycle*. The pistons move down or up four times before they repeat themselves.

Suppose that a piston in a gasoline engine starts at the top of its

Tech Talk

The word *power plant* is sometimes used in place of the word *engine*. Gasoline engines are power plants, as are diesel and jet engines. Power plants also generate power for uses other than transportation.

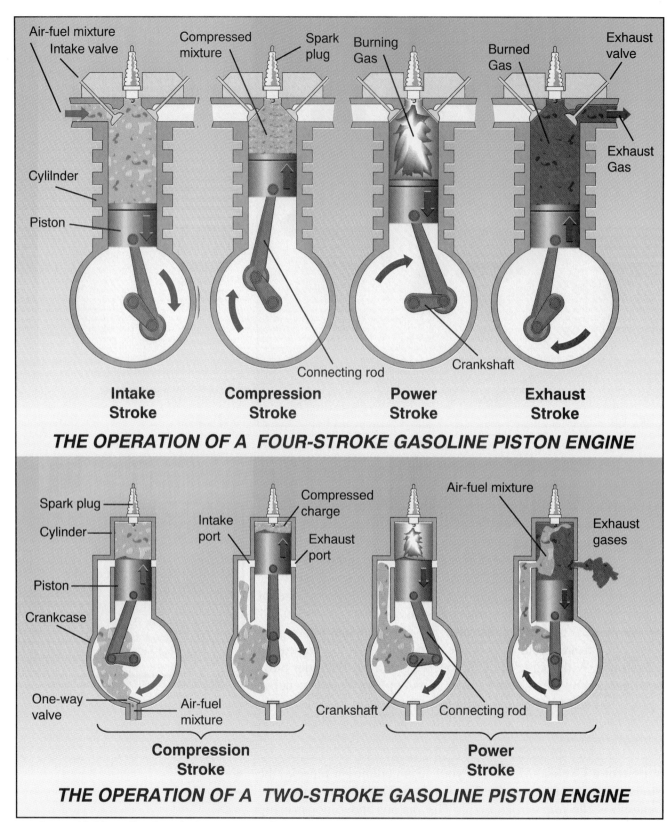

THE OPERATION OF A FOUR-STROKE GASOLINE PISTON ENGINE

THE OPERATION OF A TWO-STROKE GASOLINE PISTON ENGINE

■ **Figure 13-13** ■ The four strokes of a gasoline engine in an automobile (top) are intake, compression, power, and exhaust. The two strokes of a gasoline engine in a chain saw (bottom) are compression and power.

cylinder as in Figure 13.13. (1) The piston moves down on the *intake stroke*. It brings a gasoline and air mixture into the cylinder. (2) The piston moves up to the top on the *compression stroke*. It squeezes the mixture to about one-eighth of its original volume. (3) An electrical spark from a spark plug *ignites* the mixture, or causes it to catch on fire. The gasoline and air mixture burns very rapidly and forces the piston down on the *power stroke*. (4) The piston moves up on the *exhaust stroke*. It pushes out the exhaust gases. The four strokes then begin to repeat themselves. This means that the engine operates with a four-stroke cycle.

Two-Stroke Cycles. Some small gasoline engines operate with only two strokes. The intake and compression strokes are combined. The power and exhaust strokes are also combined. Such engines operate with the *two-stroke cycle* shown in Figure 13.13. The piston makes two strokes before it begins to repeat itself.

Two-stroke cycle engines use a mixture of gasoline and oil as fuel. They power mopeds, string trimmers, chain saws, and other devices. The Newcomen engine was a two-stroke cycle steam engine. Can you figure out why?

■ **Figure 13-14** ■ Each year, over 200 million passengers fly from one of America's 500 airports with commercial flights. Many fly in jumbo jets like this DC-10 (left). An automobile is a more common everyday piece of transportation for many people (right).

— Learning Time —

Recalling the Facts

1. How long does it take for the fuel to burn inside a gasoline engine?
2. What is the difference between a stroke and a cycle?
3. A gasoline engine squeezes the gasoline and air mixture before it's ignited. How much is it squeezed?
4. Name the four strokes of a four-stroke cycle engine.

Thinking for Yourself

1. How would an engine be affected if its piston fit too loosely?
2. Why can only rocket engines be used for space travel?
3. We don't have any three-stroke cycle engines. Do you think one could be made?
4. Which develops more pressure to push on a piston, a steam engine or a gasoline engine?

Applying What You've Learned

1. Use a tire pump and balloon to lift books with air pressure. This shows that low pressure can develop a high force.
2. Remove the cylinder head from a single-cylinder four-stroke cycle engine. See the four strokes in action.

Tech Talk

When automobile mechanics talk about the *size* of an engine, they don't mean how big it looks. To these experts, engine size means the combined inside volume of all the cylinders.

Gasoline Engines

How do gasoline engines produce power for automobiles?

There are more gasoline engines in the world than any other type. We use gasoline engines in our automobiles because they are inexpensive and start easily.

The main reason gasoline engines are so popular is because they can be inexpensively made in almost any size. There are very small gasoline engines, like those used in lawn edgers. However, they can also be very large. The largest gasoline engine ever built powered some American military airplanes. It was made in the later 1940s and named the R-4360 Wasp Major. It was more than forty times larger than the engine in most automobiles.

Automobile engines operate with a four-stroke cycle. Many modern engines have four cylinders, but others have six or even eight cylinders. See Figure 13.15. Some 1934 Cadillacs had sixteen cylinders, the most ever in a car. There are no two-stroke cycle automobile engines. Two-stroke cycle gasoline engines are less efficient and emit more pollutants.

Gasoline is a fuel that must be treated with the greatest care. When spilled on the ground, it quickly evaporates to form a potentially explosive mist. An accidental spark from a rock or piece of metal could easily cause an explosion. It may be difficult to believe, but a gallon of gasoline has six times as much energy as a gallon of dynamite. See Figure 13.16 on the next page.

The piston in a gasoline engine moves only up and down. Up-and-down, straight-line motion is called **reciprocating motion**. The word *reciprocate* means "to give in return." Unless we travel by pogo stick, reciprocating motion cannot be used for transportation. We need a way to convert it to **rotary motion**, or *circular motion*. All our transportation methods, except rocket-powered spacecraft, require rotary motion to turn wheels or propellers.

Reciprocating motion is changed to rotary motion by a *crank* or **crankshaft**. The pedals on your bicycle are part of a crank. The crank converts your reciprocating leg motion to the circular motion of the wheels. Your legs move up and down, but your bicycle wheels rotate to move you smoothly forward.

reciprocating motion
(rĭ-'sĭp-rĕ-ˌkāt-ĭng)
up-and-down or back-and-forth motion that occurs in a straight line. The pistons in an engine have this type of motion.

rotary motion
circular motion. In an engine, the reciprocating motion of the pistons is changed to rotary motion by a crankshaft.

crankshaft
the part of an engine that changes the reciprocating piston motion to rotary motion to turn the wheels

■ **Figure 13-15** ■ This photo shows a 1997 Toyota Camry V6 engine. "V6" means the engine has six cylinders arranged at about a 45° angle. There are three cylinders on each side of the V.

Figure 13-16 One gallon of gasoline has much more energy than six sticks of dynamite.

An automobile has a crankshaft inside the engine. Each piston has its own crank, and they are lined up next to each other on a shaft. See Figure 13.17. That is why the shaft is called a crankshaft. The rotating crankshaft transfers power to the driving wheels.

Diesel Engines

How are diesel engines different from gasoline engines?

Diesel engines are best suited for heavy-duty work. Although used in some automobiles, they are more important for powering trucks, buses, locomotives, and ships. They can operate smoothly under heavy loads that would cause a gasoline engine to *stall*, or stop running. Diesel engines last longer and require less maintenance than gasoline engines. Diesels are too heavy for airplane use.

The internal parts of a diesel engine are much like those inside a gasoline engine. Diesels have pistons, cylinders, and a crankshaft. They come in four-stroke and two-stroke versions. Cars and local trucks use four-stroke cycle engines. Two-stroke cycle engines are used in cross-country semitrailer trucks and locomotives. See Figure 13.18. Two-stroke cycle diesels are a bit more efficient and use less fuel than four-stroke cycle diesels.

The major differences between diesel and gasoline engines are in the diesel's fuel system and the way the fuel is ignited, or set on fire. Diesel fuel is similar to kerosene and cannot be easily ignited with a spark plug. The diesel engine uses a different type of ignition system. It uses hot air. Here's how the diesel does it.

The engine's four strokes are the same as in a gasoline engine. However, there are some operating differences, as shown in Figure 13.18. (1) On the *intake stroke*, only air enters the cylinder. It is not

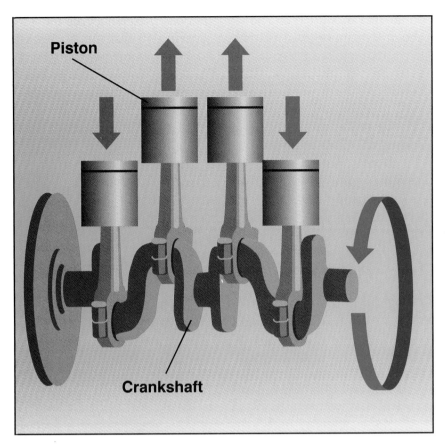

Piston

Crankshaft

Figure 13-17 A crankshaft inside an engine is a heavy and strong piece of metal that changes the up and down motion of pistons into rotary motion.

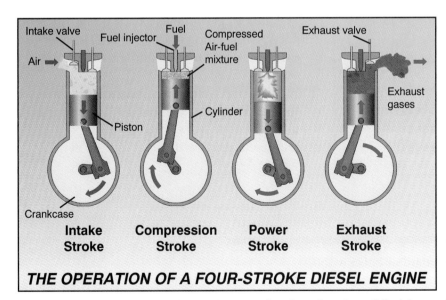

Intake valve
Fuel
Fuel injector
Compressed Air-fuel mixture
Exhaust valve

Air

Piston

Cylinder

Exhaust gases

Crankcase

Intake Stroke   **Compression Stroke**   **Power Stroke**   **Exhaust Stroke**

THE OPERATION OF A FOUR-STROKE DIESEL ENGINE

Figure 13-18 Large 18-wheelers use diesel engines to pull their heavy loads across the country (right). Many trucks use a four stroke diesel engine (left).

an air-fuel mixture. (2) The *compression stroke* squeezes the air to about $\frac{1}{22}$ of its original volume. Squeezing air that much causes its temperature to reach about 1,000° F. (3) A fuel injector squirts diesel fuel directly into the cylinder. The high air temperature causes burning to begin immediately. Pressure in the cylinder builds up very quickly and pushes the piston down on the *power stroke*. (4) The piston moves up on the *exhaust stroke*. Burned gases are pushed out of the cylinder.

Notice that the diesel engine ignites fuel with hot air. The gasoline engine ignites fuel with a spark from a spark plug.

Practically all automobiles use gasoline engines, but some have diesel engines. Why do we have both types? What are the advantages and disadvantages of each? Here is a brief comparison for automobile engines of about the same size:

Diesel	Gasoline
Less power	More power
Uses less fuel	Uses more fuel
Harder to start	Easier to start
Lasts longer	Wears out sooner

— Learning Time —

Recalling the Facts

1. What is the main reason gasoline engines are so popular?
2. Modern automobiles are sold with different numbers of cylinders. What are those numbers?
3. What are the main differences between a diesel engine and a gasoline engine?
4. What ignites the diesel fuel inside the engine?

Thinking for Yourself

1. A gallon of gasoline has six times as much energy as a gallon of dynamite. Does this mean we could use dynamite as a fuel for our cars? Explain.
2. All land and water transportation engines require rotation. Are there any animals that use rotation to travel? Name them.
3. Diesel engines are difficult to start in the winter. Why?

Applying What You've Learned

1. Make a crankshaft from cardboard or balsa wood.
2. Use rubbing alcohol to demonstrate volatility. Gasoline is highly volatile.

Gas Turbine Engines

Why do commercial airplanes use gas turbine engines instead of gasoline engines?

Gas turbine engines power all large airplanes, ships, trucks, and some locomotives. There are three basic types of gas turbine engines. They are the *turbojet*, the *turbofan*, and the *turboshaft*. A more common name for a turbojet or turbofan engine is **jet engine**. Both of them push airplanes through the air with a jet of high-pressure exhaust gas. This is known as **thrust**.

Gas turbine engines are complicated, but they are the most reliable internal combustion engine. A reliable engine is one that rarely stops running because a part breaks. If you rode in a car that frequently had flat tires, you'd say that the tires were unreliable. If the engine often broke down, you'd say that the engine was unreliable. On the other hand, if an engine kept running and running, you'd say it was reliable. Gas turbines are very reliable, and that's one reason why large airplanes use them.

Turboshaft engines produce rotational power. Propeller-operated airplanes, helicopters, and other transportation vehicles are powered by its spinning shaft. Turboshaft engines are also called *turboprop* engines when used in airplanes.

Gas turbines operate continuously like a propane torch. They don't have individual strokes like gasoline and diesel engines. However, they do have regions for intake, compression, power, and exhaust as shown in Figure 13.19: (1) Air comes in the front of the

jet engine
a type of gas turbine engine that is moved forward because of the hot air and exhaust that are shot out of the back part of the engine

thrust
the high pressure exhaust that pushes a jet engine forward

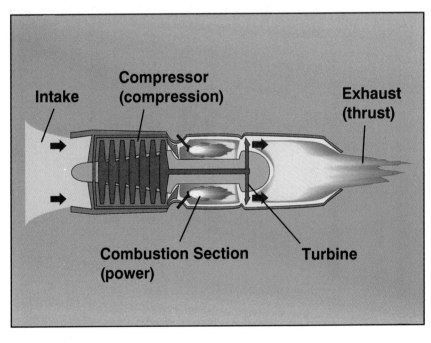

Intake

Compressor (compression)

Exhaust (thrust)

Combustion Section (power)

Turbine

Figure 13-19 Unlike gasoline and diesel engines, turbojet engines operate continuously and don't have cycles. However, they do have regions for intake, compression, power, and exhaust.

engine (intake region) and (2) passes through the compressor (compression region). The spinning compressor has hundreds of small blades that force air toward the rear, compressing it. The compressed air enters combustion chambers that look like metal food cans. (3) Fuel injectors squirt kerosene fuel that burns with the air (power region). The exhaust gases get quite hot and build up high pressure. In rushing out the engine, they spin turbines in their path. A central shaft connects the turbines to the compressor. In this way, the turbines provide power to rotate the compressor. (4) The gases continue flowing rapidly rearward and leave the engine (exhaust region). The exhaust gases produce the thrust to move the airplane forward.

Turbofan engines have large fanlike front sections. The fan blows air around the outside of the engine to provide more thrust. See Figure 13.20. Turboshaft engines have another turbine section. The exhaust gases also spin this extra turbine. A shaft connected to it provides rotational power.

Gas turbines are smaller and lighter than other engines of the same power rating. They also have a long engine life. Their biggest disadvantage is high cost. The gas turbine has parts that spin at high speeds and are kept at high temperatures. Such parts must be carefully made from special materials. This makes the engine very expensive. Turbofan engines used by airplane companies cost over $1 million each.

Tech Talk

Gas turbine engines use the prefix *turbo*. Names such as *turbojet* and *turboprop* mean that the engine uses a turbine. Some modern automobile engines have *turbochargers*. Turbochargers use a turbine to increase the power of a gasoline engine.

Putting Knowledge To Work

Metrics in Transportation

When James Watt developed the horsepower (hp) standard, he gave us a measurement unit most people could understand. Unfortunately, it's not a metric unit. Another way of measuring power is the kilowatt (kw). Although commonly used in America for electrical purposes, it can also be used for automobile engines. One hp equals 0.746 kw. Automobile manufacturers often use both units. This is called dual rating. A 120-hp engine can also be rated at about 90 kw (120 × 0.746).

Metrics can also be used to calculate speed. Kilometers per hour appears on many speedometers as kph, km/h, or something similar. One mile per hour (mph) equals 1.61 kph. A car moving at 60 mph could also be said to be traveling at about 97 kph (60 × 1.61).

Tire pressure can also be measured in metric units. The American unit used for this type of measurement is pounds per square inch (psi). One common metric unit used for tire-pressure measurement is the kilopascal (kPa), named after Blaise Pascal (1623-1662). One psi equals 6.90 kPa. A tire pressure of 26 psi could also be said to be about 180 kPa (26 × 6.90).

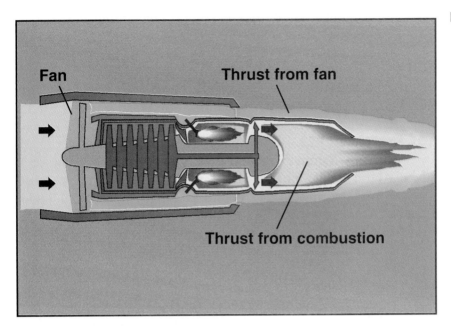

Fan

Thrust from fan

Thrust from combustion

Rocket Engines

Why are rocket engines used to send satellites into earth orbit?

A rocket engine carries its own oxygen for combustion. Oxygen and fuel are called **propellants**. The fuel burns with the oxygen to produce high-speed exhaust gas. The gas rushes out the rear to produce thrust. Jet engines also develop thrust, but they use oxygen from the air. That's why jet engines can't operate outside the atmosphere. Only rocket engines can travel in space.

The simplest rocket engines use a solid propellant. Early Chinese rockets probably used gunpowder, a solid propellant with chemicals that contain fuel and oxygen. See Figure 13.21. Model rockets also use solid propellants.

The thrust from five powerful rocket engines lifts the space shuttle off its launching pad. Two are solid-propellant rocket engines that are strapped to each side and look like long tubes. They are called solid rocket boosters (SRBs) because solid propellant in the rocket engine helps boost the space shuttle into earth orbit. Each one develops 2.6 million pounds of thrust and burns out in 2 minutes. The SRBs drop into the ocean, where a ship recovers them. They are overhauled and used on other flights. The space shuttle was the first American rocket designed to carry people in a vehicle using solid-propellant rocket engines for lift-off thrust.

The three middle engines on the space shuttle use liquid propellants. They are called space shuttle main engines (SSMEs) and are a permanent part of the space shuttle. They do not drop away like the SRBs. The propellants used in the SSMEs are liquid oxygen (LOX, pronounced "locks") and liquid hydrogen (LH_2, pronounced "L-H-2"). LH_2 is the fuel. Both oxygen and hydrogen are

Figure 13-21 The earliest rockets were solid-fuel rockets made by the Chinese many centuries ago. The first ones were used for celebrations much like those you may see on Independence Day.

gases at room temperature. Greatly reducing their temperature converts them to liquids. The supercold LOX has a temperature of $-300°$ F and LH_2 is even colder at $-425°$ F. They are stored in specially insulated tanks.

Each SSME develops a lift-off thrust of 375,000 pounds. The propellants are carried in a large external tank that's covered with insulation. After it's emptied, the tank drops away and is not recovered. See Figure 13.22.

America's Saturn 5 rocket, which put Neil Armstrong and Edwin Aldrin on the moon in 1969, remains the largest and most powerful rocket ever built. See Figure 13.23. Its engines used liquid propellants and developed 7.5 million pounds of lift-off thrust. It was 363 feet tall compared to the space shuttle's 184 feet.

Commercial airplanes aren't powered by rocket engines because rocket engines use so much propellant. They are also much more expensive than jet engines.

We use both liquid- and solid-propellant rocket engines because each has its advantages and disadvantages:

	Solid	Liquid
Easy to control the thrust?	No	Yes
Stop and restart engine?	No	Yes
Long-term propellant storage?	Yes	No
Complicated engine?	No	Yes
Made in small sizes?	Yes	No

Figure 13-23 The mighty Saturn 5 of the 1970s delivered people to the moon and is still the largest rocket ever built.

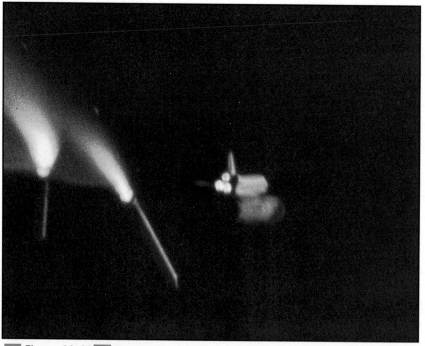

Figure 13-22 The space shuttle releases its two solid rocket boosters after they use up their propellant. Because of the way they work, some people call the SRBs *strap on solids.*

Electric Motors

What are some transportation vehicles that use electric motors?

Many transportation vehicles use electric motors. Subways and all-electric trains that obtain power from overhead wires are two examples. See Figure 13.24. However, other devices, such as elevators, escalators, and rides in amusement parks, also are powered by electric motors.

Locomotives use a diesel-electric drive system. A diesel engine turns a generator to produce electricity. The electricity is sent by wires to electric motors directly connected to the wheels. These *traction motors* turn the driving wheels. Locomotives can develop over 6,000 horsepower and pull 200 railroad cars. A few locomotives use gas turbine engines to operate the generators. You can tell if a locomotive uses a gas turbine or diesel engine by listening carefully as a train passes by. Gas turbine engines make a high-pitched sound. Diesel engines make a lower-pitched sound.

Automobile companies continually experiment with electrically powered cars. One such car has several batteries for energy storage. The solar cells recharge those batteries on sunny days. To operate any electric car, the driver presses on an accelerator pedal, just like in an engine-powered car. The pedal controls how much electricity is sent to the motor from the batteries, and the car speeds up or slows down.

Figure 13-24 Buses that use electric motors keep air pollution out of the cities and are quieter than diesel-powered buses.

— Learning Time —

Recalling the Facts

1. What type of gas turbine engine produces rotational power?
2. What are the four strokes of a gas turbine engine?
3. How is solid fuel different from solid propellant?
4. Why is the space shuttle's external tank insulated?
5. Why aren't commercial airplanes powered by rockets?
6. How much power is developed by diesel-electric locomotives?

Thinking for Yourself

1. Name two applications for turboshaft engines that were not mentioned in this chapter.
2. Which type of engine is considered the least reliable: steam, gasoline, diesel, or gas turbine?
3. The huge SRBs on the space shuttle use up all their propellant in only 2 minutes. Explain.
4. What is the difference between the power system on an all-electric train and one pulled by a diesel-electric locomotive?

Applying What You've Learned

1. Make a boat-type reaction engine with a small plastic bottle, baking soda, and vinegar. The base-acid reaction forms a gas that pushes the bottle on the water's surface.
2. Make and launch a model rocket.

——— Careers in Transportation Power ———

What kinds of jobs are there in power?

There are many jobs available at companies that manufacture engines. Many of these companies make gas turbine engines and diesel engines. They sell the engines to other companies, which then use them in the products they sell. Some companies manufacture gasoline engines for pumps, generators, lawn mowers, and other uses. They usually sell them to other manufacturers who assemble the completed product. Automobile companies usually make their own engines. See Figure 13.25 on page 412. There are only a few manufacturers that specialize in rocket engines and steam turbines.

Trained persons to service engines are always needed by automobile service facilities and the airlines. They are also needed on

A Success Story

Jill was raised in a small town and didn't have much opportunity to use different transportation vehicles. She took the bus a few times and once rode a train while on vacation, but she never rode a subway, or an airplane, or even a large boat.

She always liked airplanes and hoped to be around them someday. It was incredible to her that such a large metal box could actually fly. Jill read all about airplanes.

One day an Air Force recruiter came to Jill's high school. The recruiter's name was Lieutenant Laura Sposa, and she talked to Jill's science class. Jill was impressed with what she heard and asked some questions. She was surprised at how much Lieutenant Sposa knew about airplanes, especially since she was only four years older than Jill.

After graduation, Jill decided to join the Air Force. She knew that the military had programs to help pay for future college work. After her initial training, she was stationed in a very large city. It bothered her at first, but she adapted quickly and soon enjoyed the new surroundings. On the job, Jill was responsible for organizing spare parts for jet engines. She learned to operate many different pieces of equipment and liked her job.

After completing her military obligation, she used her savings and financial assistance from the Air Force to attend an engineering college. The courses were difficult, but Jill knew what she wanted. She was a good student and graduated with a degree in mechanical engineering.

She now works for a large manufacturer of gas turbine engines. She designs and installs instruments for testing engines before the engines are delivered to an airline. Jill really enjoys her job, and she even operates a million-dollar engine once in a while. She works with people who share her interests, and she has many close friends. She sees her family several times a year, and her parents are quite proud of her.

Jill does a lot of traveling in her job. She flies around the country and has ridden on many subways. She even took a boat trip.

---

large cargo boats, ships, and railroad lines. Factories hire people to maintain the engines in the equipment they own.

Many good jobs require education after high school, but not necessarily college. Training at a trade or vocational school is often satisfactory. High school counselors have information on employment.

Your school librarian can help you find the names of nearby companies that manufacture engines. Here is a short list of some large engine-building companies.

Large Gasoline Engines
Chrysler Corporation
Highland Park, MI 48288

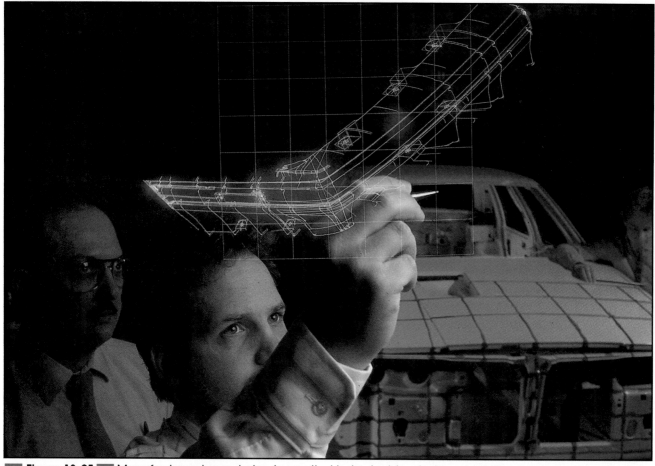

Figure 13-25 Manufacturers have design teams that help decide what products the company should make.

Ford Motor Company
Dearborn, MI 48121

General Motors Corporation
Warren, MI 48090

Small Gasoline Engines
Briggs and Stratton Corporation
Wauwatosa, WI 53222

Tecumseh Products Company
Tecumseh, MI 49286

Diesel Engines
Caterpillar, Inc.
Peoria, IL 61629

Cummins Engine Company, Inc.
Columbus, IN 47202

Detroit Diesel Corporation

Detroit, MI 48239

Gas Turbine Engines
Garrett Engine Division
Phoenix, AZ 85010

General Electric Flight Propulsion Division
Cincinnati, OH 45215

Pratt & Whitney Aircraft
East Hartford, CT 06108

Rocket Engines
Rocketdyne Division of Rockwell International Corporation
El Segundo, CA 90245
(Liquid propellants)

Thiokol Corporation
Brigham City, UT 84302
(Solid propellants)

Steam Turbines
General Electric Steam Turbine Division
Lynn, MA 01904

Summary

Directions

The following sentences contain blanks. For each numbered blank, pick the answer that makes the most sense in the entire passage. Write your answer on a separate sheet of paper.

Power is a measure of how quickly energy is changed into useful work. Much of our useful power comes from engines. The energy used by an engine is ___1___. Engines that use gasoline are found in cars. Rocket engines power space shuttles and ___2___ engines are used in jumbo jets.

The world's first engine was a steam engine made by Thomas Newcomen. It was used to pump ___3___ from coal mines. James Watt made many improvements to the engine. He made it small and powerful enough to be used in ___4___. The wood or coal used by steam engines was burned outside the engine. That's why the steam engine is a(n) ___5___ combustion engine.

The next power plant to come on the scene was the gasoline engine. Nikolaus ___6___

1.
  a. air
  b. power
  c. fuel
  d. force

2.
  a. gas turbine
  b. electric
  c. diesel
  d. steam

3.
  a. gasoline
  b. milk
  c. oil
  d. water

4.
  a. cars
  b. factories
  c. homes
  d. schools

5.
  a. hot
  b. external
  c. internal
  d. burning

6.
  a. Goddard
  b. Fruehauf
  c. Otto
  d. Drake

(Continued on next page)

Summary

(Continued)

made the first successful one in 1876. It was used in a car manufactured by Karl Benz. Like modern gasoline engines, that early one used a piston that could slide up and down inside a ___7___ . The piston moved up and down only a few inches. Each upward or downward motion of the piston was a ___8___ . The up and down motion was converted to more useful rotary motion by a crank or crankshaft. Because the fuel burned inside the engine, the gasoline engine was called a(n) ___9___ combustion engine.

Rudolph Diesel made an engine that also used pistons. One difference from a gasoline engine was that its kerosene fuel was ignited by ___10___ . Modern diesel engines are best suited for ___11___ . Currently, only a few automobiles are powered by diesel engines.

Gas turbine engines use a jet of high-pressure exhaust gas to push airplanes through the air. They are more expensive than gasoline engines but are also ___12___ . That's why all large commercial airplanes use gas turbine engines. Some gas turbine engines are connected to ___13___ . They are called turboprop engines.

Unlike gas turbine engines, rocket engines carry both oxygen and fuel. That's why only rocket engines can travel ___14___ . Space shuttles have two solid propellant boosters strapped to each side. The propellants used by the three middle engines are ___15___ . One advantage of the three SSMEs is that the engines can be ___16___ . The space shuttle is a big rocket but it is only about half as tall as was the Saturn 5 moon rocket of the 1970s.

A great many trains are powered by electric motors. Long-haul locomotives use diesel engines to turn ___17___ that produce electricity. The electricity is sent to electric motors connected to the driving wheels. Some locomotives can develop over ___18___ horsepower and pull 200 railroad cars. A few locomotives use gas turbine engines instead of diesel engines.

7.
  a. cylinder
  b. crack
  c. tank
  d. pipe

8.
  a. cycle
  b. rotation
  c. series
  d. stroke

9.
  a. a hot
  b. external
  c. internal
  d. burning

10.
  a. hot air
  b. sparks
  c. fire
  d. electricity

11.
  a. cars
  b. heavy work
  c. light work
  d. airplanes

12.
  a. easier to service
  b. bigger
  c. simpler
  d. more reliable

13.
  a. disks
  b. chargers
  c. propellers
  d. fans

14.
  a. in space
  b. fast
  c. straight up
  d. above the clouds

15.
  a. Gasoline and air
  b. CLOX and LH_3
  c. Gasoline and kerosene
  d. LOX and LH_2

16.
  a. removed
  b. restarted
  c. overhauled
  d. made in small sizes

17.
  a. gearboxes
  b. wheels
  c. shafts
  d. generators

18.
  a. 60
  b. 600
  c. 6,000
  d. 60,000

Key Word Definitions

Directions

The column on the left contains the key words from this chapter. The column on the right contains a scrambled list of phrases that describe what these words mean. Match the correct meaning with each word. Write your answer on a separate sheet of paper.

Key Word

1. **Motor**
2. **Engine**
3. **Boiler**
4. **External combustion engine**
5. **Combustion**
6. **Vacuum**
7. **Condenser**
8. **Turbine**
9. **Internal combustion engine**
10. **Piston**
11. **Cylinder**
12. **Cycle**
13. **Stroke**
14. **Reciprocating motion**
15. **Rotary motion**
16. **Crankshaft**
17. **Jet engine**
18. **Thrust**
19. **Propellant**

Description

a) Up-and-down motion

b) Contains less pressure than atmospheric pressure

c) Oxygen and fuel in a rocket engine

d) Converts energy into motion

e) Engine with an inside fire

f) Converts reciprocating to rotary motion

g) Changes water to steam

h) Hole in which a piston slides

i) Force of gas turbine exhaust gas

j) Uses heat to produce motion

k) Engine with an outside fire

l) An event that repeats itself

m) Burning

n) Type of gas turbine engine

o) Changes steam to water

p) Circular motion

q) Disk that continually spins

r) Slides inside a cylinder

s) Movement in one direction

Building a Steam Turbine

Did You Know?

In a steam power plant, steam is directed toward a fan-shaped turbine. The pressure from the steam causes the turbine to spin. The rotary motion is used to turn an alternator that generates electricity for use in your home and school. Turbines in steam power plants sometimes operate nonstop for years before they are shutdown for inspection or repair.

Problems To Solve

The first modern steam turbine was made in 1884. Electric companies now use them to produce electricity. Their huge turbines are expensive and have many parts. You can make your own small steam turbine with just eight simple parts.

A Procedure To Follow

1. Review the assembly drawing in Figure A to get an idea of how the steam turbine will be made.

2. Use scissors and the template shown in Figure B to cut out an octagon-shaped disk from the beverage can. Punch a

Resources You Will Need

- Copper tube, 1 inch in diameter and 4 inches long
- Two corks to fit tightly into the ends of the copper tube
- Aluminum beverage can
- Plastic bead
- Small nail
- Two metal clothes hangers
- Propane torch
- Scissors
- Razor knife
- Heavy wire cutters
- File
- Safety goggles

A.

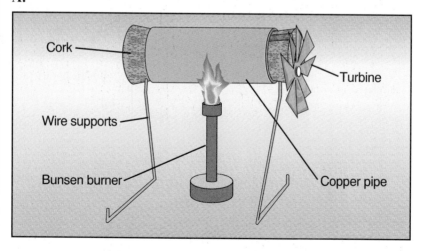

Cork

Turbine

Wire supports

Bunsen burner

Copper pipe

hole in the center with a small nail and bend the blades as shown in Figure B. This part will be your spinning turbine disk. Color one blade with a permanent marker to help you judge how fast the disk will spin.

3. Use the razor knife to cut a notch in one of the corks as shown in Figure B. The notch will direct steam from the copper tube to the turbine disk and cause the disk to spin.

4. Slide the turbine disk onto the nail. Then slide on the plastic bead, which will act as a bearing. See Figure B. Push the nail into the notched cork, and place the cork in one end of the copper tube. Place the other cork in the other end of the tube. Your copper tube is now a completed boiler.

5. Use the heavy wire cutters to make two supports out of clothes hanger wire. See Figure B. Sharpen the ends with a file, and stick one support into the edge of each cork. Your steam turbine is now ready to operate. See Figure B.

6. Remove a cork and fill the copper tube about two-thirds full of water. Replace the cork.

7. Put on the safety glasses and light the propane torch. Carefully and evenly heat the copper tube. Steam will soon come out through the notch you cut in the cork. The force of the steam will spin the turbine.

8. *Safety note:* Never allow all the water in the turbine to be converted to steam. The copper tube would overheat, and the corks could catch on fire.

What Did You Learn?

1. Could you tell how fast the disk was rotating? Would you expect it to spin faster when the boiler (copper tube) was two-thirds full of water or only one-third full of water?

2. Would the disk spin faster or slower if the notch you cut in the cork were smaller?

3. Try twisting the turbine disk blades to greater or lesser angles. What effect, if any, does this have on the rotating speed?

B.

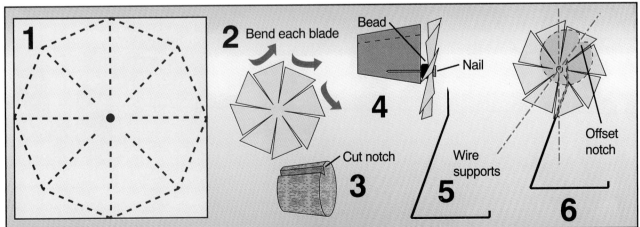

Transportation Systems

Key Words

In this chapter you will learn and study the meanings of the following important technology terms.

TRANSPORTATION SYSTEM
VEHICLE
TRANSPORTATION TERMINAL
AXLE
MASS TRANSPORTATION
TRANSMISSION
REAR-WHEEL DRIVE
FRONT-WHEEL DRIVE
FOUR-WHEEL DRIVE
MPG REQUIREMENTS
TRACTOR TRAILER
DISPLACEMENT
SUPERTANKER
COMMERCIAL AIRPLANE
JUMBO JET
HELICOPTER
LIGHTER-THAN-AIR CRAFT
NATIONAL AERONAUTICS
   AND SPACE
   ADMINISTRATION (NASA)

You probably don't give much thought to a trip across town or a weekend visit to a relative's home in another city. Nor do you think much about how that banana you had for breakfast made its way into your kitchen all the way from some faraway country. You don't think about these things because of the advances people have made in developing transportation systems.

A **transportation system** is an organization of the components (parts) or processes necessary to move people or products from one place to another. See Figure 14.1 on the next page. One of the major components is some type of vehicle. A **vehicle** is a container for moving people or products. A car is a vehicle, and so is a truck.

It's easier to understand transportation systems if we group them by *land, water, air, space, conveyor,* and *pipeline.* Something all these groups have in common is a transportation terminal. **Transportation terminals** are buildings or locations where people or products enter or leave the system. An airport is a terminal. Bus stops, railroad stations, and seaside docks are also terminals.

Transportation systems are all *interrelated.* That means that each one depends on the others. Buses and cars, for example, take passengers to airports and ship docks. Ships deliver fuel for airplanes. Loaded trailers are sometimes transported on railroad cars or ships. Can you think of other examples?

■ **Figure 14-1** ■ The railroad yard and ship loading area at Tacoma, Washington is an example of interrelated transportation systems.

The History of Transportation

What did the first transportation carts look like?

Until the invention of engines, transportation was always slow and difficult. Prehistoric people traveled mostly by foot. When necessary, they moved goods from place to place by carrying them or dragging them along the ground.

A large improvement in land transportation came with the invention of the wheel about 5,000 years ago. The first wheels were made of two or three boards held together by two crosspieces. A hole in the middle allowed the wheels to be attached to an axle. An **axle** is a circular rod that connects wheels. The first vehicles with wheels were carts that used four solid wheels and were pulled by oxen. See Figure 14.2.

Road Vehicles

What was the first American automobile?

The first *self-propelled* land vehicle was built in France by Nicolas Cugnot in 1769. *Self-propelled* means that the engine is part of the vehicle. A horse-drawn cart, for example, is not self-propelled. Cugnot's vehicle was powered by a two-cylinder steam engine. It had three wheels, traveled at 2 mph, and was very difficult to steer. The first self-propelled land vehicle in America was built

axle
('ăk-səl)
a rod or shaft on which one or more wheels turn

■ **Figure 14-2** ■ The earliest wheeled vehicles included carts (for carrying people and goods) and chariots (for military uses). The picture of the chariot (above) was drawn on a Sumerian clay tablet around 4,500 years ago. High-wheeled "bone shaker" bicycles with solid tires were very popular in the late 1800s (lower left). This replica (lower right) of the 1831 *De Witt Clinton* locomotive was operated in 1892.

by Oliver Evans of Philadelphia in 1805. It was a steam-powered 20-ton boat with wheels. He drove it from his workshop to the river, about $1\frac{1}{2}$ miles away. Neither of these two vehicles were automobiles as we know them. However, they marked the beginning of powered land transportation.

Charles and Frank Duryea built the first U.S. car offered for sale to the public. The factory opened in Springfield, Massachusetts, in 1895. Automobiles have been an important part of our society ever since. The Duryea automobile soon had competition from other manufacturers of cars.

The Duryea automobile had a gasoline engine, but at the turn of the century, cars were also made with steam engines and electric motors. The best-known *steamers* were built by the twin brothers F. E. and F. O. Stanley in Newton, Massachusetts. In 1906, a Stanley race car with a steam engine set a world speed record of 127.659 mph at Daytona Beach, Florida.

Electrically powered cars had heavy batteries under the seat. After about 30 miles of traveling, they had to be recharged at home with electricity from the power company.

Around 1900, people began to realize that the gasoline engine was the best power choice for automobiles. It developed more and more power as the years passed. A gasoline-powered car was lightweight in comparison to electric cars and their heavy batteries. Gasoline engines were smaller than steam engines with their bulky boiler and water storage tank. By 1917 there were few steam and electric cars being manufactured. Some advantages and disadvantages are summed up in Figure 14.3.

The first semitrailer truck used a wagon hitched to a Ford Model T automobile. It was designed and built by Detroit blacksmith Augustus Fruehauf in 1914 for the owner of a lumber company. Fruehauf called it a *semitrailer* because it had two wheels instead of the four that other trailers had.

As cities grew, officials saw the need for **mass transportation**, an organized passenger service available to the general public. During the 1870s they built covered wagons called *streetcars* with enough benches to seat about thirty people. Horses pulled the wagons on steel rails.

There is some disagreement over where the first electrically powered streetcar operated. However, the first *successful* system was built by Frank Sprague in 1887 for Richmond, Virginia. It had forty cars, 12 miles of track, and an electrical generating plant in which a steam engine turned a generator to make electricity.

Locometives

What country made the first locomotives?

James Watt built practically all of his steam engines in England, and that country soon became the leader in steam engine tech-

Duryea
(door'yā)

Fruehauf
(frü'höf)

mass transportation
public passenger service used to move large numbers of passengers from one place to another

Sprague
(sprāg)

	Advantages	Disadvantages
Steam	• Engine started easily • Cars were fast	• Wintertime temperatures froze water in the engine and broke parts • Water tank had to be refilled every 50 miles • Had to wait 20 minutes after lighting the boiler before drive-away pressure was built up
Electric	• Motor started easily • Quiet and clean • Little servicing required	• Driving range was limited to 30 miles or less • Slower than others • Only good for in city use
Gasoline	• Smaller engine • More power • Unlimited driving range (no stops for water or recharging batteries)	• Engine breakdowns were common • Had to be started with a crank

Figure 14-3 Cars of 1900 were powered by steam engines, electric motors, or gasoline engines. Each one had advantages and disadvantages.

nology. The first successful steam railroad began service in 1825 between Manchester and Liverpool, about 30 miles. The *Rocket* was the first locomotive used on the railroad line.

America's first locomotive made specifically for a railroad line was *The Best Friend of Charleston*. Built in late 1830, it could pull four loaded cars. The locomotive operated out of Charleston, South Carolina, for only a few months because it was destroyed by a boiler explosion the following year.

Boats and Ships

What's the difference between a boat and a ship?

Wind-powered sailboats were first used by the ancient Egyptians about 3200 B.C. Before this, canoes and rafts were powered by people using paddles, oars, or poles. Sailing vessels permitted civilization to spread throughout the world as sea voyagers took ideas and inventions to less developed societies.

Mediterranean shipbuilders of the mid-1400s combined all the best features of sailing ships to obtain one with more power and better handling. They developed a ship with three masts to hold the sails. The main mast was in the middle of the ship. The *foremast* was in front. And the *mizzenmast* was in the back. It was a full-*rigged* ship. To rig means to attach sails, ropes, and braces to the masts. Such explorers as Christopher Columbus, Vasco da Gama, Sir Francis Drake, and Ferdinand Magellan used ships of this type.

Tech Talk

The *Rocket* locomotive reached 29 mph during a test run. However, speed had nothing to do with its name. A British engineering journal said that passengers would be safer riding a rocket than a train. Father and son inventors George and Robert Stephenson took up the challenge, built a successful locomotive, and named it *Rocket*.

Mediterranean
(ˌmĕd-ə-tə-ˈrā-nē-ən)

423

Tech Talk

When you see a floating craft carrying people or cargo, are you looking at a boat or a ship? A *boat* is a small watercraft propelled by oars, sails, or an engine. A *ship* is larger than a boat and navigates in deep water.

The most beautiful of all sailing ships were the clippers of the mid-1800s. The name came from the word *clip*, which means to move swiftly. Some clipper ships had as many as thirty-five sails rigged and could travel over twenty mph. They were used to carry cargo from New York, around the tip of South America, to San Francisco. Such a voyage usually took 100 days.

The *Thornton* was the first successful steam-powered boat. It used large oars that dipped into the water on each side. It was built by John Fitch in 1790. The *Thornton* operated regularly between Burlington, in New Jersey, and Philadelphia, about 20 miles. Unfortunately, it never made a profit.

Robert Fulton built a steamboat in 1807 that he named *Clermont*. It operated on the Hudson River between New York and Albany. The *Clermont* was the first commercially successful steamboat.

The first ocean crossing by a steamship was the SS *Savannah* in 1819. It sailed from Savannah, Georgia, to Liverpool, England, in thirty days. It wasn't under steam power for the entire distance because it couldn't carry enough coal. It also had three masts on which to hang sails. However, the *Savannah* is credited as the first steamship to cross an ocean.

Balloons, Airplanes, and Helicopters

Who first flew across the Atlantic Ocean in an airplane?

Balloons are also called lighter-than-air craft, or LTAs. The first LTA to carry people was built in 1783 by Joseph and Jacques Montgolfier in Annonay, France. It was a 75-foot-high cloth and paper bag filled with hot air and smoke from a large fire. It rose about 6,000 feet.

The toss of a coin allowed Orville Wright to pilot the first engine-powered heavier-than-air craft. He and his brother Wilbur had spent several years completing the design of their *Flyer*. Its first flight was off the sands of Kitty Hawk, North Carolina, in 1903. It flew for 12 seconds and covered 120 feet.

The first ocean crossing by airplane was made by Lieutenant Commander Albert C. Read and his flight crew in the U.S. Navy's NC-4 flying boat. The four-engined airplane shown in Figure 14.4 completed a flight from Rockaway, New York, to Lisbon, Portugal, in 1919. It made the trip in stages, landing several times to refuel from U.S. Navy ships in the ocean.

Igor Sikorsky was a Russian-born, naturalized American citizen. Wearing a favorite hat, as he always did during test flights, he flew the VS-300 in 1939. It was the world's first successful single-rotor helicopter. See Figure 14.5. When he was only twenty-four, Sikorsky designed, helped build, and piloted the world's first four-engined airplane.

Montgolfier
(mänt-'gäl-fē-ər)

Sikorsky
(sĭ-kŏr'skē)

Figure 14-4 NC-4 was the first airplane to complete a transatlantic flight. This photograph was taken as the flying boat taxied in Lisbon, Portugal harbor in 1919.

Figure 14-5 Russian immigrant Igor Sikorsky designed, built, and flew the first successful helicopter.

Spacecraft

Why are Russian space travelers called cosmonauts?

The space age officially began in 1957 when Russia launched *Sputnik I*. It was the first artificial satellite to orbit the earth. The Russians sent the first person into space in 1961 when cosmonaut Yuri Gagarin orbited the earth in the *Vostok 1* spacecraft. See Figure 14.6. A few weeks later, Alan Shepard made America's first space flight from Cape Kennedy, Florida. John Glenn was the first astronaut to orbit the earth when he made three revolutions in 1962.

Tech Talk

We call our space travelers *astronauts. Astro* means "a star," and *naut* means "sailor." Russians use the word *cosmonaut. Cosmos* means "universe." An astronaut is a star sailor, and a cosmonaut is a universe sailor.

Figure 14-6 Russian cosmonaut Valentina Tereshkova was the first woman in space in 1963 (left). Cosmonaut Yuri Gagarin was the first person in space. He orbited the Earth in 1961 (center). Astronaut Alan Shepard was America's first astronaut. He made a suborbital flight in 1961 in a capsule he named *Freedom 7* (right).

425

— Learning Time —

Recalling the Facts

1. How were the first wooden wheels made?
2. Who invented the semitrailer, and in what year?
3. Why was England a leader in steam engine technology for locomotives?
4. The SS *Savannah* was the first steamship to cross an ocean. Why didn't it use steam power for the entire journey?
5. Why did Orville Wright make the first airplane flight and not Wilbur Wright?
6. Who was the first person to orbit the earth in a satellite?

Thinking for Yourself

1. Give examples of trips you've made in which you've used more than one type of transportation system.
2. If you had only the tools available 5,000 years ago, how would you make a wheel? Make a step-by-step list of procedures.
3. Many people don't know that Albert Read piloted the first airplane across the Atlantic Ocean. Why do they think it was Charles Lindbergh?
4. The first artificial satellite was Russia's *Sputnik I*. What does the word *sputnik* mean?

Applying What You've Learned

1. Design a complete railroad system between your hometown and a distant city. Select a route that does not already have a railroad.
2. Make wheels like they did 5,000 years ago. Use as few modern tools as possible.
3. Build a model of an early U.S. vehicle that many people don't know about. Examples include Oliver Evan's dredge, Augustus Fruehauf's semitrailer, *The Best Friend of Charleston* locomotive, John Fitch's steamboat, the SS *Savannah*, the NC-4, and the VS-300.
4. Build a conveyor system with Legos. Use a Lego computer interface and Lego sensors to convert your system into an automated system that can be programmed to sort materials.

—— Land Transportation ——

What are some examples of land transportation vehicles?

When you travel in a car, bus or train, you are using a land transportation vehicle. Land transportation also includes travel by bicycle, motorcycle, and subway. Perhaps you can think of other land vehicles.

An automobile is an important part of our land transportation system. However, you need more than just a car to get from place to place. Your car drives on roads and over bridges. Without roads and bridges, a car wouldn't be very useful.

When the car is not running well, someone in your family takes it to a service center. The center has the proper repair parts on hand because someone ordered them ahead of time.

Roads, bridges, and service centers are just a few of the parts of a land transportation system that allow you to go wherever you want in your family car. Can you think of other parts to the system?

Automobiles

Why do we have so many automobiles in the United States?

The United States is one of the world's largest countries. It is expensive to develop a mass transportation system that can serve such a large country. That is one reason why the automobile has become such an important part of our way of life. Automobiles are for personal transportation, not mass transportation. Only 6 percent of the world's population live in the United States. However, about 50 percent of all the world's automobiles are used in the United States. Do you think that's too many? What do you think we should do about it?

▬▬ Putting Knowledge To Work ▬▬

Henry Ford's Model T

What would you call the greatest land vehicle ever made? Do you think it could possibly be the spindly-looking Model T that Henry Ford started building in 1908? During a time when automobiles were built for wealthy people, Ford made a car aimed at the average person. It cost as little as $260, and his company built 15 million of them.

Ford and two engineers designed the car over a three-year period. Remembering his farming background, Ford made sure that it would be suitable for rural America. It was lightweight to keep the car from easily sinking into muddy roads. The simple four-cylinder engine could be repaired by almost anyone. The axles and frame were made of special steel that was strong enough to hold up to the poor roads. The car's high wheels easily rolled over deep ruts.

Its name was as simple as the vehicle. Ford called it the Model T. His first car in 1903 had been called Model A, his second was Model B, and so on. Ford didn't actually manufacture car models A through S. He used the letters for cars he was thinking of making.

Ford wanted to provide durable and inexpensive transportation for ordinary people. His Model T shown in Figure 14.7 on the next page did just that. It was in such demand that at one point, more Model T's were made than all other American automobiles combined. It brought farmers and city dwellers together and united the nation like no other vehicle before it. The Model T was used worldwide and in 1957 was labeled the greatest single vehicle in the history of world transportation. That is quite a title.

Figure 14-7 The Ford Model T united America like no vehicle before or after it. This 1920 *Runabout* sold for $550.

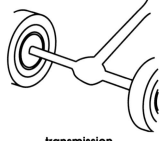

transmission
the parts and gears that cause the power to be sent from the engine to the axle

Modern automobiles are quite different from early cars. However, your family car has at least two things in common with the Model T. Both have **transmissions** to transmit power from the engine to the driving wheels, and both have front-mounted engines.

Transmissions. All cars have transmissions with gears that are similar to those on a ten-speed bicycle. Your legs work best when they turn the bicycle pedals once every 1 or 2 seconds. When you pedal up a hill, you shift into a *low gear* so that your legs can still move at one revolution every 2 seconds. The rear wheel moves more slowly because it takes more effort to pedal up a hill.

You shift your bicycle into a *high gear* for travel on flat land. Your legs still move at one revolution every 2 seconds, but the rear wheel turns more quickly. It takes less effort to pedal on level ground. Notice that you pedal at about the same speed, but the bicycle's speed changes. A car's transmission works the same way. See Figure 14.8.

A car's engine operates best if you use a low gear while climbing a steep hill. If you don't, the engine might stall. It's similar to what happens to you when you decide to hop off your bicycle and walk it up a steep hill. You hop off because you don't have a low enough gear. Even at your lowest gear, your legs can't provide enough power to keep the bike moving, which means you have stalled. Low gear is also used to start a car moving from a dead stop. It's similar to pedaling a bicycle from a dead stop. Isn't it easier to start pedaling in a low gear than in a high gear? Why do you think that's true?

You use a car's high gear when driving on a flat highway. Some cars have a three-speed transmission, some have four-speed

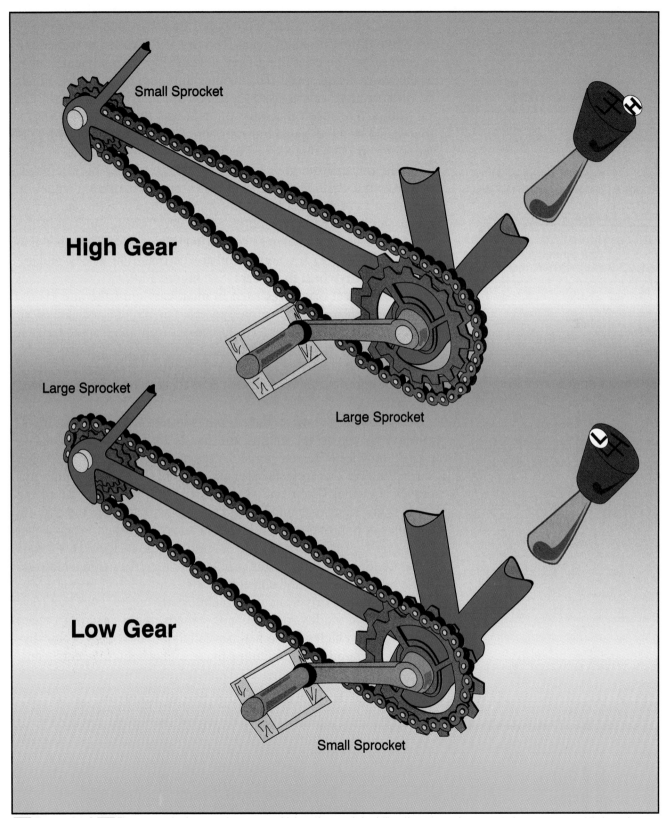

High Gear

Small Sprocket

Large Sprocket

Low Gear

Large Sprocket

Small Sprocket

■ **Figure 14-8** ■ The sprockets on a ten-speed bicycle work just like the gears in a car's transmission. High gear (top) combines the larger crank sprocket with the smallest sprocket at the rear wheel. Low gear (bottom) combines the smaller crank sprocket with the largest sprocket at the rear wheel.

and some have five-speed. Five-speed transmissions, for example, have five different sets of gears. The first set is called low gear and is used to get the car rolling from a dead stop. Second gear is next as the car picks up speed. Then comes third and fourth gears. Finally, fifth, or high, gear is used for steady-speed driving. Generally, a car with a five-speed transmission uses less fuel than a car with a four-speed or three-speed transmission. Can you figure out why? Is it easier to pedal a three-speed bicycle or a ten-speed bicycle?

The power from your legs is transferred to your bicycle's rear wheel with a chain. The rear wheel is a bicycle's driving wheel. A car doesn't use a chain. Instead, it transfers power with one or two metal shafts called *drive shafts*. Some cars transfer power to the rear wheels and are known as **rear-wheel drive** cars. There are also **front-wheel drive** cars. Some send power to all four wheels and are known as **four-wheel drive** cars. See Figure 14.9.

Rear-wheel drive was used in most cars until the late 1970s. Rear-wheel drive requires more space for the engine, transmission, and drive shafts. These parts are spread out from the front to the rear of the car. The transmission is located below a big hump in the floor next to the driver's feet. Over the past several years, the most popular vehicle sold in America has been a rear-wheel drive, full-sized pickup truck.

In front-wheel drive, the engine is turned sideways. The transmission is next to the engine and has a drive shaft connected to each front wheel. The transmission does not extend into the passenger compartment. The front floor is almost completely flat, which improves leg room. The front wheels carry more of the weight of the engine and transmission. This additional weight over the driving wheels can improve traction in wet or snowy weather. Front-wheel drive cars are easier to make and assemble at the factory. This means there are fewer manufacturing problems, which results in a more reliable automobile for the consumer.

Four-wheel drive cars can go almost anywhere. Cars that travel on unpaved muddy roads or snowy roads might need four-wheel drive. These vehicles are sometimes called four-wheelers or four-by-fours. Automobile manufacturers make four-wheel drive cars, vans, and trucks. Some companies call them all-wheel drive. A popular four-wheel drive vehicle is called a sport utility vehicle (SUV). This vehicle resembles a box on wheels and is made by almost all automobile manufacturers.

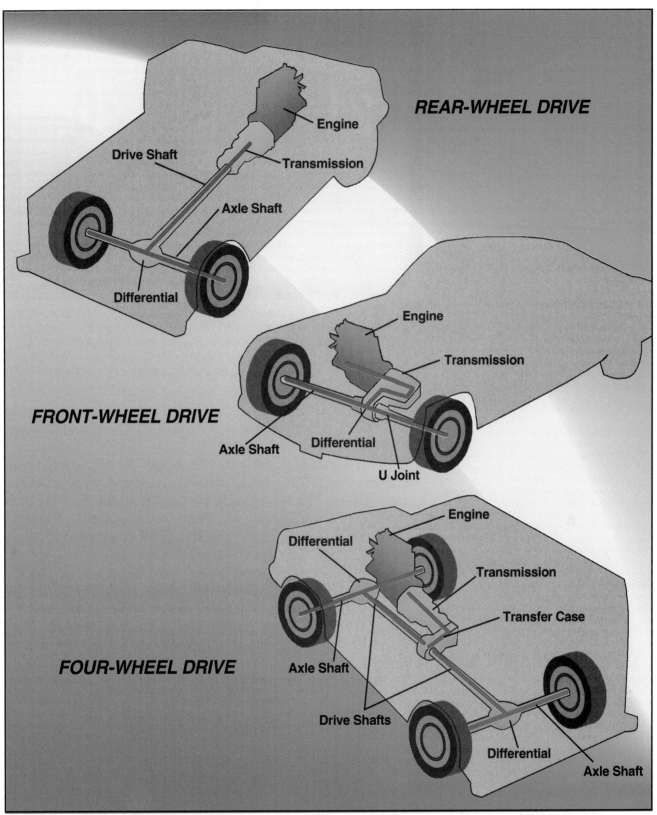

REAR-WHEEL DRIVE

Engine

Drive Shaft

Transmission

Axle Shaft

Differential

FRONT-WHEEL DRIVE

Engine

Transmission

Axle Shaft

Differential

U Joint

FOUR-WHEEL DRIVE

Engine

Differential

Transmission

Transfer Case

Axle Shaft

Drive Shafts

Differential

Axle Shaft

■ **Figure 14-9** ■ Not too many years ago, all cars and trucks were rear wheel drive (top). Most current automobiles use front wheel drive (center). Some specialty vehicles use four wheel drive (bottom).

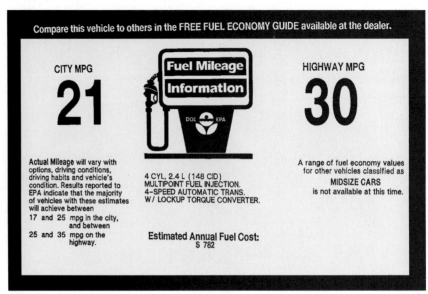

Figure 14-10 This new-car sticker shows how many miles per gallon this vehicle will get in the city and on the highway. To extend fuel supplies, the government requires automobile manufacturers to meet annual mpg requirements. These figures are also very important to people when they are trying to decide which vehicle they want to buy.

Fuel. Although there are some diesel-powered automobiles, they are few in number. Most cars use gasoline for fuel. Gasoline, like diesel fuel, is a nonrenewable resource.

Some brands of gasoline are mixed with alcohol, which is a renewable resource. The mixture is usually 90 percent gasoline and 10 percent alcohol. The pollution from engines using the mixture is lower than with straight gasoline. That is because alcohol is a chemical compound containing oxygen. Extra oxygen is released during combustion and helps to burn the fuel more completely. Also, because alcohol comes from grains such as corn and sugarcane, the mixture allows us to extend our supplies of gasoline.

To further extend fuel supplies, the government requires automobile manufacturers to meet annual **mpg requirements.** This means that cars must get more miles per gallon than they used to. Miles per gallon is a way of measuring how much gasoline an engine uses. A high-mpg car uses less gasoline than a low-mpg car because a high-mpg car travels more miles on a gallon of gasoline. See Figure 14.10.

mpg requirements
the government regulations that set the minimum number of miles per gallon a vehicle may produce during city driving and highway driving

The government conducts tests, operating each engine as if it were in a car driving in the city. They measure a *city mpg*. Then they conduct a driving test as if the car were on a highway. They measure a *highway mpg*. A car uses more gasoline in stop-and-go city driving. The city mpg rating is always lower than the highway mpg rating.

A small car might have a city rating of about 28 mpg and a highway rating of about 33 mpg. A medium-size car might get about 17 mpg in the city and 25 mpg on the highway. Four-wheel drive automobiles usually get the lowest ratings. Some are as low as 11 mpg in the city and 13 mpg on the highway.

You should always find out the mpg rating for any car your family might want to buy. The government estimates that cars are driven an average of 15,000 miles each year. A car whose overall fuel usage is 30 mpg will consume 500 gallons of gasoline a year. A car with an overall fuel usage of 12 mpg will consume 1250 gallons of gasoline in a year.

— Learning Time —

— Recalling the Facts —

1. What is one reason why the automobile has become an important form of transportation in the United States?
2. What does a transmission do?
3. Some cars have a front floor that's almost flat. Which are the driving wheels in that type of car?
4. What are two advantages of a fuel that is 90 percent gasoline and 10 percent alcohol?
5. What do mpg requirements mean?

— Thinking for Yourself —

1. In a ten-speed bicycle, is high gear a *small*-diameter gear at the rear wheel or a *large*-diameter gear at the rear wheel? Explain.
2. The federal government estimates that cars are driven an average of 15,000 miles each year. For a gasoline that costs $1.50 per gallon, what would be the annual gasoline cost for a 30 mpg car? What would be the cost for a 12 mpg car?

— Applying What You've Learned —

1. Design and construct a clay model of a futuristic automobile.
2. Use a city map to set up a sequence for traffic signal operation.
3. Make a car powered with a rubber band or mouse trap.
4. Use a ten-speed bicycle to determine overall gear ratios. Some ratios are so similar that most ten-speed bicycles have only six significantly different gear ratios.

■ Figure 14-11 ■ Large buses have always been an important part of intercity mass transportation.

tractor trailer
a two-part truck that includes a tractor, or engine and cab, and a trailer that holds the load to be hauled

434

Buses

How did school buses help provide better schools?

Large and boxlike, buses usually carry thirty or more passengers. They are used for mass transportation between cities, called *intercity* transportation. See Figure 14.11. Buses are also used for local urban transportation within cities and for school transportation. We have about 20,000 intercity buses in the United States and 50,000 buses used for local urban transportation. School systems use over 230,000 buses. More buses are made in the United States than in any other country.

The first regular urban motor bus service was established in London, England, in 1904. New York City quickly followed in 1905. The first scheduled intercity bus line in America opened in 1914 in Hibbing, Minnesota. Carl Eric Wickman converted an automobile to seat ten persons and scheduled hourly trips to nearby cities for 15 cents.

School buses were used as early as 1920. They made it possible to combine many small rural schools into one improved school. About 18 million students ride buses to school each day.

Intercity buses are powered by diesel engines. The buses carry up to sixty-four seated passengers and provide cheaper transportation than do railroads or airlines.

Urban buses can carry more people than intercity buses because they have standing passengers. Use of urban buses eases traffic congestion and saves fuel. They use about one-third as much fuel per passenger as do automobiles. However, they account for only 15 percent of all passenger miles traveled in the United States. These are some of the reasons why the federal government provides money to improve a city's bus fleet.

Trucks

How many different kinds of trucks are there?

From coast to coast, trucks move much of our country's freight. Over half the cities in America rely on trucks to supply them with food, fuel, furniture, and other products.

No one knows who invented the first truck, but trucks began appearing in the 1890s. It was not a popular form of hauling because the roads were so poor. In 1904, there were only 700 trucks in the United States. However, roads soon improved, and by 1918, there were over 600,000. Today there are over 30 million trucks on the roads of America.

There are hundreds of different kinds of trucks. Most are diesel powered, but gasoline engines are also used. Some commercial trucks are as small as pickup trucks, while others are as large as the eighteen-wheel semitrailer trucks that carry cargo on interstate highways. These large trucks are also called **tractor trailers** and have two main parts. The tractor is the front part and includes the engine and the cab. The trailer is the rear part and holds the cargo. Trailers can be specially designed to be placed on a railroad car. They can also be made to fit on ships without having to remove the cargo.

There is no such thing as a standard truck, but there are three general types: light duty, medium duty, and heavy duty. Panel and pickup trucks are examples of light-duty trucks. Medium-duty trucks are used locally and include sanitation trucks, soft drink trucks, and fuel trucks. Heavy-duty trucks carry large loads, and an eighteen-wheel tractor trailer is one type. Only about one in every twenty trucks is an eighteen-wheeler. See Figure 14.12.

The flat front and square shape of many trucks creates a lot of air resistance. At 45 mph, one-third of the engine's power is used just to overcome the resistance. This means that the trucks waste fuel. Manufacturers have tried two ways to streamline trucks. One way is to change the shape of the tractor and the front of the trailer. The other way is to place a wind deflector on the tractor's roof. See Figure 14.13 on the next page. A properly designed deflector can reduce air resistance by 20 percent. Next time you take a ride in a car, count the number of trucks you see with deflectors and the number without deflectors.

■ **Figure 14-12** ■ Heavy duty trucks carry large amounts of cargo. A semi-trailer like this one from the Fruehauf Trailer Company in Detroit, can go wherever there's a road.

Figure 14-13 Many large trucks have wind deflectors like this one on the roof over the driver. Deflectors reduce air resistance and improve fuel mileage.

Zephyr
('zĕf-ər)

Locomotives

☰ *Why did diesel engines replace steam engines in locomotives?*

If you lived during the 1800s, the faraway sound of a steam locomotive whistle would have caused your ears to perk up. Young people who saw trains roar by wondered where they came from, or where they were going, or who was on board. Perhaps one of those long-ago youngsters was an ancestor of yours. Trains fired the imagination of everyone, both young and old.

America's east and west coasts were connected by steel rails in 1869. One railroad company started construction east from Sacramento, California, where a river flowed to the Pacific Ocean. Another railroad company started west from Omaha, Nebraska, which was already linked by rails to cities along the Atlantic Ocean. After about six years of construction, the two railroad companies met at Promontory, Utah, where railroad officials drove the last nail. It was a golden spike.

Practically all locomotives were powered by steam engines until the 1934 diesel-electric *Zephyr* shown in Figure 14.14 began service in the United States. It gained public attention after making a 13-hour nonstop run from Denver to Chicago while using only $17 worth of diesel fuel. The steam locomotive it replaced took 27 hours and $225 worth of fuel for the same trip. By 1960, diesel-electrics had completely replaced steam locomotives on all mainline railroads. There are now about 27,000 diesel-electric locomotives in America.

Railroads earn most of their money by hauling freight. They carry bulky items like coal and iron ore. They also carry automo-

Figure 14-14 The streamlined two-stroke-cycle diesel-powered 1934 *Pioneer Zephyr* restored the public's interest in train travel. This specific train carried a total of one million passengers and is now on display at Chicago's Museum of Science and Industry.

biles, television sets, and anything else you could name. About 10,000 freight trains roll over the tracks of America each day. Some are over 200 cars long.

Trains also carry passengers, and one of the busiest lines connects New York City with Washington, D.C. The train averages 80 mph and reaches a top speed of 100 mph. However, trains carry less than 1 percent of all U.S. intercity passengers. That is a very low number. In some other countries, trains carry up to 50 percent of the intercity travelers.

Many of today's locomotives are all-electric. They don't have a diesel engine or an electrical generator. They get power from overhead wires carrying electricity or from a third rail placed near the tracks. All-electric locomotives are quiet and produce no exhaust gas. Commuter trains—those that carry commuters to work—and subways are all-electric. See Figure 14.15. Each working day, they carry hundreds of thousands of people to and from work in large cities.

There is a new modern train that doesn't roll on rails. It doesn't even touch the ground. The train is called a *maglev* for "magnetic levitation." Strong magnets allow the train to float, or levitate, about 1 inch over its *roadbed*. A roadbed is the surface over which the train travels.

A 25-mph maglev train has operated at England's Birmingham airport for several years. However, maglevs can also be fast. World speed records of over 300 mph have been achieved by trains in Japan. Maglev trains are very quiet and have almost no vibration.

Tech Talk

The word *locomotive* really means a steam, diesel, or electric motor on wheels. It's the power unit at the front of a train of railroad cars. *Loco* comes from the word *locus*, which means "place." *Motive* means "moving." The word *locomotive* was probably used to describe the power unit because it moves from one place to another. Locomotives were the first self-propelled land vehicles seen by the general public.

Figure 14-15 Train yards used by commuter and cargo trains like this one in Geissen, Germany appear confusing. Actually, trains are carefully controlled and accidents are quite uncommon.

Figure 14-16 Steel roller coasters are fun because they can safely turn cars upside down. This one is at Six Flags in Valencia, California.

Thrilling Technology

Some railed vehicles are built just for fun, to thrill you and make you want to come back for more. They're called roller coasters, and they can cost as much as $5 million to build. See Figure 14.16 above. Roller coasters actually coast. They're only under power while being pulled to the top of the first hill, about 140 feet high. The older ones with wooden frames are exciting, but most new ones are made of steel. Steel roller coasters are the only kind that can turn you upside down. The *Shockwave* at Six Flags Great America in Gurnee, Illinois, does it seven times. The *Iron Dragon* at Cedar Point in Sandusky, Ohio, is a suspended coaster. It hangs from a track, so you don't feel hemmed in by braces and supports.

Roller coasters thrill their passengers because the ride seems to be dangerous. It rarely is. The manufacturers make the equipment as safe as humanly possible. Coaches are locked to the tracks with wheels both above and below the rails. Computers lock the controls until all seat restraints are fastened. Those same computers apply the brakes if a roller coaster goes too fast. The structure is inspected every day, and critical parts are regularly X rayed. The government's Consumer Product Safety Commission calculated that riding a roller coaster is eighty-four times safer than riding a bicycle.

Many people still enjoy the wooden-framed roller coasters. One of the all-time favorites is the 1927-built *Cyclone* at Astroland in Coney Island. The *Texas Cyclone* at AstroWorld in Houston is modeled after Coney Island's. The one in Texas is slightly bigger and faster than the original.

— Learning Time —

Recalling the Facts

1. How is an intercity bus used? How is this different from an urban bus?
2. Why can urban buses carry more passengers than intercity buses?
3. Why do some tractor-trailers have air deflectors on top of the tractor?
4. Do railroads earn more money hauling passengers or freight?
5. How is an all-electric train different from one pulled by a diesel-electric locomotive?

Thinking for Yourself

1. There are many advantages to traveling to work by bus. From a passenger's standpoint, name three advantages.
2. Name three reasons why some people might prefer to drive their own car to work rather than ride the bus.
3. The federal government estimates that trucks carry about 25 percent of all intercity ton miles of freight. What does *ton mile* mean?

Applying What You've Learned

1. If you can see a highway from your school, count the number of tractor trailers with and without air deflectors. Determine a percentage.
2. Build a simple truck model out of cardboard. Include a removable air deflector. Direct a fan at your model and evaluate. Which student created the most aerodynamic model?

—— Water Transportation ——

How can metal boats float?

Ships deliver practically all the overseas cargo leaving or arriving in America. To transport by ship, we need docks with special loading and unloading equipment, and we need properly trained people to operate the ship. We also need good communications for weather data and other information. All these components are part of a water transportation system.

Water has provided transportation routes for centuries. Rivers, lakes, and oceans have made natural routes between cities, between states, between countries, and between continents. Of the twenty largest U.S. cities, for example, nineteen are located on an ocean or a navigable waterway. A *navigable waterway* is a lake or river that

displacement
the weight of the water that is moved out of the way by a floating ship

supertanker
a very large ship that is fitted with tanks for carrying oil across oceans

is deep and wide enough to allow ships and boats to pass. The five Great Lakes are navigable waterways, as are the Mississippi and Ohio rivers.

For centuries, people used sailing ships to haul cargo and passengers between countries. The ships of 1800 had displacements of about 1,200 tons (1 ton = 2,000 pounds). **Displacement** is the weight of water displaced, or occupied, by the floating ship. It's generally the weight of a fully loaded ship. The wind limited how much cargo those older ships could carry. A heavy ship would move much more slowly under wind power. A heavy ship is much more difficult to push than a lighter one.

Today's ships are pushed by powerful engines. It is not unusual for a modern ship to displace 100,000 tons. That's as much as could be carried by eighty ships in 1800. Very large ships called **supertankers** transport oil across oceans in storage tanks. Their displacements are as high as 500,000 tons. See Figure 14.17. Does this tell you why they are called *super*tankers?

■ **Figure 14-17** ■ Most early wind jammers had three masts: foremast, main mast, and mizzen mast (left). Modern supertankers transport oil and can carry as much weight as about 400 three-masted ships (right).

Steamships first used paddles to move through the water. Some people argued that a propeller would produce more power. A tug-of-war contest took place between two British ships in 1845. The *Rattler* had a propeller, and the *Alecto* had paddles. Although both had 200- horsepower steam engines, *Rattler* easily won the contest.

Early boats were made out of wood. Wood floats. Metal can also be made to float if it is shaped properly. A piece of metal shaped into a leakproof box will float as long as the metal box weighs less than the water it displaces. This is called *buoyancy*, the ability to float. England launched the first all-iron ship, named *Vulcan*, in 1818.

Like trucks, water transportation vessels are of three general types. *Passenger vessels* carry people. See Figure 14.18 on the next page. Large ocean liners used for vacation cruises are one type of passenger vessel, but there are others. *Cargo ships* transport oil, grain, iron ore, automobiles, and many other products. Like passenger vessels, they use steam turbines for power. *Specialty craft* include everything else, such as river barges for transporting coal and other goods, tugboats for pulling large ships into dock, icebreakers, and many other boats. Almost all of them use very powerful steam turbine, gas turbine, or diesel engines.

It usually costs less to transport goods by water than by rail, road, or air. Whenever possible, people try to save as much money as they can by transporting products the cheapest way.

Air Transportation

▽ *What are the main types of air transportation vehicles?*

Transportation by air takes place in airplanes, helicopters, and lighter-than-air craft. Hang gliders and sailplanes are transportation vehicles primarily used for recreation. Can you think of other air transportation devices?

Airplanes are the most important part of our air transportation system. However, many other components are necessary for safe air travel. We need airports, training programs, and radar, for example. Many airplanes are in the air at the same time. This is why safe air travel is our most complex transportation system. Can you think of other components of the air transportation system?

Airplanes

▽ *Why can jumbo jets stay up in the air for a long time?*

Many important airplanes were built and flown after the Wright brothers' first flight in 1903. Few were as important as the 1935 DC-3 airplane that marked the beginning of profitable air

Tech Talk

You may have seen a movie about ships and heard an actor say something like, "We're making 15 *knots*." In the 1500s and later, sailors at sea threw a small piece of wood overboard with a thin rope attached to it. The rope had knots tied in it about every 50 feet. The number of knots passing through the sailor's hand in 30 seconds was a measure of the ship's speed. One knot equals 1.151 mph. It is incorrect to say "knots per hour."

Figure 14-18 This hydrofoil boat has small wings hidden under the surface of the water. In port, the boat floats on its hull. As it speeds up when leaving port, the boat lifts out of the water and literally flies on the underwater wings.

commercial airplane
an airplane that carries passengers or freight in order to earn money

jumbo jet
a very large airplane that carries several hundred passengers at one time. They can make very long flights before needing to refuel.

Figure 14-19 The remarkable DC-3 was designed and built in just ten months in 1935 by the small Douglas Aircraft Company. The plane is affectionately named the *Gooney Bird* and many are still flying today.

transportation. See Figure 14.19. Using two gasoline engines, it was the first airplane to make money just carrying passengers. By 1939, the twenty-one passenger DC-3 carried 75 percent of all air passengers.

Commercial airplanes are those used to make a profit. They once used gasoline engines to turn propellers. Gas turbines, or jet engines, first appeared on the 1952 four-jet British *Comet*. It could only carry thirty-six passengers, so it wasn't as successful as the Boeing 707, the first American jet, which came out in 1958. The first Boeing 707s carried 179 passengers, a huge number at that time.

The number of airline passengers doubled during the early 1960s. Manufacturers decided to build very large passenger airplanes they called **jumbo jets**. Boeing's 747 was the first; it could carry about 500 passengers. See Figure 14.20. Other U.S.-built jumbo jets are the McDonnell-Douglas MD-11 and the Boeing 777.

Because jumbo jets have such powerful engines, they can lift more weight than other planes. This allows them to carry a lot of fuel. As a result, they can stay up in the air for a long time. Jumbo jets fly non-stop from Cincinnati to London and from Detroit to Tokyo, for example.

There are smaller jet aircraft, like the two-engine Boeing 737 and the two-engine DC-9. They are used on shorter flights between smaller cities. The smaller airplanes use less fuel and don't need a

Figure 14-20 The Boeing 747 was the first jumbo jet. With four powerful turbofan engines, it can lift more weight than practically any other airplane in the world.

helicopter
an aircraft that is lifted straight up off the ground by a rotor, or horizontal propeller-like device

long runway to take off. Without such airplanes, smaller U.S cities might not have good air service.

Airplanes mainly fly passengers, but some carry only cargo. However, even the biggest airplanes can carry only a fraction of what a ship or train can haul. A Boeing 747 can carry 100 tons of cargo, about the most of any airplane in the world. This is much less than any cargo ship and makes air transportation very expensive. Usually only lightweight items, such as mail and electronics, are shipped by air.

Helicopters

Why do some helicopters have a small tail rotor?

A **helicopter** is an aircraft with one or two rotors that allow it to lift straight up. Helicopters can be as small as a one-person machine with a low-powered gasoline engine. They can also be as large as cargo-carrying helicopters that easily lift 10 tons. These larger helicopters are powered by one or two gas turbine engines. Some helicopters have a small tail rotor to keep them from spinning around. See Figure 14.21.

Passenger-carrying helicopters connect downtown New York City with the three major airports in the area. Such services are also available in Chicago, San Francisco, and other cities. Some helicopters specialize in industrial operations, such as pipeline inspection and delivering parts to construction sites. Others are used to check on automobile traffic and transport people to hospitals.

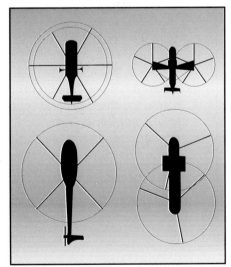

Figure 14-21 Helicopters use three ways to keep from spinning around in the air. They are: two rotors, one over the other, each spinning in opposite directions (top left); two rotors, side by side, each spinning in opposite directions (top right and bottom right); or one small tail rotor (bottom left).

443

Figure 14-22 A blimp is a large gas bag, just like an ordinary balloon. Filled with nonflammable helium, the Miami-based Goodyear *America* is the world's best-known modern airship.

Lighter-Than-Air Craft

What makes a hot-air balloon lift into the air?

Lighter-than-air craft, or LTAs, are also known as *dirigibles*, *zeppelins*, *blimps*, and *airships*. They use helium to lift into the air and gasoline engines turning propellers to move forward. The engines are located in *gondolas*, or cars, suspended from the craft. The passenger compartment is located in a separate gondola.

A hot-air balloon is an LTA. It uses large torches to heat the air inside a huge nylon bag. Hot air weighs less than cooler air, so the balloon rises. Hot-air balloons have no way to control direction. They travel where the wind pushes them.

You may have heard of one of the better-known airships. It's the cigar-shaped Goodyear blimp named *America*. See Figure 14.22. It's about 200 feet long and travels at 35 mph. The *America* is used for advertising and for television cameras. Future LTAs might be used for rescue missions, to patrol ocean shores, or to lift very heavy commercial loads.

— Learning Time —

Recalling the Facts

1. Why are the ships that transport oil called *super*tankers?
2. Which is better for moving a ship through the water: a paddle or a propeller?
3. Name the three general types of water transportation vessels.
4. Why do some helicopters have a small tail rotor?
5. List four names for lighter-than-air craft.

Thinking for Yourself

1. A flat barge is 20 feet wide and 80 feet long. It settles 10 feet into the water when loaded. What is its displacement in tons? Water weighs 62.4 pounds per cubic foot.
2. How many Boeing 747s would it take to transport all the students, teachers, and staff employees in your school?
3. Twin-rotor helicopters don't need a tail rotor. Why?

Applying What You've Learned

1. Using easily shaped metal like aluminum or brass, demonstrate buoyancy. Shape it to float and to sink. Determine displacement by measuring with a ruler and knowing that water weighs 0.0361 pound per cubic inch.
2. Make a glider, and see how it flies with different control settings for the rudder, elevator, and ailerons.
3. Make a hot-air balloon from a paper bag. Have it lift a payload.

Space Transportation

≡ *What is a typical space shuttle mission like?*

Space transportation has been a dream of all people who ever looked up at the stars. Today's spacecraft are the most complicated transportation devices ever made. However, not many rockets leave the earth's surface every year. They lift off only in Florida and land only in California or Florida. Many airplanes take off and land every day from hundreds of cities. Thousands of people are always in the air. Their safety is of utmost importance. This is why our air transportation system is more complex than our space transportation system.

The highest airplane flight was by the Russian Alexander Fedotov in 1977. He flew his MIG-25 fighter to a height of 123,523 feet, about 23 miles. However, even that altitude is not considered *space*. Space is usually said to begin 100 miles above the earth's surface. At that height, a satellite can orbit for months. At lower altitudes, air in the atmosphere would slow it down and cause the satellite to fall.

Space transportation is probably the most exciting way to move people and cargo. The **National Aeronautics and Space Administration (NASA)** is responsible for directing the entire U.S. space program. This includes the space shuttles. NASA called the first space shuttle flight STS-1, for space transportation system. The second flight was STS-2. And so on.

John Young and Roger Crippen flew *Columbia* into space in 1981. It was the first space shuttle to go into orbit. Each flight carries cargo, called a *payload*, in the large cargo bay behind the pilots. The payload can weigh up to 65,000 pounds and is usually a satellite.

On a typical mission, the space shuttle with four astronauts orbits 115 miles above the earth. Its speed is about 17,000 mph. An astronaut opens the doors to the huge cargo bay, which contains a communications satellite. A 50-foot mechanical arm, controlled by the astronaut, removes the satellite and releases it a safe distance from the space shuttle. See Figure 14.23 on the next page. The arm can also grab satellites already in orbit that aren't working properly. Sometimes these satellites can be repaired and returned to orbit.

The space shuttle can also carry a complete scientific laboratory in the cargo bay. The laboratory permits such experiments as surveying the earth's resources and doing research on processing materials in weightless space. Studies are also made on plants and animals carried in the orbiting laboratory.

After being in orbit for about a week, the astronauts fire small rockets to slow down the space shuttle. It re-enters the earth's atmosphere and glides toward a landing strip. The space shuttle usually touches down at Edwards Air Force Base in California at 220 mph.

Fedotov
(fĕd'ō-tŏv)

National Aeronautics and Space Administration (NASA)
the government agency that regulates and direc's the space program for the United States

Figure 14-23 These astronauts are practicing satellite construction and repair in the weightlessness of space. You can see the vertical fin of the shuttle in the middle of the photograph.

An orbiting space station is a long-term NASA goal. The United States is working with Russia, Canada, and other countries on this project. Although it is uncertain when the space station might be completed, it is being designed to be serviced by American space shuttles. The space station will probably hold eight astronauts.

Pipeline Transportation

What kinds of materials are transported by pipeline?

Pipeline transportation is used to move gases and liquids over long distances. Pipelines transport natural gas, water, and such fuels as gasoline, kerosene, and diesel fuel. Pipelines also carry industrial waste and sewage.

Usually buried 3 feet underground, pipelines are quiet, unseen, and do not contribute to traffic congestion. They transport gases and liquids by using pressure to push them along. Natural gas moves at 15 mph and liquid fuels travel at 3 mph, about a brisk walking speed. The pressure is developed by large pumps powered by gas turbine or diesel engines. The engines are located in pumping stations every 30 to 150 miles along the pipeline.

Pipelines continuously deliver large amounts of gases or liquids directly from a supplier to a user. They are a very important means of transportation, but because we don't see them, it's easy to

Figure 14-24 American pride was felt by everyone in the world on July 20, 1969. Astronaut Neil Armstrong, the first person on the moon, placed the American flag on the surface and had his photograph taken by Edwin Aldrin, the second person on the moon.

446

Messages from the Moon

When it sat on its Cape Kennedy launch pad, people said its top was the highest point in all of Florida. The gleaming white Saturn 5 rocket was as tall as a thirty-six-story building. It developed 7.5 million pounds of thrust at lift-off. No rocket was ever bigger or more powerful.

On a summer day in 1969, astronauts Edwin Aldrin, Neil Armstrong, and Michael Collins strapped themselves into the command module of a Saturn 5. Five mighty rocket engines sent the three men into orbit around the earth. After two revolutions, they fired another rocket engine to begin the first journey to the moon's surface by a human being.

Near the moon, the astronauts fired still other rocket engines to slow down and put themselves into orbit around the moon. Collins remained in the command module while Aldrin and Armstrong flew the lunar excursion module to the surface. Immediately after the lunar excursion module safely touched down, mission commander Armstrong radioed back the first words ever sent from another world. "Tranquility Base here. The Eagle has landed."

The lunar excursion module was named *Eagle*, and they had landed on an area of the moon named the Sea of Tranquility.

On July 20, 1969, Armstrong set the first human foot onto the moon's surface. He said, "That's one small step for man, one giant leap for mankind." A TV camera recorded the event, which was seen by millions around the world. See Figure 14.24. Aldrin and Armstrong stayed on the surface for a few days, collecting 48 pounds of soil and rock samples, making measurements, and setting up scientific equipment. Aldrin described the view as "magnificent desolation."

They left a plaque behind. It reads: "Here, men from the planet Earth first set foot upon the Moon. July, 1969 A.D. We came in peace for all mankind." They also left medals and shoulder patches in memory of three Russian cosmonauts and three American astronauts who had died.

Saturn 5 rockets made six flights to the moon. The last lunar landing was in 1972.

---

forget they exist. In the United States, 222,000 miles of pipelines carry oil products. That's enough pipe to wrap eight times around the world. Natural gas pipelines are even longer, at 980,000 miles. How many times would that go around the world?

The Trans-Alaska pipeline first started pumping oil in 1977. It stretches 775 miles from Prudhoe Bay in the north to Valdez in the south. Much of the 4-foot-diameter pipeline is aboveground. It crosses 20 rivers, 300 streams, and 3 mountain ranges.

The oil in an oil pipeline system comes from oil wells, where it's pumped to holding tanks or storage tanks. See Figure 14.25 on the next page. It's sometimes combined with oil from a supertanker and pumped to a refinery. Refined petroleum products are then pumped to storage tanks, railroad tank cars, trucks, or another section of the supertanker.

Prudhoe Bay
('prüd-(ˌ)(h)ō)

Valdez
(văl-dēz')

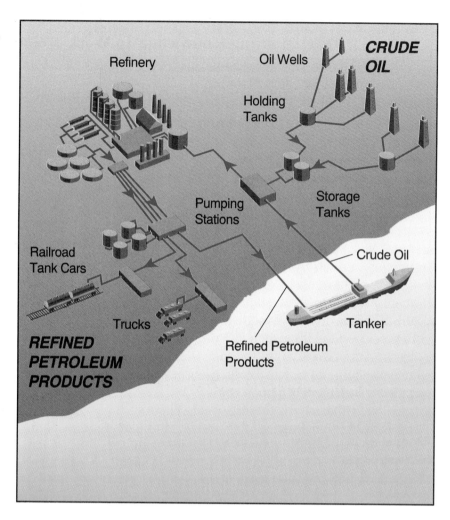

Figure 14-25 An oil pipeline system takes crude oil and converts it to gasoline, diesel fuel, asphalt, and many other useful products.

Conveyor Transportation

How do airports use conveyors?

A conveyor is an endless belt that carries large amounts of material from one place to another. Just like pipelines, conveyors work continuously. Most are wide belts powered by electric motors. When used in deep mines or quarries, they can be 1 mile in length or longer.

Some belts are flat and wide. Others form a slight V shape for transporting bulk materials like salt and sugar without spilling them. Some factory conveyors are chains up high with hooks for hanging items, such as automobile bumpers.

Ships, trucks, and railroad cars are loaded and unloaded with conveyors. See Figure 14.26. Airports use them to move baggage. Even an escalator is a conveyor. It's an endless belt that forms stairs.

—— Transportation Careers ——

Figure 14-26 Conveyors are part of a ship unloading system that also includes heavy duty cranes and cargo packaged in large containers.

What are some transportation careers you might enjoy?

Few jobs are more rewarding than those in the transportation industry. The career possibilities are almost as endless as your imagination.

Pretend for a moment that you're a flight attendant on a jumbo jet flying between Dallas and Seattle. A planeload of passengers depends on you and your co-workers for special assistance and meals. Your workplace is friendly, professional, and 30,000 feet in the air. As passengers leave the plane after landing, you get a true sense of accomplishment as each one smiles and says, "Thank you."

Maybe you'd rather be a truck driver carrying vegetables from the West Coast to Nashville. You pull into a truck stop and meet other drivers. Some of the faces are familiar, and you exchange stories over good food with good friends. Each day brings new acquaintances and new experiences. No two days are ever alike.

Perhaps you'd like to stand evening watch on a large Great Lakes iron ore boat as it plows through Lake Michigan between Duluth and Chicago. The huge vessel hits the waves and sprays water over the bow. You have an important job and have to stay alert. Excitement and responsibility come your way every day.

You might enjoy working at the rapid transit office in a large city. The large glass building is bright and airy. Lots of sunlight

Duluth
(də-ˈlüth)

449

streams in. Your work area is a large desk with a computer, a telephone, and two green plants. Your job is to help make sure that buses and subways stay on schedule. Thousands of commuters depend on your judgment to help them get to work on time.

Careers in the transportation industry are as varied as the people who have them. Each and every career is a vital link that keeps our transportation system functioning. If you decide to become a part of it, you will reap benefits and satisfaction that comes from a job well done.

You may want some detailed information about specific careers. Here are the names and addresses of some organizations that may be able to help you.

Automobiles
Society of Automotive Engineers
400 Commonwealth Drive
Warrendale, PA 15096

Buses
American Public Transit Association
1225 Connecticut Avenue, NW
Washington, DC 20036

Trucks
American Trucking Association
2200 Mill Road
Alexandria, VA 22314

Railroads
Association of American Railroads
1920 L Street, NW
Washington, DC 20036

Ships
Maritime Administration
U.S. Department of Transportation
400 Seventh Street, SW
Washington, DC 20590

Aircraft
Air Transport Association of America
1709 New York Avenue
Washington, DC 20590

Spacecraft
NASA Headquarters
Washington, DC 20546

Pipelines
American Petroleum Institute
1220 L Street, NW
Washington, DC 20005

Conveyors
Conveyor Equipment Manufacturers Assn.
932 Hungerford Drive
Rockville, MD 20850

— Learning Time —

Recalling the Facts

1. What do the letters NASA stand for?
2. What is the usual type of payload carried by a space shuttle?
3. How fast do liquid fuels travel in a pipeline?
4. Do we have more natural gas pipelines or oil pipelines?
5. Why are some conveyor belts formed into the shape of the letter V?

Thinking for Yourself

1. Astronauts use a mechanical arm to grab satellites already in orbit. Why don't they float out to get them?
2. Explain how firing small rockets can slow down an orbiting space shuttle.
3. The Trans-Alaska Pipeline requires a lot of maintenance and was expensive to build. Explain.

Applying What You've Learned

1. Build a model of the space shuttle or other space equipment. With some of your classmates, produce a videotape documentary showing a typical mission.
2. Pump fluids down a line and measure flow rate. Use a 1- or 2-liter plastic soda pop bottle. Fill with water and attach a tube to the top. Invert and pierce with an inflator needle valve connected to a tire pump (the kind used to inflate soccer balls). Operate the tire pump and measure flow rates.

Summary

Directions

The following sentences contain blanks. For each numbered blank, pick the answer that makes the most sense in the entire passage. Write your answer on a separate sheet of paper.

Nicholas Cugnot of France built the first self-propelled land vehicle in 1769. Charles and Frank Duryea built the first ___1___ car offered for sale. Later on, Augustus ___2___ built the first semi-trailer offered for sale. We now call them eighteen wheelers. The first part of an eighteen wheeler is called the ___3___ .

Trains are also part of a land transportation system. America's first locomotive built for a railroad line was *The Best Friend of* ___4___ . Modern railroads earn more money by hauling

1.
a. British
b. Canadian
c. U.S.
d. worldwide

2.
a. Evinrude
b. Fruehauf
c. Ford
d. Evans

3.
a. tractor
b. semi
c. truck
d. front

4.
a. Savannah
b. Charleston
c. Dallas
d. New York

(Continued on next page)

Summary

freight instead of passengers. The freight is usually ___5___ items like coal and iron ore. However, some trains are built to carry only passengers. All-electric commuter trains get their power from overhead wires or ___6___ .

No drawings of Christopher Columbus's three ships have ever been found. So no one is exactly sure what they looked like. However, it is known that each had ___7___ masts to hold the sails. Some fast ___8___ ships of the 1800s had as many as thirty-five sails and carrried cargo on the open sea between New York City and San Francisco. Sail power gave way to steam power when the SS (Steam Ship) ___9___ crossed the Atlantic Ocean in 1819. Very large ships are used today to transport products and oil between countries. Their ___10___ can be as high as 100,000 tons and more. It usually costs ___11___ to transport goods by water than any other way.

Airplane flight became a reality when Orville Wright was lifted into the air by the airplane he and his brother built. Lieutenant Commander ___12___ and his flight crew made the first flight over an ocean. Their NC-4 flying boat started from New York and ended at ___13___ . Many years later, the first jet-powered passenger plane was the British-built *Comet*. It was less successful than the American-built ___14___ that came out in 1958. Because passenger travel increased so much, the Boeing Co. built a large plane they called the 747. It became the first of several ___15___ , which can carry up to 500 passengers. The jet airplanes that provide service at smaller airports are not so large. They don't need a long runway and use ___16___ .

The space age began when Russia launched the first artificial satellite. America's first astronaut was ___17___ who flew a sub-orbital flight in 1961. John Young and Roger Crippen flew the first space shuttle into orbit in 1981. During a typical space shuttle mission, astronauts

5.
a. packaged
b. expensive
c. lightweight
d. bulky

6.
a. a third rail
b. batteries
c. the wheels
d. solar cells

7.
a. one
b. two
c. three
d. four

8.
a. mizzenmast
b. clipper
c. rigged
d. steam

9.
a. Washington
b. Charleston
c. Savannah
d. Philadelphia

10.
a. displacements
b. sizes
c. lengths
d. volumes

11.
a. more
b. less
c. about the same
d. a lot

12.
a. Amelia Earhart
b. Charles Lindbergh
c. Wilbur Wright
d. Albert Read

13.
a. France
b. England
c. Ireland
d. Portugal

14.
a. Boeing 707
b. Douglas DC-3
c. Curtiss NC-4
d. Ryan NYP

15.
a. zeppelins
b. jumbo jets
c. sky trains
d. cargo planes

16.
a. fewer passengers
b. less material
c. less fuel
d. less time

17.
a. Alan Shepard
b. John Glenn
c. Neil Armstrong
d. Sally Ride

(Continued on next page)

Continued

orbit ___18___ miles above the earth. The cargo bay usually carries a communications satellite. The satellite is positioned in orbit with a 50-foot ___19___ . The astronauts return to earth after about a week in orbit.

18.
a. 55
b. 115
c. 1,000
d. 25,000

19.
a. door
b. line
c. mechanical arm
d. rocket

Key Word Definitions

Directions
The column on the left contains the key words from this chapter. The column on the right contains a scrambled list of phrases that describe what these words mean. Match the correct meaning with each word. Write your answer on a separate sheet of paper.

Key Word	*Description*
1. Transportation system	a) Another name for blimp or airship
2. Vehicle	b) Government regulations for fuel use
3. Transportation terminal	c) The engine's power goes to rear wheels
4. Axle	d) Airplane used to make a profit
5. Mass transportation	e) Container for moving people or products
6. Transmission	f) Aircraft that moves straight up
7. Rear-wheel drive	g) The engine's power goes to all wheels
8. Front-wheel drive	h) Place to enter a transportation system
9. Four-wheel drive	i) Directs the U.S. space program
10. MPG requirement	j) An organized way of moving people or products
11. Tractor trailer	k) Large ship that transports oil
12. Displacement	l) Gears for transmitting engine power
13. Supertanker	m) Very large passenger airplane
14. Commercial airplane	n) Circular rod connecting wheels
15. Jumbo jet	o) Passenger service available to the public
16. Helicopter	p) Large truck with two main parts
17. Lighter-than-air craft	q) The engine's power goes to front wheels
18. National Aeronautics and Space Administration (NASA)	r) Weight of water occupied by a ship

Rubber-Band-Powered Vehicle

Did You Know?

Modern land transportation vehicles have many basic parts that are the same as those used hundreds of years ago. For example, they all have *wheels*, a *body*, and a way to provide *power*. You can learn more about land vehicles by making your own.

Problems To Solve

You will make a simple land transportation vehicle powered by a rubber band. You can then experiment with the design to see if you can make it go faster or farther.

A Procedure To Follow

1. Look at the drawing in Figure A to get an idea of what your vehicle will look like.
2. Drill a ⅜-inch hole at one end of the body. See Figure

B. Use the saw to cut notches at the front and rear of the body.
3. Put the dowel rod through the hole. Cut it to length and attach the 2¼-inch

Resources You Will Need

- **One wooden block, 1 inch thick, 2 inches wide, and 6 inches long**
- **Two wooden wheels, about 2¼ inches in diameter; ready-made wheels from a toy are okay.**
- **Two wooden wheels, about 1 inch in diameter; ready-made wheels from a toy are okay.**
- **One dowel rod, ⅜ inch in diameter and about 3 or 4 inches long**
- **Three heavy rubber bands, ¼ inch wide and about 2 inches in diameter**
- **Nails**
- **Staples**
- **About 18 inches of fishing line**
- **Hammer**
- **Punch**
- **Saw**
- **Wire cutters**
- **Hand drill with ⅜-inch drill bit**
- **25-foot tape measure**

A.

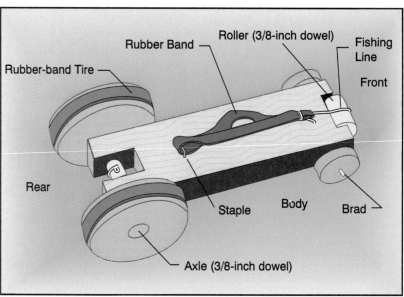

Rubber Band

Roller (3/8-inch dowel)

Fishing Line

Front

Rubber-band Tire

Rear

Staple

Body

Brad

Axle (3/8-inch dowel)

wheels. Place a rubber band around the outside of each wheel. These will be your car's drive wheels, the wheels that provide the driving power. Mark the center of the dowel rod with a punch and hammer a small nail. See Figure B. Leave about ¼ inch sticking out. Later on, you will tie one end of the fishing line to the nail.

4. Attach the front wheels with small nails. Make sure all four wheels can spin freely.

5. Make the front roller from a ⅞-inch-long section of the ⅜-inch-diameter dowel rod. Hammer small nails into each end and cut off the heads of the nails. Attach the roller to the body with staples, as shown in Figure B. Don't drive the staples too deep. The roller must turn easily.

6. Attach a rubber band to the bottom of the body by hammering in a staple. See Figure B. Tie one end of the fishing line to the rubber band. Pass the line over the front roller and under the body of the vehicle. Tie the other end of the fishing line to the small nail on the dowel rod connecting the large wheels.

7. To provide power for your car, wind the larger wheels backward. See Figure B. Continue until all the fishing line is wound onto the axle and the rubber band is stretched over the front roller.

8. Place your vehicle on a smooth floor and let go. It will quickly accelerate and then coast to a stop.

9. Use the 25-foot tape measure to measure how far your car traveled. Do it a few times to get a good average.

What Did You Learn?

1. Did your car travel as far as you expected it would?
2. How could you change your vehicle to make it go faster or farther?
3. What would happen if you used bigger drive wheels? What would happen if you used smaller drive wheels?

B.

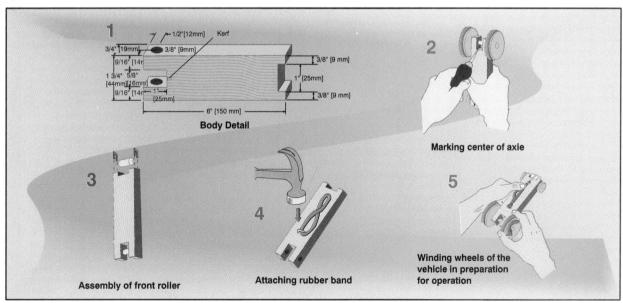

1 Body Detail

2 Marking center of axle

3 Assembly of front roller

4 Attaching rubber band

5 Winding wheels of the vehicle in preparation for operation

What Does the Future Hold?

In 1900 a baby born in North America had a life expectancy of 45 years. Today, a newborn's life expectancy is 75 years. In this section you will learn how technology is helping us lead longer, healthier lives. You'll also learn what our lives may be like in the future.

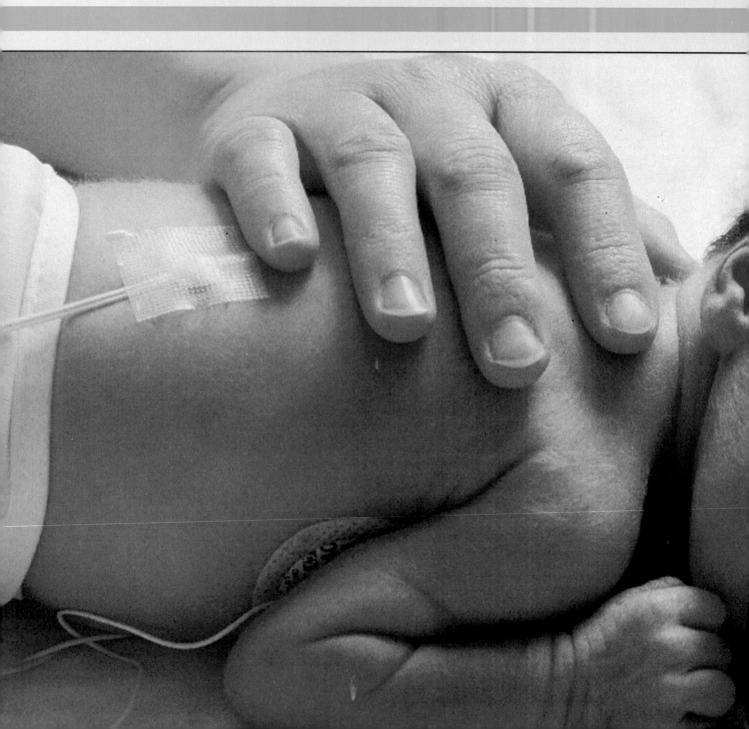

Section V

Special Topics

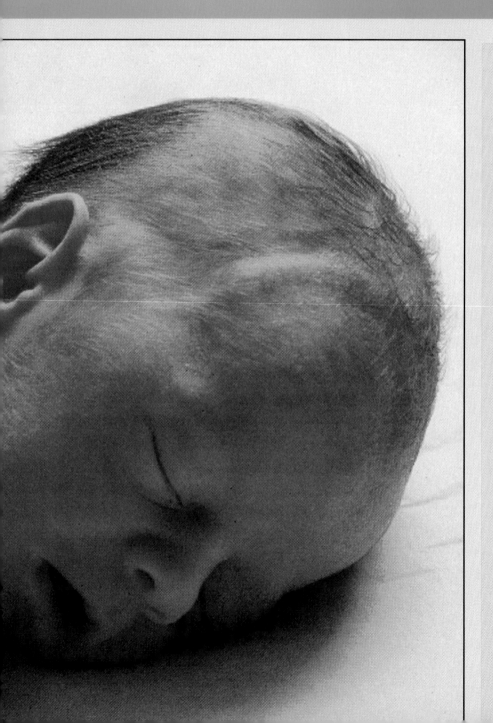

457

Biotechnology

— **Key Words** —

In this chapter you will learn and study the meanings of the following important technology terms.

BIOTECHNOLOGY
FERTILIZATION
IRRIGATION
HYDROPONIC FARMING
MONOCULTURE
LIVESTOCK
RADIOLOGY
ERGONOMICS
BIOENGINEERING
BIONICS
BIOMETRICS
ACID RAIN
GENETIC ENGINEERING

Did you ever find yourself cold, hungry, and frightened? Perhaps you forgot your key, and no one was home. Or you might have become separated from your friends or parents when suddenly you realized you were lost. When that uncertain feeling of danger clamped down upon you, what did you think about?

Food, clothing, shelter, and safety are very basic needs. People who lack food, clothing, and shelter think constantly about surviving. Early technology helped our ancestors meet these needs.

Biology is the branch of science that seeks knowledge about living things. **Biotechnology** is the branch of technology that uses this knowledge to develop new systems to improve agriculture, health care, and the environment. Biotechnology is also responsible for the development of all the products and processes that are made from the resources that come from living things.

In this chapter you will explore devices and processes that have increased food production, improved health care, furthered our understanding of living things, and affected our environment. You will learn about agriculture, health care, bioengineering, and biotechnology's effect on the environment—all subdivisions of biotechnology.

Agriculture

How does modern agriculture compare to agriculture of the past?

The development of agriculture—including the taming of wild animals and invention of the plow—helped release people from their never-ending search for food. This freedom gave people the opportunity to think, discover, and create.

Agriculture today is part of biotechnology because all of its products come from living things. It has become the most mechanized, science-based technology that we currently have. Agriculture is the oldest of the different areas of biotechnology.

Agriculture is the laboratory where biotechnology was born. At first science and technology set out to improve farming. Many of the discoveries that were made evolved into new areas of technology. Eventually, this expansion left agriculture as one part of the bigger picture that became known as biotechnology.

A 1990 report on the American work force indicated that less than 3 percent of all working Americans are involved in farming. In 1987 it took 650,000 large farms to produce 75 percent of our nation's food supply. In the year 2000 it is expected that 50,000 farms will produce 75 percent of our food supply. Never before have so few been able to produce so much. American farmers produce enough food to meet our own country's needs. Then they export any extra food to other countries.

The modern-day farmer is meeting our agricultural needs through the use of very specialized machines, the scientific breeding of plants and animals, and the control of plant and animal disease and insect infestations.

Farm Tools and Machines

What were some early farm tools and machines?

In the beginning, sticks were used to scratch the soil to plant the crops, and crude stone tools were used to harvest them. By 10,000 B.C. the mortar (container you grind in) and pestle (tool for grinding things) were already being used to grind wheat and barley crops. See Figure 15.1. Eventually, people developed special tools to help with the planting and harvesting. These tools included shovels, saws, axes, wine presses, ovens, pots, and the plow.

Originally, these tools were made of stone or wood. When people learned to work with metal, many of these tools were improved and made out of bronze, iron, and finally steel. The invention of the plow greatly increased our ability to cultivate the land. The earliest plows and metal sickles made it easier to plant and harvest crops. Figure 15.2 shows you what some of these early farm tools looked like.

■ **Figure 15-1** ■ The mortar and pestle have been used since 10,000 B.C. to grind materials into a fine powder.

■ **Figure 15-2** ■ Early farm machinery that was used to help cultivate the land was very different from the machinery used today.

Mules, horses, and oxen were used for food, transportation, and to power farm machines. The plow was pulled by farm animals and guided by the walking farmer.

The 1900s saw the introduction of complex farm machinery that was capable of replacing large numbers of unskilled laborers. See Figure 15.3 on the next page. As animal power was replaced by machine power, the animals on the farm became livestock or pets.

Current Agricultural Problems and Solutions

≡ *What issues are confronting modern-day agriculture?*

Modern techniques of farming have created bumper crops and some dangers. The constant **fertilization** and **irrigation** has started to cause a buildup of minerals and salts in the soil. If not controlled, these deposits can cause smaller crop yields in the future. New experimental farms now exist where farmers grow plants without soil. The roots of the plants receive their nutrients from a water-fertilizer mix. This is called **hydroponic farming.**

Most farmers plant the same crop for as far as the eye can see. See Figure 15.4 on the next page. This is called **monoculture**, and it is a very cost-effective way of farming. It does, however, leave an area more susceptible to insect destruction. This is especially true when the entire crop has been bred to have the same characteristics—which unfortunately include the same weaknesses.

fertilization
the process of applying different substances, such as manure or chemicals, to the soil as food for plants

irrigation
the process of supplying water to fields by use of pipes, canals, or sprinklers

hydroponic farming
the process of growing plants in a mixture of water and fertilizer without the use of soil

monoculture
the cultivation, or growing, of only one crop

To reduce this problem, biotechnologists are now trying to breed back some of the genetic inherited variations that once existed in the crops. They are also introducing some new traits, creating crops that need less water, that are less affected by bad weather, that grow faster, and that have natural defense against insects and disease. **Livestock** meanwhile, are being bred to have less fat content, live on crop residue, resist disease, and reach market weight sooner. Figure 15.5 shows how farmers use airplanes to spray their crops.

livestock
animals raised or kept for pleasure or for use and profit

Figure 15-4 This farm practices monoculture. The same crop has been planted on this large parcel of land.

Figure 15-5 This crop of wheat s being sprayed with chemicals that vill protect the wheat from insects.

Figure 15-6 A satellite 500 miles above the earth can now tell farmers how to improve their crops. This technology combines GPS, computers with special farming programs, and other technologies called precision farming.

Putting Knowledge To Work

Precision Farming

In *precision farming,* a global positioning system (GPS) satellite positioned 500 miles above the earth creates images that locate the farm and divide the land into a series of plots on a grid. See Figure 15.6. This is the start of the process necessary to do a full computer map of the farm. The next step is to acquire soil samples for each plot in the grid. This is done by hand, except that the farmers track their location using handheld GPS receivers. The information is added to the computer database for each grid point on the map.

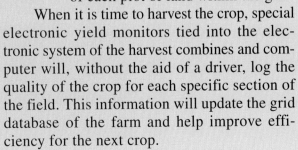

Now problem areas in the middle of a field are registered as specific points on the grid. The satellite images can also register certain types of soil problems directly from space. By placing a computer overlay of the grid onto the computer image from space, special software can tell the farmer moisture conditions, pest infestations, and the need for fertilizer in specific plot areas of the grid.

Farm land is the perfect place to test a fully GPS-controlled driving system. Within the air-conditioned cab of this precision-farming tractor, computers will communicate with the satellite for GPS positioning. The computers will then use the grid database of the farm to control the release of seed, fertilizer, and pesticide that match the soil conditions of each plot of land within the grid.

When it is time to harvest the crop, special electronic yield monitors tied into the electronic system of the harvest combines and computer will, without the aid of a driver, log the quality of the crop for each specific section of the field. This information will update the grid database of the farm and help improve efficiency for the next crop.

463

— Learning Time —

Recalling the Facts

1. What is the difference between biology and biotechnology?
2. Why is agriculture considered a part of biotechnology?
3. Describe some of the current agricultural problems that are facing biotechnology. Suggest some solutions to these problems.

Thinking for Yourself

1. Why was the agricultural revolution important to the development of technology?
2. Describe some of the major developments that changed our early farm-based society into a technological society.
3. Why is monoculture farming cost-effective and at the same time dangerous for the farmer?

Applying What You've Learned

1. Construct a small greenhouse in the window of your room or on the school grounds. Grow two plants from seeds. Test different fertilization methods on these plants.
2. Research an early farm tool or machine and construct its model. Write a report on the effect that this machine or tool had on agriculture, and prepare a chart that shows how this tool or machine worked.

— Health Care —

What effect has modern health care had on life expectancy?

The ancient search for the fountain of youth was, in a sense, an attempt to find an instant health care system. Our ancestors tried to develop chemical processes and develop machines that would cure all disease. Although these discoveries were never made, special medicines were found that can cure or prevent certain diseases. Since the start of the Industrial Revolution, advances in health care have doubled the life expectancy of our world's population.

Our understanding of the human body could never have reached its present level without the development of the microscope, which made it possible to see the physical characteristics of life. The development of the X-ray machine and other scanning devices made it possible to see inside a living organism.

The development of various monitoring machines made it possible to analyze the electrical impulses of the heart, brain, and nervous system. People also developed many laboratory tests that made it possible to determine if a person's chemistry was normal.

The Stethoscope

What is the purpose of the stethoscope?

Medical professionals use stethoscopes to hear the sounds of the heart and lungs. Doctors can tell just by listening if your heart valves are opening and closing properly.

The stethoscope works on the same principle as a child's tin can string telephone. In a stethoscope, a hollow tube connects the earpieces to a diaphragm. The diaphragm is placed against your chest. The sound vibrations are picked up by the diaphragm and then transmitted to the health professional's ears. See Figure 15.7.

The Electrocardiogram (EKG)

What does a doctor learn from an EKG?

An electrocardiogram (EKG) picks up your heart's electrical impulses from twelve leads attached to your body. The information from these sensors are amplified and used to print a paper chart that shows the condition of your heart. See Figure 15.8.

A relative of the EKG, the electrocardiographic body surface potential map registers the heart's electrical impulses from thirty-two points on the human body. By using almost three times as many locations as an EKG and the information-generating power of a computer, this machine can create a visual picture of your heart's size, shape, and tissue condition. See Figure 15.9 on the next page.

electrocardiogram (EKG)
(ī-,lĕk-trō-'kärd-ē-ə-,grăm)

Tech Talk

The electrocardiogram is usually referred to as an *EKG*. Can you see why? If you worked with heart patients in a hospital or doctor's office would you want to say *electrocardiogram* several times a day, or just *EKG*?

Figure 15-7 These children are being examined with a stethoscope as part of a preventative medicine program.

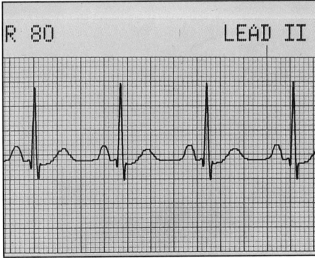

Figure 15-8 A trained medical professional can use this EKG readout to determine the condition of a person's heart.

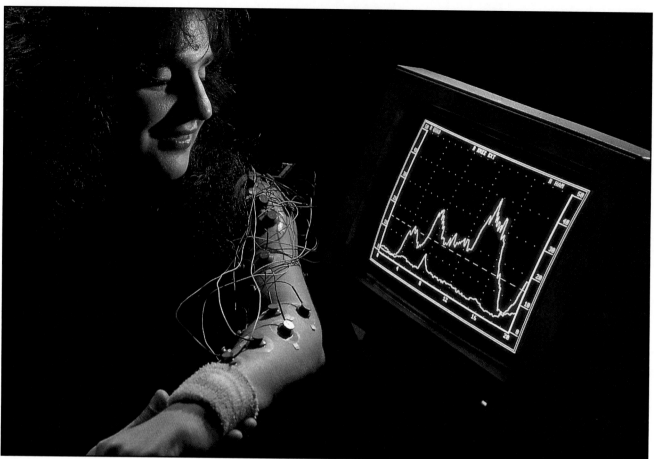

Figure 15-9 Electrocardiographic machines use a computer to convert 32 body signals into a map of the human heart.

Microscopes

What is the purpose of a microscope, and how does it work?

Most of the enemies of good health are too small to be seen with the naked eye. They are microscopic in size. The development of the microscope made it possible to see even the smallest particles that make up our body tissue.

Doctors routinely send blood and tissue samples for microscopic examination. From these samples it is often possible to determine what is wrong with the patient. Once diagnosed, proper medical treatment can be started.

The development of more powerful microscopes opened doors to smaller and smaller worlds. The most powerful microscopes may help us understand and develop treatments for diseases now considered incurable.

How the Microscope Works. Did you ever notice that a glass filled with water can magnify objects? Water-filled magnifying glasses were used 3,000 years ago. Glass magnifying lenses weren't introduced until the thirteenth century.

The microscope that you use in science is a compound microscope based on the magnifying properties of lenses. The light shinning through the sample is spread by the lens, causing the object to appear larger than it actually is.

The power of these optical (lens) microscopes has now been pushed up to 2,000 times the size of the object. This is believed to be the maximum magnification that can be reached using daylight or artificial white light. Image magnifications of 3,000 times actual size have now been achieved through the use of ultraviolet light.

The Electron Microscope. A number of different types of electron microscopes exist. These machines use a beam of electrons instead of a beam of light to magnify objects. Just as light is spread by a glass lens, electrons can be spread and the resulting picture recorded on a television picture tube. Since each stream of electrons carries a small spot of the picture of the specimen, the spreading of the streams causes the picture to grow like a projected movie image.

The scanning-tunneling electron microscope produced the first picture of a single silicon atom in January, 1983. With this electron microscope, it is possible to view a single atom of any material that will conduct electricity. An atom is less than one hundred-millionth of an inch in diameter.

The atomic force microscope was invented in 1985. See Figure 15.10. It allows this same powerful viewing of nonconducting materials. It uses a fine laser beam to measure surface changes. A picture of the surface is created by recording and displaying millions of height changes measured by the laser beam. This microscope can record images of nonconducting living tissue.

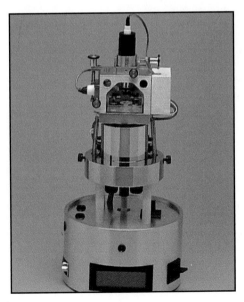

Figure 15-10 Atomic force microscopes allow for the viewing of single atoms in living tissue. An atom is less than one hundred-millionth of an inch in diameter.

X Rays

How are X rays used in health care?

The most portable X-ray equipment ever devised belongs to the fictional character Superman. But his ability to see through solid objects is no match for today's medical imaging machines.

The branch of medicine that uses radiation for the diagnosis and treatment of disease is called **radiology**. The first radiology tool was the X-ray machine. X-ray and newer imaging machines make it possible to see inside a living person without cutting the person open.

The X-ray beam was discovered by the German physicist Wilhelm Roentgen. See Figure 15.11. His 1893 discovery was an accident. Roentgen was experimenting with a *Crookes Tube*, which was an early form of a cathode-ray tube. The tube was hooked up to a high-voltage electric current. When his wife called him to lunch, he set the unit down on a book that had a key sitting between its pages. This book was sitting on top of a photographic plate. When he developed the plate, he discovered the picture of the key.

Roentgen announced his discovery to the world in 1895, and X rays were soon being used to photograph bones of the human body. Roentgen built a device that generated fast-moving electrons that crashed into a metal surface inside the tube. These collisions caused the tube to give off electromagnetic X-ray energy. Figure 15.12 shows how your dentist uses X rays to photograph your teeth.

To photograph internal organs using X rays, it is necessary to add a contrast medium to the body. Physicians usually give their pa-

radiology
the branch of medicine that uses radiation, as in X rays, to diagnose and treat diseases

Roentgen
(rĕnt′gən)

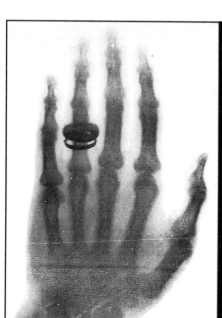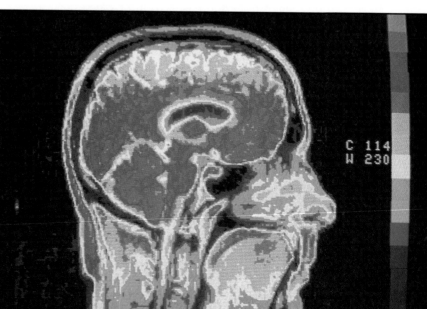

■ **Figure 15-11** ■ Roentgen's X rays, exampled on the left, provide a view of the living human skeleton. The modern cat scan, viewed on the right, combines Roentgen's X-ray technology, dyes, and the computer to create detailed pictures of the human body.

tients barium sulphate, which causes the organs to show up on the photographic film.

Computerized Axial Tomography (CAT)

What does a cat scan show, and how does it work?

CAT scans combine the X ray's ability to see inside the human body with the computer's ability to visually display and process information. In this process, thousands of measurements are taken with low-voltage X rays. To create these pictures, the X-ray tube rotates completely around the part of the body being scanned. See Figure 15.11 again.

This information is then processed by a computer and turned into three-dimensional representations of what is going on inside the body. The health technician can create different images of the body part by changing the viewing angle, enlarging the image, and enhancing tissue color. All of these changes are done by the computer.

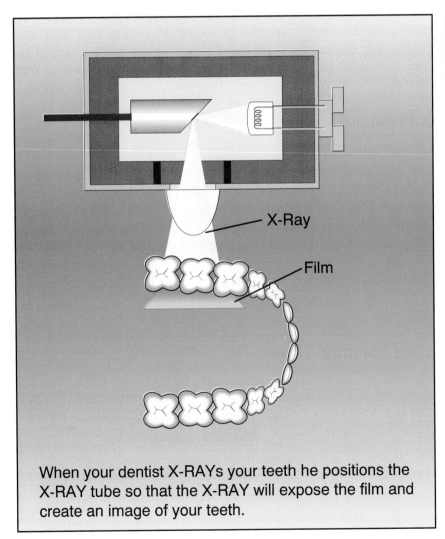

Figure 15-12 When your dentist X rays your teeth he positions the X ray tube so that the X ray will expose the film and create an image of your teeth.

When your dentist X-RAYs your teeth he positions the X-RAY tube so that the X-RAY will expose the film and create an image of your teeth.

469

Magnetic Resonance Imaging (MRI)

What's the difference between a CAT scan and an MRI?

Magnetic resonance imaging creates images that are very similar to the images from a CAT scan. In the MRI scan, however, the patient isn't exposed to X rays. The image is created using measurements that were created by magnetic waves.

For this imaging to be done, you are placed into a machine that resembles a very large donut. You can't be wearing any metal jewelry or have a pacemaker or any metal plates inside your body. Metal fillings in your teeth are okay.

A very powerful magnetic field is then generated by a supermagnet. This field causes all the hydrogen atoms in your body to line up with the field. This lineup is a little like the flags around the United Nations building on a very windy day. They are all blowing and pointing in the same direction.

At this point the MRI machine hits your body with a burst of radio waves. Waves of a certain frequency cause the atoms of each tissue of your body to drop out of alignment for a second. When they drop out, they transmit a radio frequency signal of their own. Each body tissue has its own radio signature that is picked up and interpreted by the computer. These computer images are so clear that they are as good as the view that would be seen by a surgeon.

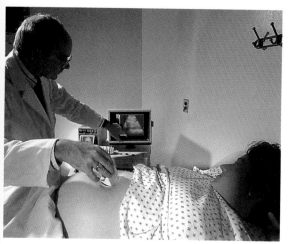

Figure 15-13 Ultrasound technology is often used to view and examine an unborn baby.

Ultrasound

What is the most common use of an ultrasound?

An *ultrasound* image is created by directing an ultrahigh-frequency sound against the part of the body that is going to be viewed. This sound energy will be altered by the internal structure of the body and bounce back as an echo. The created image is shaped like a triangle and rendered in black and white or color. See Figure 15.13. Ultrasound is often used to view an unborn baby. Ultrasound is completely harmless to the mother and her unborn baby.

— Learning Time —

Recalling the Facts

1. Explain how a stethoscope works.
2. How did Roentgen accidentally discover X rays?

Thinking for Yourself

1. Of the many advances in medical technology that you just studied, which one do you think was the most important?

(Continued on next page)

3. Explain how an ultrasound image is created.

Give reasons for your answer.
2. Explain the difference between a CAT scan and an MRI scan.

Applying What You've Learned

1. Construct a stethoscope.
2. Research a medical diagnostic machine that has not been covered in this chapter. Write a report that covers the machine's development and use. Include a large poster that shows how the machine works. Build a rough model of the machine.

— Bioengineering —

≡ *What are some of the benefits of bioengineering?*

The bionic man and the bionic woman were fictional characters whose physical capabilities were improved because of high-technology implants. Many of the replacements that they received as fictional characters are now created by scientists and engineers and implanted in real people. These devices are only capable of restoring lost physical ability. They do not, as yet, give their recipients superhuman power.

Ergonomics is another name for **bioengineering**. The scientists and engineers that work in this field are called ergonomists or bioengineers. They often design products around the limitations and comfort needs of people. They develop car instrumentation panels, comfortable seating, safety devices, people-friendly machines and appliances, and protective clothing. See Figure 15.14 on the next page.

Bioengineers are also responsible for the creation of life-sustaining environments for aquanauts, astronauts, and airplane passengers. Building life-sustaining units in alien environments makes it possible for people to work deep underground, under the oceans, in airplanes flying high in our atmosphere, and in outer space.

Bionics: Spare Parts

≡ *What are some of the body parts being replaced by bionics?*

The human body is a perfectly crafted machine that contains thousands of living parts. Bioengineers are developing systems that

ergonomics
the branch of technology that designs products with the comfort and needs of people in mind. Also known as *bioengineering.*

bioengineering
the branch of technology that designs products with the comforts and needs of people in mind. Also known as *ergonomics.*

471

Figure 15-14 A sneaker is an example of a product that has been designed to support and protect the bones and muscles of the body.

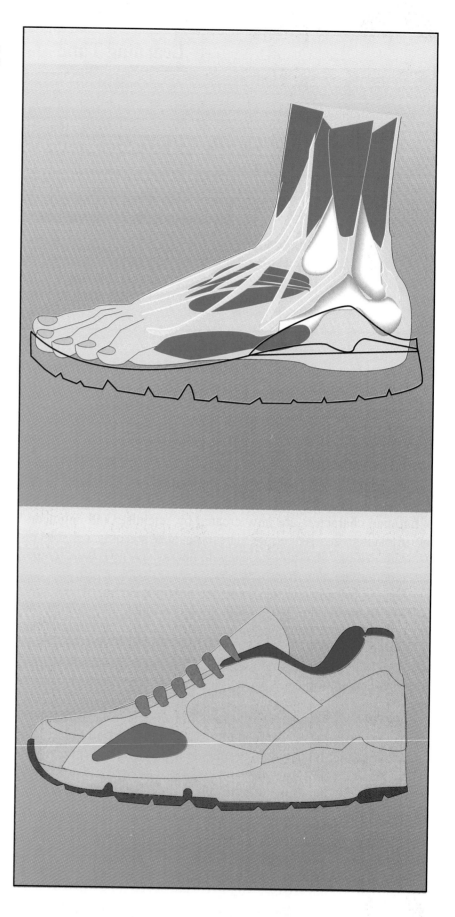

can replace these parts if they get worn out, are destroyed in an accident, or are defective at birth. This area of bioengineering is called **bionics**.

Creating replacement parts for the human body involves a number of problems. All construction materials must be biologically *inert* (chemically neutral), so that the person's immune system won't send antibodies to attack what the body sees as foreign.

The new part must be able to work in the environment of the human body. It should be able to duplicate the function of the part it replaces for many years to come. Having to replace this part again would mean putting the person through surgery again.

The new system must perform the body function of the part it is replacing well enough that it doesn't cause a breakdown in another body system. If a replacement heart valve causes blood clots, it can't be used. If a replacement bone joint causes other body bones to break, then it can't be used. Figure 15.15 shows a Robodoc ready for surgery.

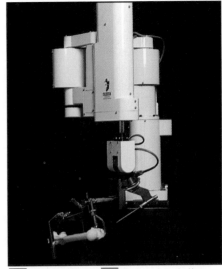

Figure 15-15 Robodoc is the marriage of CAD technology, CAM technology, and biotechnology.

Putting Knowledge To Work

Robodoc

In the fiction movie *Robocop*, a cyborg (part human and part robot) performs law enforcement activities better than the human police officers that it works with. *Robodoc* isn't science fiction. It is a three-foot, 250-pound robot that started performing surgery on people on November 7, 1992. Just like the fictional character Robocop, this machine is capable of performing its programming tasks better than humans. Its job has nothing to do with law enforcement and everything to do with operating on people.

Robodoc isn't the first robot to work in an operating room. It is, however, the first robot to do more than just guide surgical instruments. Until Robodoc, the surgery was always under the full control of a human surgeon. In hip replacements human surgeons cut the bone with a 20% accuracy. The robot performs its surgery by following the information contained in a CAT scan. Robodoc prepares the bone for the installation of the artificial replacement with a 96% accuracy.

It performs this job without seeing or understanding what it is doing.

To start performing its operation, Robodoc needs to be positioned at exactly the right location. This is done by the surgeon, who installs three steel pins into the bone in a preliminary operation that takes place before the CAT scan is made. During the surgery the surgeon exposes the bone and the three pins. Three sensors from Robodoc are connected to these pins. Robodoc then carves out the cavity for the joint replacement without any further instructions. After the carving is complete, the surgeon is able to install the replacement. Robodoc lets human surgeons who have been assisting with the operation also close the incision.

Robodoc is now performing surgery in Europe and Japan and waiting for final FDA approval in the U.S. Robodoc will probably soon be performing many of the 750,000 hip replacements that take place each year.

The *Jarvik-7* artificial heart passed all kinds of animal tests before it was placed into the body of a dying man. The surgery was performed because no human heart was available.

To work, this bionic heart had to be connected to an outside system that gave it the power to pump 7 to 12 quarts of blood per minute. The heart kept Barney Clark alive for 112 days, but the system started to cause blood clots. These blood clots caused the man to have a number of strokes and die.

Biometrics: Your Eye, Hand, or Voice Is Your Key

What are some of the common uses of biometrics?

The technology of **biometrics** is replacing keys, credit cards, and combination locks with the unique pattern of a person's voice, finger, palm, or retina eye scan. With such systems, you can't lose your key or credit card or forget your combination number.

When banking systems start using a biometric system, you won't have to worry about anyone getting into your account unless they take you along on the robbery. If a retina eye scan were necessary to start your car, few cars would be stolen.

At this time the most common biometric reader uses a person's fingerprint. You have used similar, though less sophisticated, readers if you have ever bought a soda from a machine that took dollar bills.

The biometric reader and its cousin, the dollar bill reader, both convert what they see into a mathematical image. The dollar bill reader only has to compare the paper that it receives with its memory of what a dollar bill should look like. The biometric reader has to compare your fingerprint, let's say, with all the fingerprints that it has on file and identify you as being okay for access.

Biometric security devices are currently being used in high-security government facilities. In December of 1997, biometric eye scans were introduced for banking security at ATMs. Here, three cameras locate you, zoom in on your eye, read your eye like a fingerprint, and then approve your banking transaction.

Some cars in England and Japan have biometric key systems. You should soon find biometric security systems on your family car.

Protection of People and the Environment

What are some of the major environmental concerns facing the world today?

Do you remember a nursery rhyme about an old woman who swallows a fly? At each point in the rhyme, she swallows something else that is capable of swallowing the last thing she ate. Each

biometrics
the branch of technology that uses voice prints, finger prints, and retina eye scans as positive forms of identification. Used for security and other purposes.

474

Insect in Amber

Tyrannosaurus Rex

Dolly the Sheep

■ **Figure 15-16** ■ If scientists can clone a sheep, will they ever be able to bring back extinct animals or dinosaurs?

▬ **Putting Knowledge To Work** ▬

Cloning

The two *Jurassic Park* movies dealt with the cloning of dinosaurs from ancient DNA. After seeing these movies, did you wonder if ancient DNA can be recovered? Can dinosaurs or Dolly the lamb be cloned? See Figure 15.16.

A microscopic blueprint exists in the DNA that is in the cells of every living organism. In nature, this blueprint is used to build each individual plant, insect, bird, reptile, and mammal. When bees pollinate a plant, they help nature draw a new set of plans by joining the DNA from two different flowering plants. Cloning uses the DNA from only one plant or animal to create an offspring that is an exact duplicate of the one parent.

The *Jurassic Park* movies were based on an amazing technological achievement. At this very moment, people are recovering ancient DNA from amber in laboratories around the world. Amber is tree sap after it hardens into a resin. See the first photo in Figure 15.16.

These movies switch to science fiction when the Jurassic scientists create dinosaurs by "filling in the gaps" of the incomplete dinosaur DNA because the ancient DNA that has been found in amber is only one percent intact. To understand just how much of the information is missing, imagine trying to solve a hundred-billion-piece jigsaw puzzle with just a tiny segment from one single piece.

Would the creation of multiple twins in prize livestock be cloning? Scientists have been able to duplicate a natural process that causes the birth of identical twins. The calves or ponies are exact duplicates (clones) of each other, but they have the DNA of both parents.

How was Dolly the lamb cloned? Dolly the lamb is a genetic duplicate of a single parent. Scientists removed the nucleus from an unfertilized egg and replaced it with the nucleus from an adult cell. A spark of electricity caused the nucleus to fuse with the egg and sparked a change in the nature of the genetic material that was in this nucleus. The egg then started to behave like an ordinary fertilized egg, even though it had the nucleus of an adult ewe. The egg was implanted into a surrogate ewe, and after birth the DNA of the baby only matched the adult cell donor. Dolly cloning has not yet been verified by the scientific community.

Cloning can help keep endangered species from becoming extinct. It will continue to be used to help breed prize livestock that have less fat content and are more resistant to disease. In time, it might help bring back species that have recently become extinct. All technology can have negative as well as positive effects on our society. It is up to our society to make certain that cloning doesn't become an ethical emergency.

episode ends with "I think she will die." Finally, she swallows a horse, and the rhyme ends with "She's dead, of course."

Many people worry that our technology has our environment swallowing all kinds of terrible things. Everyone today wants to reduce the pollution that presently exists in our atmosphere, oceans, and rain forests.

Biotechnology must find solutions to many environmental problems that our past technology has created. **Acid rain** is a by-product of our use of fossil fuels. It is killing our rain forests and the life that lives within them. It is also killing aquatic life, damaging farm crops, corroding steel and stone structures, and endangering human health.

Smog is produced when hydrocarbons combine with nitrogen oxides and sunlight. The chemicals needed to make smog are produced by the burning of fossil fuels. Smog has become a major world problem. See Figure 15.17.

Almost half the population of the United States is now breathing substandard air. Our automobiles and factories are slowly poisoning the air we breathe. Catastrophes such as the oil fields being set afire during the 1991 war with Iraq can do great harm to the world's environment. What effect do you think the burning of over 5 million barrels of Kuwaiti oil per day for several months will have on our environment?

Our technology is feeding our waterways with untreated sewage, treated sewage, industrial waste, and agricultural drainage. All of these impurities affect the animals and plants that live in or near these waterways. When we poison our water, we poison ourselves. What steps can biotechnology take to reverse this current

acid rain
polluted rain caused by the use of fossil fuels such as coal, oil, and gasoline. It causes much damage to the environment.

■ **Figure 15-17** ■ Smog is a major environmental concern that has led to many medical problems.

ecological disaster?

What can we do to produce less solid waste? The most common method of solid waste removal is open dumping. Most cities are running out of open dumping areas. They are now trying to reduce the amount of solid waste by recycling. New facilities are being designed and built to convert solid waste into usable energy.

Environmental aspects of biotechnology also include the development of adequate safeguards for new pesticides and medical drugs. Are enough safeguards in effect to guarantee that a new life form won't threaten our very existence? Can the current research that seeks to replace human genes and thereby correct defects that were present at birth, lead to unplanned and unwanted human evolution?

genetic engineering
the process of changing genes that contain information for developing certain characteristics or traits in plants and animals

Putting Knowledge To Work

Turning Bacteria and Plants into Plastic-Making Machines

In the mid-1970s, a British chemical company started producing 25 tons a year of biopolymers—plastics from living organisms. They began this program to find a way of making plastic without the use of petroleum.

They found a natural bacteria that produced a poor-quality plastic. Then they changed the bacteria's diet to include organic acids. The new foods they fed their bacteria produced a plastic that could be used to make plastic bottles.

This plastic is completely biodegradable; that is, it will decompose. Any plastic that is made by bacteria can be eaten by bacteria.

At this time, all plastics made by people, called synthetic plastics, are not fully biodegradable. The bacteria at the dump or treatment plant that try to break these plastics down can only convert the large plastic items to very tiny plastic balls.

The problem with bacteria-made plastics is that they cost $15 a pound to make, while more conventional plastics only cost 50 cents a pound to make. Technological advances will, in time, make this process cost-effective.

Researchers at MIT wanted to determine just how these bacteria made plastic. They ended up using **genetic engineering** to improve the process. They were soon making different types of plastic by altering the genes of the bacteria.

The great jump forward is now under way. Bioengineers are trying to get plants to grow plastics. Plants have the same ingredients from which bacteria assemble plastics. Once the bacteria's gene for making plastics is successfully transferred to plants, we could have plant seeds that will grow into plastic-making machines.

This process won't give us plastic flowers or bottles attached to plant stems. It will give us small plastic pellets inside the cells of the plant. These pellets would be harvested and processed into more useful forms.

Biotechnology must address these issues. Systems must be developed to protect all life on our planet. When it comes to developing products that affect the food we eat, the air we breathe, and the health care that we receive, the profit motive of our free-market system must be kept under careful scrutiny.

Environmental safety affects the relationship between living things and technology. Since biotechnology forms a bridge between the study of living things and technology, the environment demands biotechnology's full attention.

The hope is that scientists and engineers working together will be able to find ways of reducing the negative effects of human inventions. The current problems are easy to identify. Many of the solutions will need to be found by you, the future work force of our world.

Biotechnology Careers

What are some of the career choices in biotechnology?

The occupational outlook in the biotechnology field is very good. Career ladders exist in all the occupational areas. This means that you can qualify for some jobs straight out of high school, while other positions will require an associate degree or more.

Some of the careers mentioned in the area of agriculture cross over into the other areas of biotechnology. Botanists, for example, study plant life. They use this knowledge to improve agriculture and to develop new medicines for plants, animals, and people.

Agriculture-Related Careers

What are some careers in agriculture?

In the area of agriculture, the trend of more food being produced by fewer people is expected to continue. If you want a career in farming, you will need to learn about modern farming practices.

Depending on your area of specialization, you can expect to take some of the following courses. You might study crop production and science, livestock management, animal science, plant and animal production, soil conservation, natural resource management, business management, and marketing.

Your career choice will determine how much education you actually need for entry-level positions. You can prepare for your future occupation in a high school or community college studying landscaping, ornamental horticulture (decorative flowers and plants), farming, forestry, conservation, or farm-related sales of supplies and equipment.

A four-year college program will meet minimum requirements if you are interested in becoming a teacher of agriculture, botanist, agricultural engineer, animal pathologist (studying animal disease), animal scientist (doing research in the selection and breeding of animals), biologist, industrial microbiologist (studying food-related disease and bacteria that produce toxins), or plant pathologist (researching the cause and control of disease in plants).

If you decide to go into veterinary medicine, you will need to complete at least seven years of higher education. You will become a Doctor of Veterinary Medicine (D.V.M.) upon graduation. There are only twenty-eight veterinary medical programs in the United States, and their admission requirements are more demanding than those of medical school. See Figure 15.18.

Health Care and Bioengineering Careers

What are some careers in bioengineering and health care?

What do people mean when they talk about the graying of America? This statement means that the largest percentage of our population is middle-aged. As our population continues to age, there will be an even greater need for more workers in health-related fields. According to the Bureau of Labor Statistics, the fastest- growing occupations are health-related occupations.

Building machines to protect and compensate for human frailties has become a design necessity for today's manufacturer. Engineers and scientists who specialize in ergonomics will find plenty of projects to work on in the future.

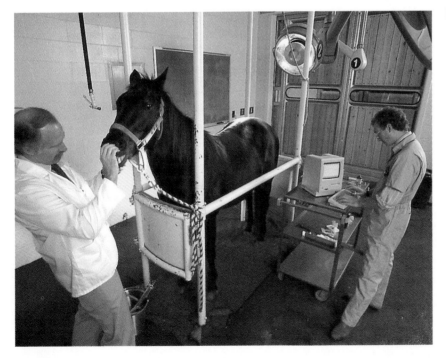

Figure 15-18 In veterinary medicine you learn how to provide medical care to large and small animals.

Some high school and community college programs teach the basics of nursing care and rehabilitation. They can prepare you for a career as a nurse's aide, home health aide, practical nurse, dental assistant, medical assistant, dental laboratory technician, biomedical equipment technician (running and maintaining high-tech medical equipment), cytotechnologist (working in a laboratory with a pathologist), or dental hygienist (cleaning patients' teeth and gums in a dental office).

A four-year college program is the minimum requirement for a career as a biochemist (studying the chemical makeup of living things), industrial hygienist (studying and working to prevent occupational disease and injuries), registered nurse, health inspector, or occupational therapist (planning and directing recreational and vocational programs for physically or mentally disabled persons). Many other occupations exist within this area at this level of education.

A four-year degree plus a master's degree is the minimum requirement for a career as an audiologist (hearing examiner), biomedical engineer (designing medical instruments and machines), ergonomist (bioengineer), or geneticist (studying and experimenting with genes). If you wish to design pacemakers, artificial kidneys, and artificial hearts, then you might want to become a biophysicist. You will need to complete a four-year degree, master's degree, and doctorate.

If you want to become a doctor, be prepared to complete three or four years of college, medical school, an internship, a residency, and an area of specialization. You might want to specialize as an allergist-immunologist (working with people who have allergies), anesthesiologist (making patients insensitive to pain so that surgery can be performed), cardiologist (heart specialist), dermatologist (skin specialist), or ophthalmologist (specializing in eye injuries and diseases).

The *Dictionary of Occupational Titles* can be found in your school or public library. It contains information about many other careers. Your guidance counselor or computer teacher might have a computer program that can help you explore different careers.

— Learning Time —

Recalling the Facts

1. What three conditions must a spare part meet in order for it to be used in the human body?

Thinking for Yourself

1. Spare body parts are currently needed in three basic situations. Identify these situations. Then give other reasons why

(Continued on next page)

2. Name and describe three types of systems that bioengineers design and build.
3. Why is our environment a technology issue?
4. What part of the body did the *Jarvik-7* replace? Is it still being used?
5. What is a biometric reader used for? What other device uses the same technology?

people might want bionic parts installed in the future. (What the future holds is anyone's guess. So be creative.)
2. How much blood would an artificial heart have to pump per minute, per hour, per day, per month, and per year?
3. If you could design the perfect bioengineered food supply, what would the animal or crop look like? What special characteristics would it have?

Applying What You've Learned

1. Redesign something you own so that it is more comfortable to use. Your new design can change the size, shape, location of controls, and even the material that the device is made of. Your completed project can be a model, mechanical drawing, computer drawing, or written report. Be sure to include a picture of the original object. You can work on something as complex as a television or as simple as a paper clip.
2. Design a self-contained environment that can meet the needs of six astronauts during a trip to Mars.
3. Research the career that you find most interesting. Determine the educational requirements, salary, working conditions, job responsibilities, and future outlook for that career.

Summary

Directions

The following sentences contain blanks. For each numbered blank, pick the answer that makes the most sense in the entire passage. Write your answer on a separate sheet of paper.

Biotechnology is the branch of technology that uses knowledge about living things to develop new technology for agriculture, health care, and the environment. All products that are made from once-living resources are produced by ____1____ industries.

Today less than 3 percent of the American work force is involved in farming. So few people can ____2____ so much because of automated

1.
a. manufacturing
b. communication
c. biotechnology
d. medical

2.
a. produce
b. buy
c. sell
d. destroy

(Continued on next page)

Continued

machines and specialized breeding of ___3___ . Because of biotechnology's selective breeding programs, we now produce larger crops from the land and more ___4___ from our livestock than ever before.

The chemical treatments developed in biotechnology laboratories ___5___ farm products from most diseases and insect infestations. These ___6___ guarantee that most of what is ___7___ will make it to market.

The monoculture system of ___8___ has very large areas of land dedicated to growing one crop. Selective breeding has produced plants with many strengths. Selective breeding, however, has left these same plants with the same ___9___ . When ___10___ or diseases are not stopped by spraying, the entire crop is at risk. New programs are now under way to reintroduce some of the inherited plant and animal variations that once naturally existed. By altering plant and animal genes, genetic engineers are developing new ___11___ that give these plants and animals natural protection against insects and disease.

Our health care system is also a part of ___12___ . The CAT scan, ultrasound, and MRI make it possible to see inside living organisms. By combining these scanners with ___13___ , scientists have been given a window through which they can see the workings of a single cell, DNA, and viruses.

Bioengineering is also a part of biotechnology. ___14___ design bionic replacement parts, protection systems, and artificial environments. All bionic parts must do the work of the part that they replace. These parts must be invisible to the immune system, reliable, and not cause other body systems to ___15___ . Many bionic parts are now in use. Even an artificial heart was developed and tested. The heart was ___16___ because it caused life-threatening blood clots.

Bioengineers have also developed ___17___ gear for sports and bulletproof vests for the police and military. Bioengineers have de-

3.
a. cats and dogs
b. plants and animals
c. machines and tools
d. weeds and fertilizer

4.
a. plants
b. fertilizer
c. pets
d. beef, pork, and poultry

5.
a. protect
b. inoculate
c. deliver
d. take

6.
a. tools
b. machines
c. people
d. chemicals

7.
a. sold
b. grown
c. purchased
d. frozen

8.
a. manufacturing
b. farming
c. freeze drying
d. pasteurizing

9.
a. weaknesses
b. diseases

c. colors
d. farmers

10.
a. fires
b. drought
c. hunger
d. insects

11.
a. products
b. colors
c. traits
d. faces

12.
a. manufacturing
b. communication
c. agriculture
d. biotechnology

13.
a. computers
b. telescopes
c. machines
d. tools

14.
a. Artists
b. Bioengineers
c. Mechanics
d. Nurses

15.
a. work
b. fail
c. move
d. become unnecessary

(Continued on next page)

veloped automobile restraint systems and artificial environments for people traveling underwater, in airplanes, and in outer space.

Until recently, people ignored the negative effects of our technology on our _____18_____. After years of neglect, our environment is now receiving the attention that it deserves. New technology has to meet new environmental _____19_____. Special groups are pressing for new laws that will require old technology to meet newer, stricter environmental standards in the future.

16.
a. used
b. successful
c. redesigned
d. installed

17.
a. protective
b. invisible
c. dangerous
d. heavy

18.
a. roads
b. homes
c. cars
d. environment

19.
a. biotechnology
b. standards
c. treatments
d. packaging

Key Word Definitions

Directions

The column on the left contains the key words from this chapter. The column on the right contains a scrambled list of phrases that describe what these words mean. Match the correct meaning with each word. Write your answer on a separate sheet of paper.

Key Word	Description
1. **Biotechnology**	a) Growing plants without soil
2. **Fertilization**	b) Technology of health care, agriculture, and environment
3. **Irrigation**	c) Farm animals
4. **Hydroponic farming**	d) Same crop for as far as the eye can see
5. **Monoculture**	e) The process of supplying water to the fields
6. **Livestock**	f) The process of feeding plants
7. **Radiology**	g) Designing products for comfort and needs of people
8. **Ergonomics**	h) Voice print, fingerprint, or retina eye scan
9. **Bioengineering**	i) Artificial parts
10. **Bionics**	j) Changing the traits of future generations
11. **Biometrics**	k) Another name for bioengineering
12. **Acid rain**	l) Radiation for diagnosis and treatment of disease
13. **Genetic engineering**	m) Killing our rain forests

PROBLEM-SOLVING ACTIVITY

Constructing a Hydroponic Planter

Resources You Will Need

- One 6-inch diameter plastic pipe that is 2 feet long
- Two ¼-inch-thick sheets of plastic, 6½-inches square
- One 4-foot by ⅜-inch flexible tubing
- Two tubing/container adapters with rubber washers and nuts
- Two ⅝ clamps
- 32 ounce plastic pail
- Drill set
- Electric drill
- 2¼-inch hole saw
- Magic marker
- Drill press
- All-purpose plastic cement
- Hot glue gun
- Hot glue
- PH testing kit
- Baking soda
- Vinegar
- Sterilized garden sand
- Fine plastic screening
- Plant food
- Four small potted plants
- Wood blocks
- 18" x 10" ¾" thick plywood sheet.

Did You Know?

Only 5 percent of the earth's surface is suitable for soil farming. To grow plants, you need seeds or young plants, water, sunlight, air, nutrients, and a material to hold the plants in place. Did you know that you can grow plants without soil?

The ancient Aztecs of Central America discovered this technology long before any Europeans landed on our shores. In their marshlands the Aztecs built large floating islands that they farmed. The plants received water and nutrients from their roots which grew under the islands.

The modern technology associated with hydroponic farming didn't develop until the 1920s, when William Gericke, from the University of California, used the technology to grow 25-foot tomato plants. He chose the name hydroponic from the two greek words *hydro* ("water") and *ponus* ("work").

Hydroponics is a good method of farming for those regions of the world where water is scarce, soil is poor or doesn't exist, insects are difficult to control, or land is in short supply. In hydroponic greenhouses plants grow faster, take up less space, and have larger crops. What's more, their water and nutrients are recycled.

A.

Problems To Solve

In the future, farmers will rely more on hydroponic technology to meet the food demands of the increasing world population. Your problem is to build a hydroponic planter, like the one shown in Figure A, that can grow plants and vegetables without soil.

A Procedure To Follow

1. Stand the pipe on end in the center of the plastic sheet.
2. Draw a circle at the center of the plastic sheet using the pipe as a template. See Figure B.
3. Mark the location for the tubing adaptor. See Figure B.
4. Drill the hole for the tubing adaptor through the plastic sheet and the pail. Figure C.
5. Slip one tubing adaptor through the plastic sheet and one through the pail.

Make certain that you place rubber washers on both sides of the fitting to prevent water leaks.
6. Tighten the nuts to hold the tubing adapters in place.
7. On the inside of your plastic sheet, place a circle of hot glue around the tubing adaptor.
8. Press your screen guard into the hot glue. See Figure D on the next page. This protector will prevent sand from escaping from your planter when you remove

extra nutrients from your planter.
9. Place the pipe on a ¾-inch piece of plywood so that its ends are raised off the work table. For best results, place stop blocks into the plywood to keep your pipe from rolling. See Figure E on the next page.
10. The plastic sheets that you will attach to the ends of your pipe will seal the ends so that the pipe can hold water and also serve as your stand. Check the

B.

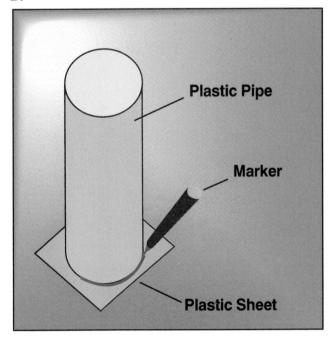

Plastic Pipe

Marker

Plastic Sheet

C.

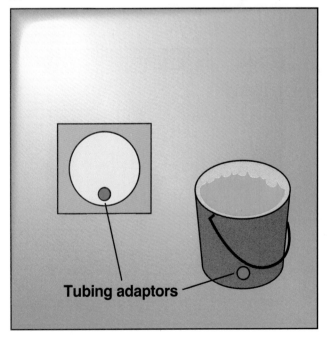

Tubing adaptors

alignment of your parts before applying glue. Ask another student to help you glue these parts together.

11. Coat both pipe ends with plastic glue. Attach both plastic pieces so that they are straight and even. Make certain that one edge of the plastic is resting squarely on the table so that your finished planter won't rock.

12. Hold or clamp the parts together until the glue sets.

Carefully set the project aside to dry until the following session.

13. Mark the location for your plant holes. See Figure F.

14. Drill through the plastic pipe using the hole saw mounted in the drill press. See Figure G.

15. Attach flexible tubing to the planter and the pail. Use small hose clamps to lock tubing to adapters.

16. Fill the planter with water to check for leaks, and repair where necessary.

17. Partially fill the planter with sand.

18. Remove plants from pots and carefully wash away soil.

19. Carefully place your plants into the planter; filling in sand where needed.

20. Tap water contains chlorine. Let the water stand for two days prior to use.

D.

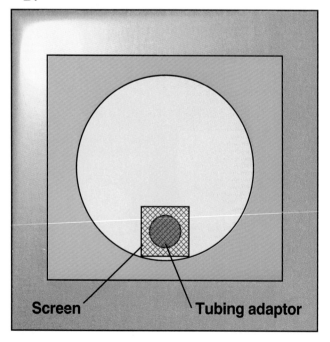

Screen   Tubing adaptor

E.

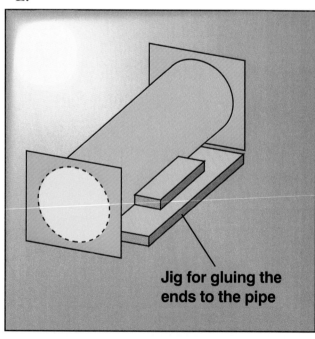

Jig for gluing the ends to the pipe

21. Test the pH of the water. It should read 6 or 7 for best plant growth. Add a few drops of baking soda if pH is low or a few drops of vinegar if pH is high. Then retest your solution.

22. Partially fill the pail with a nutrient-water mixture. Place the pail higher than the planter so that the mixture fills the planter. Remember that it is better to have too little solution and have to add more than to flood the room.

23. Lower the pail so that the extra solution drains from the planter.

24. Check daily to see that the sand is moist. Add water to the pail to replace water lost from evaporation.

25. Refer to a reference book such as *Hydroponics Soilless Gardening* by Richard Nicholls to learn how to care for your plants. This book should be available at your school or public library.

What Did You Learn?

1. Why is hydroponic gardening important?
2. To glue the ends to the pipe, you used a wood board. What purpose did this board serve?
3. In what countries will you most likely find large commercial factories engaged in hydroponic farming?

F.

G.

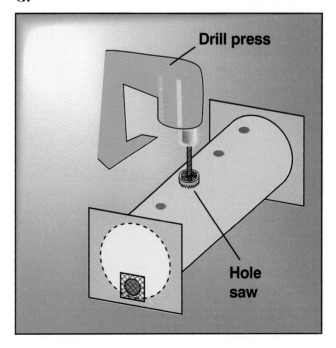

Drill press

Hole saw

Chapter 16

Tomorrow's Technology

Throughout history, people have tried to predict the future. Most people expect the future to be pretty much like the present—only much more advanced. The future, however, is often completely changed by the development of totally new ideas. For example, when drums were used for communication, people expected that future technology was going to improve the drum rather than invent the telegraph, telephone, radio, and television.

Futurists are people who predict the future. Let's play the futurist's role and try to predict what our technology will look like in the future. But remember history's lesson. At any time, unexpected breakthroughs can create the same kind of revolutionary change that was brought about by the wheel, plow, printing press, telephone, automobile, airplane, television, and computer. Also remember, if technology continues to grow at its present rate, our print information alone will double every eight years. This means that what we now know will only be 1 percent of our knowledge in the year 2050.

Let's look at some of the changes on the computer-aided drawing boards of the present. They represent many of the devices that you will be using in the future. This means that you can expect to see some of these new-tech achievements by your next birthday, others by the first part of the 21st century.

Fuzzy Thinking Computers

Computers of the past have worked on a *go, no go* technology. Everything has been seen as a yes or no. Fuzzy Logic computer systems now exist. These machines are programmed to mimic human thinking.

Current uses of this new technology allow you to shake your Panasonic video camera and still get perfect movies that look like they were taken on a tripod. The autopilot on the B1 bomber uses Fuzzy Logic to control the plane. Fuzzy Logic systems are also being used by alarm control systems to screen out false alarms, by cameras to focus your pictures, and by IBM's Deep Blue to win chess matches.

Today, Fuzzy Thinking systems will slow down the warming or cooling of your car. In the future, they will determine washing machine running time based on the computer's interpretation of dirt levels in clothes. They will also make transportation systems carry you without feeling a sensation of motion, and iron out the bumpy, angled lines on your computer screen drawings and video games.

The Future of Communication

What inventions will be used for communication in the future?

The printing press, telephone, television, computer, camera, video camera, VCR, and CD player are but a few of the many communication devices of the present that you have already studied. These devices are now evolving into one integrated super communication machine that you will use in the future.

To complete this evolution, each of these communication systems must process and transmit data in the same way. The graphic arts industry today produces a lot of its text and illustrations on computer-based digital machines. The telephone, television, VCR, video camera, and still camera are all evolving into digital systems.

Telephones of the Future

What will the telephone of the future look like?

To move your phone conversation on fiber-optic cables, phone signals must be converted from electrical analog signals to digital pulses of light. At this time these fiber-optic signals must be converted back into electrical signals. The company that invented the transistor in 1947, Bell Labs, is now Lucent Technologies. In 1997, they developed new Digital Signal Processor (DSP) chips that translate video and speech into digital and analog signals. These chips can convert the signal back and forth between digital and analog to meet equipment needs.

Cellular telephone manufacturers are now building new digital

Personal Computer at the Cutting Edge

What do you like least about using a computer? Most people answer "Typing," and "Remembering computer commands." For almost the last 30 years technologists have been trying to develop a computer that works without a keyboard.

If you could design the ideal computer, how would you communicate with your machine? In the *Star Trek* movies, the captain and crew communicate with their computers using a keyboard, talking directly to the machine, or using a clipboard with a special pen. The clipboard computer that uses a special electronic pen and accepts handwritten notes is now on sale along with Personal Digital Assistants (PDAs), voice recognition software, and very tiny watch-size computers that can receive beeper messages.

The introduction of the keyless computer doesn't mean that you should no longer learn how to use a computer keyboard. Futurists expect the computer keyboard to co-exist with verbal and handwritten instructions on the computers of the 21st century.

How does the new clipboard PC operate? The keyboard has been replaced by a large liquid crystal display and an electronic pen or plastic stylus.

How do these computers work? One PDA manufacturer covers its liquid crystal display with a clear glass sheet. The sheet is coated with an electrical conductor that runs as fine horizontal and vertical lines across the screen. A small electrical charge is sent across the glass vertically and horizontally. This has the effect of covering the glass with a fine electronic graph paper. The voltage varies at every point on the graph that covers the screen. When the pen touches the screen, the pen acts as a volt meter, reading the voltage at the pen's exact location. As the pen is moved across the screen it sends the voltage readings to the CPU, which tells the computer the location of the pen. The CPU lights up each spot that the electronic pen has touched.

The location of the pen is checked hundreds of times per second, and the information is quickly updated on the screen. To convert what has been drawn on the screen into useful information, the computer has been programmed to look for points where the pen is lifted off the screen for at least $\frac{1}{3}$ of a second. The writing between each lift is a printed letter, number, or punctuation mark.

The computer software now has the job of recognizing the pattern drawn and converting this information into a digital signal. This software contains hundreds of thousands of different versions of our English alphabet and it uses all of these variations to identify which letter, number, or punctuation mark was drawn.

The Palm Pilot PDA has special games. These games help you learn how to draw its graffiti style of on-screen lettering. PDAs also have an on-screen keyboard. Here, you just touch the letters that you want.

networks that are operating alongside the analog system. See Figure 16.1 on the next page. The old cellular system will continue to exist for current analog phone owners. The two systems will not, however, be compatible, and in time the analog system will be dropped. Many digital phones can operate as analog phones where digital service isn't available.

Figure 16-1 ■ (A) This Seiko watch is also a pager, an e-mail receiver, and a sports-score reporter, and it gives up-to-the-minute weather reports. (B) Today's smallest cellular phone weighs around 3.1 ounces and comes in both analog and digital versions. (C) "A" and "B" will soon be combined to create this product, which will have a small video screen for real videophone and Internet communication.

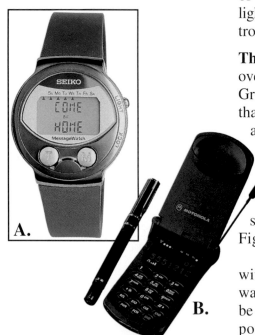

A.

B.

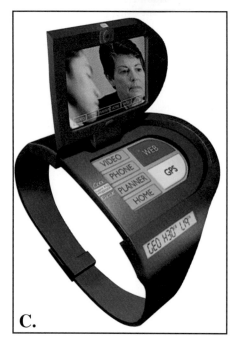

C.

The Optical Computer. The first digital **optical computer** has already been built by AT&T. An optical computer uses laser signals and fiber-optic cables instead of electricity. At the present, optical computers are still very experimental. In theory, optical computers will be able to process information at 1,000 to 10,000 times the speed of our current electronic computers.

The race to build optical computers now exists because researchers have shown that the electron that carries messages in modern computers can be replaced with photons. **Photons** are tiny units of light. They carry more information than electrons at the speed of light. Optical computers can recognize visual patterns faster than electron-based supercomputers.

The Future of Wireless Communication. It is now estimated that over 200 million people use some sort of wireless communicator. Growth in this area of communication is so rapid that it is estimated that 800 million people will use a wireless communicator by 2000 and over one half of all communication will be wireless by 2005.

One of the newest innovations in video display technology is a pair of very light Virtual Retinal Display (VRD) glasses that use a beam of colored light to paint a video image directly on the retina of your eyes. In this case, your eyes are the screen and the VRD is the equivalent of a movie projector. See Figure 16.2.

Your palm-size computer cellular videophone will provide you with instant wireless communication and information. The wristwatch telephone, first introduced in the Dick Tracy cartoon, will be built and include a video image. In time this unit will contain the power of your palm-size computer and still be small enough

Figure 16-2 ■ The Virtual Retinal Display (VRD) paints images directly onto the retina of your eye using a color beam of light.

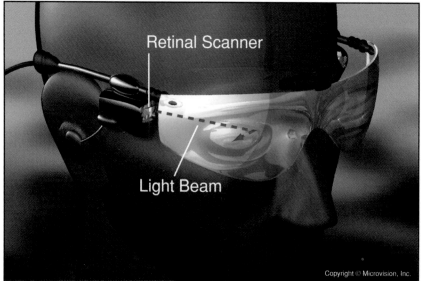

Retinal Scanner

Light Beam

Copyright © Microvision, Inc.

492

to wear as a watch or pin. All computer commands and responses will be verbal, and watches will automatically reset their times.

You will no longer have a separate telephone number for your home, office, and car. One telephone number will serve you at all times. The time, date, and name of connected parties will be automatically logged for your phone bill.

You will be able to use any phone in any office, restaurant, or home and have individual calls billed directly to you. In time your voice print and other biometric measurements will be your only needed identification.

The Future of News Media

What will your future newspaper look like?

Your newspaper of the future will be a personal edition containing only the features that you want to read. Your paper could appear in a number of formats, depending on your own personal preferences.

In the video format, a *computer-generated* person with the personality and characteristics that you like will present the news. Your news show can be a two-way conversation. You ask questions or make comments, and the computer news reporter responds to your statements.

If you want to read your news, it can be fed to your computer screen in a text and illustrations format. You will also be able to ask the computer for more information whenever you want. All illustrations will appear as miniature motion pictures. This newspaper as well as books will be delivered from cyberspace through an Internet II provider.

— Learning Time —

Recalling the Facts

1. What prevents us from making accurate long-range predictions about the future of technology?
2. When the telephone was introduced, why did people think that a million or more telephone operators would be needed to handle telephone switchboard operations?
3. Name and describe two different newspaper formats that will exist in the future.

Thinking for Yourself

1. Dramatic change can be brought about by the development of new technology. Describe some of the changes that were brought about by the development of the wheel, plow, printing press, telephone, automobile, airplane, television, and computer.
2. What do you believe was the most important technological development of the twentieth century? Why?

(Continued on next page)

The Future of Transportation

What will transportation be like in the future?

Trains, planes, and automobiles are the transportation vehicles of our recent past, present, and future. See Figure 16.3. All of these methods of transportation will continue to evolve. The quickest evolution will occur in the areas of guidance systems, fuel, and speed.

Figure 16-3 Trains, planes, and automobiles will continue to provide our transportation for the foreseeable future.

The Smart Systems of Tomorrow

What is a transportation smart system?

The automatic pilot of the past kept a plane on a particular heading at a preset speed. Today and tomorrow's *autopilots* are called **fly-by-wire systems**. In these systems the computer gets its say in almost every move the pilot makes.

The new Boeing 777 airliner has the newest, most automatic fly-by-wire commercial system. See Figure 16.4. The system on this plane can actually refuse to carry out a pilot's command. The pilot can't turn off the system.

In most planes the pilots use a large control wheel or stick to control the plane. In new fly-by-wire systems, pilots use a joystick that resembles a game machine controller, and their controls are all television monitors. See Figure 16.5 on the next page.

In the 1960s, 10 percent of the cost of an F4 fighter airplane was due to its electronic equipment. The newest military aircraft is pilotless. In combat, the plane can be shot down, but the pilot can't be hurt because he or she is located in a special remote flying simulator far from the battlefield.

The next move in fly-by-wire commercial airline technology will be the reduction of flight crew—and finally no flight crew at all. By then, total control of the aircraft will be given over to its computers and on-the-ground flight controllers.

This same technology is destined to be incorporated into your family automobile. In 1997, Buick demonstrated a car that could drive itself. This was done on a special section of road on the San Diego Freeway. This system is called *hands-off driving*. The Buick's **object avoidance system** kept the car very close to the other smart cars that

fly-by-wire system
a computer controlled method of flying aircraft

object avoidance system
an advanced automobile system that enables a vehicle to avoid hitting objects

Figure 16-4 Through the use of computer monitoring and control, new technology will simplify the controls that are shown in this Boeing 777 commercial jet cockpit.

495

magnetic levitation
a system based on the principle that same poles of a magnet will repel, or push, the other away. This repulsion is used to lift objects so that they rest on a cushion of magnetism.

maglev train
a rail system in which the vehicle rides suspended on a cushion of magnetism

Maglev—Moving Beyond the Test Track

The Japanese are now testing a fully functional 11.5-mile **magnetic levitation** system. As this system passes operational tests, the line will be extended so that it eventually connects Osaka and Tokyo. The Japanese Railway Technical Research Institute's maglev vehicles are designed to travel in U-shaped magnetic repulsion guideways (Figure 16.6) with no physical contact with the guideway walls.

To accomplish this task, levitation coils have been placed on the sidewalls of the guideways. The train is the superconducting magnet, and as it passes these induction coils, it induces the levitating magnetic field. The electromagnetic forces then push and pull the train in the middle of the guideway. The repulsion and attraction forces of magnetism are so reliable that these trains will be able to travel at speeds that exceed 300 miles per hour.

In 1997, the train shown in Figure 16.7 set a world train speed record, with passengers, of more than 300 miles per hour. It is assumed that a maglev train will, in time, be capable of breaking the sound barrier.

To build this **maglev train**, it was necessary to build a relatively light superconducting magnet and to find a cost-effective method of lowering the magnets' working temperatures into the cryogenic range (for this maglev system, -269°C). The magnet is made out of material that loses all electrical resistance at a temperature that can be sustained by liquid helium. Once energized, the superconducting magnet doesn't lose its electrical charge for a very long period of time. How long would the batteries in your CD player last if the entire device worked without any electrical resistance? Did you consider indefinitely as a possible answer?

The engine-and-braking system for this maglev train also uses magnetic repulsion and attraction. Alternating current energizes the propulsion coils that are located on the walls of the guideway. These coils set up a shifting field that actually propels the maglev. The computers that control the system can use this same force to accelerate or stop the train.

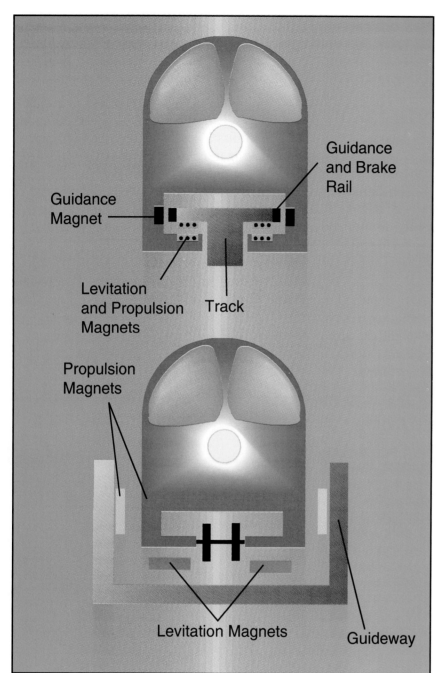

Figure 16-6 Maglev trains are suspended in air using magnetic attraction (top illustration) and magnetic repulsion (bottom illustration).

Guidance and Brake Rail

Guidance Magnet

Levitation and Propulsion Magnets

Track

Propulsion Magnets

Levitation Magnets

Guideway

were traveling together. The system instantly reacted to the slowing down or speeding up of the other cars.

Are you now looking forward to learning how to drive? Don't worry. You will reach driving age before the car completely drives itself.

One of the first *drive-by-wire car systems* (called **smart highways**) is an electronic navigation system that will direct you to your destination. General Motors has recently introduced such a system called OnStar.

The next system for the automobile will be an accident avoidance system. The Honda Automated Cruise Control uses a laser to judge distance to the next vehicle. At first, these systems will only adjust the car's speed if an object is getting too close. Honda expects this avoidance system to be a new-car option in 1999. In time, object avoidance will become part of a navigation system that will automatically handle all driving functions. At that time, your destination will be your only driving command. Once given, the car's computers will determine the route of least traffic by communicating with GPS satellites. Then your smart car will take you where you want to go.

Starting in 1998, new air bags will release with less force and have a shut-off switch for the passenger seat. Beginning in 1999, some new cars will have smart air bags. These air bags will determine the weight and size of the occupant in the seat. A child in the seat will create a slow release and an empty seat will shut off the air bag.

■ **Figure 16-7** ■ The fastest trains in the world are now maglev. The train pictured set a new world's speed record of 341 miles per hour on December 24, 1997.

Future Fuels

What fuels will power the vehicles of the future?

Because of our current energy crisis and growing concern for the environment, we will surely increase our efforts to find and develop alternate fuels. During your lifetime you can expect to see a number of developments grow out of stricter government standards and a wealth of current research into alternate fuels.

The Solar-Powered Vehicle. Do you own a watch or calculator that runs on solar batteries? Why can't auto manufacturers build solar cars for everyone? To manufacture solar cars commercially we need a breakthrough in solar panel research.

Electric Vehicles. Saturn dealerships started selling the first true production model electric vehicle (EV1) sports car in 1996. This electric car has a range of 90 miles before its lead-acid batteries need a 15-hour charge using a standard 110-volt outlet or a 3-hour charge using a 220-volt outlet. EV1 owners will save two-thirds of their current gasoline dollars, but they will have to plan on replacing the $1,500 battery package every three years.

Gas/Electric Hybrid Cars. Wouldn't it be nice to own a car that gets 70 miles to the gallon, even if all of your driving is in rush-hour traffic? Toward the end of 1998, Toyota is planning to start selling a gas/electric hybrid car in Japan. They don't plan on selling this car in the U.S. until at least 1999.

In this vehicle, the 1.5-liter gasoline engine uses its power to drive the wheels, to power the electric generator to run the motor, and to recharge the battery system. When the car is in stop-and-go traffic, all the power to the wheels is supplied by the electric motor, which is still run by the engine's electric generator. During deceleration, energy is recovered from the turning wheels as they take their turn running the electric generator to recharge the battery system. American car manufacturers are also working on a hybrid gasoline engine/electric motor vehicle. See Figure 16.8.

The Hydrogen-Powered Vehicle. Hydrogen and oxygen are created when an electric current is passed through water. When hydrogen is burned, the by-product of the burning process is water and a little nitrogen oxide. Hydrogen would be a perfect fuel if there was a safe and convenient way to store it in the car. Figure 16.10 (p. 501) shows an experimental hydrogen-powered Mercedes.

■ **Figure 16-8** ■ In this vehicle, the gasoline engine generates electricity to run the electric motor. When needed, they both turn the wheels to make the car go. Energy is recovered when you apply the brakes. This energy is used to recharge the battery.

Wireless Electric Power

What if we launched very large solar-powered satellites into space. Couldn't we then transmit their power back down to earth for our own consumption?

The idea of sending electric power through the air over long distances has been a focus of experimentation since Nikola Tesla demonstrated a remote (wireless) electric transmission tower in 1899. The Tesla coil that he invented did transmit electricity into the air.

The National Aeronautics and Space Administration (NASA) is now working together with Texas A&M University and twenty-five companies. Their research and development aims to improve current transmission technology that will lead to the transmission of power from space.

By 2023, they hope to have a working system that can be used to power satellites and a space station by remote transmission of electricity. See Figure 16.9. Currently, each satellite sent into space carries its own batteries or solar panels. Once energy is ready to be beamed through space, individual satellites would get their power from these space utility companies rather than from their own electric generators.

Canada has invested $20 million to design an experimental drone (radio-controlled) airplane. This plane is to fly in a circle 70,000 feet off the ground, powered by a remote transmitter.

Raytheon, the company that makes the Patriot Missile, successfully powered a helicopter using remote transmission of electricity in 1964. They discontinued this research when the first geosynchronous orbiting satellite was launched.

A utility company in space now seems practical. The technology for remote transmission of electricity is now possible. Your utility company will buy and then sell electricity to you if it is cheaper than the power they could make themselves.

This technology is not being developed as quickly as possible. The U.S. Congress worries about cost. Other people worry about birds that might fly through the microwave beam that sends the electric power down to earth.

Space Cannons and Space Vehicles

How will future satellites be launched into space?

You just can't beat the fiery image of a rocket being launched into space. See Figure 16.11 on page 502. The energy needed to reach the breakaway speed, freeing the rocket from the earth's gravity, demands very large engines. The heavier the ship, the more powerful and fuel-consuming the engine must be.

Figure 16-9 This futuristic space station will someday transmit electric power from space.

Figure 16-10 This experimental vehicle is powered by hydrogen, a promising fuel of the future.

The Jules Verne Gun is named after the fictional gun described in the 1865 novel entitled *From the Earth to the Moon*, by Jules Verne. This NASA project will use a gas gun with multiple explosions of power as the projectile moves through the gun tube. The goal is to reach 4.35 miles per second, which is enough to launch a 10-ton vehicle into low earth orbit. The Jules Verne Gun will be so powerful that it will subject a satellite to 1,000 times our earth's gravity within one second of firing. This means that all satellite parts must be able to withstand the pressure, and human launches would be impossible.

The space cannon in the future will launch vehicles into space without the need of conventional fuels or large rocket engines. The cannon itself will be used over and over again.

— Learning Time —

Recalling the Facts

1. Describe how a fly-by-wire system works.
2. What is the purpose of an automobile avoidance system? How would it work?
3. What fuel might future cars run on?
4. What is a space cannon?

Thinking for Yourself

1. What would be some of the advantages and disadvantages of pulling out the flight crew in future fly-by-wire systems?
2. Why can't automakers manufacture solar-powered cars for today's auto market?

(Continued on next page)

Applying What You've Learned

1. Construct a small solar-powered electronic device.
2. Construct a rubber-band-powered vehicle or wind-powered vehicle to race against others in your class. The racers will be judged on appearance, speed, and distance.

The Future of Biotechnology

What new discoveries can you expect in the area of biotechnology?

You can expect to see many advances made in the fields of health care and bioengineering. Studies have shown that bone healing can be accelerated up to 25 percent by exposing the break to ultrasound. By the turn of the century, small ultrasound devices will be placed into the cast of a broken limb and programmed to stimulate bone healing. Your dentist will use a laser to painlessly remove cavities from your teeth. See Figure 16.12.

To protect our food supply, the FDA has approved gamma radiation of fruits, vegetables, pork, and chicken. See Figure 16.13. Beef irradiation was approved in 1997 to kill harmful bacteria now found in beef.

New machines are now being designed that will help doctors diagnose medical problems. The computers in these machines will compare the data supplied by your body against its own memory. If something is wrong, the computer will suggest possible illnesses.

New machines have been developed that can locate people through walls and other obstructions. These machines can help rescuers find survivors in collapsed buildings. Other new systems can search people for weapons and illegal substances as they pass by a wall-mounted detector.

A new artificial heart's design is shown in Figure 16.14. Its power system is an electric motor that drives two plastic sacs. The battery pack feeds power to the orange-size artificial heart through electric induction.

Why build an artificial heart when real heart transplants are so successful? There just aren't enough real hearts and other organs to go around. In time, most of nature's body parts will be duplicated through bioengineering. The artificial heart in Figure 16.14 will begin FDA-approved testing in people in the year 2000.

■ **Figure 16-11** ■ Current technology uses very large rocket engines to provide the thrust needed to break free of earth's gravity.

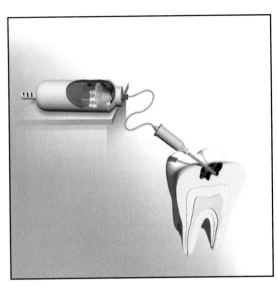

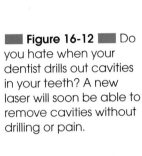

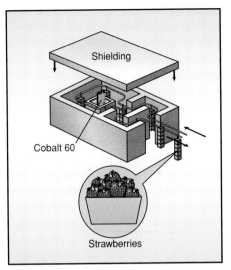

Figure 16-12 Do you hate when your dentist drills out cavities in your teeth? A new laser will soon be able to remove cavities without drilling or pain.

Figure 16-13 Many people are afraid of food that has been exposed to gamma radiation. This process is often called *cold pasteurization*. It leaves the food free of bacteria and radiation.

New materials are making it possible for replacement parts to be constructed so they give back feelings of touch, pressure, and temperature. See Figure 16.15. This means that future artificial hands and legs will move under the direct control of a person's brain. Not only will these bionic parts move more naturally, but they will look and feel real to the touch. New artificial skin, muscles, and bones will duplicate the look and feel of real body parts.

Amputees will have permanently bonded replacements that will give them the ability to experience temperature changes, tickling sensations, and even pain. The technology of the future will allow the approximately 2 million amputees in America complete use of their bodies.

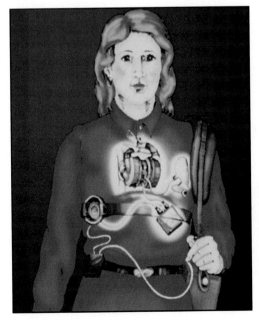

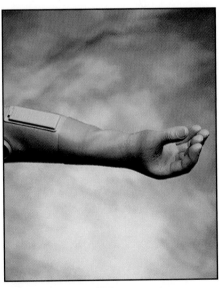

Figure 16-14 This artificial heart will be ready for human trials in the year 2000. If these tests are successful, it will be ready for general use in the year 2005. At this time, this artificial heart is seen as a stop-gap measure until a human heart can be found.

Figure 16-15 New artificial limbs can provide the feelings of touch, pressure, and temperature.

503

The Future
of Manufacturing

How will manufacturing in America change in the future?

Automated mass production will provide the goods and services that we need and want in the foreseeable future. For U.S. manufacturers to compete in our global economy, they will have to produce higher-quality products that can't be produced in low-wage nations.

Manufacturers will survive in America by creating new automated systems that operate at the cutting edge of technology. These systems will keep manufacturing costs below the level of low-wage nations by building the products faster and using less labor. See Figure 16.16.

As this new technology spreads to other parts of the world, the same manufacturing systems will be used in low-wage nations. Therefore, future American manufacturers will need to continually

Figure 16-16 This automated factory has computers guiding the operation of the entire factory. Can you find the people who are monitoring this entire system?

speed up the introduction of new products and services, or Americans will have to work for low wages.

The future will continue to see the displacement of many low-skilled workers by robotic machines that can manufacture products in the dark. This is called *lights-out manufacturing*. In the future, manufacturers will continue to create joint ventures with their competitors. This is called *virtual manufacturing*.

Manufacturers will also build their factories in outer space. Without the earth's gravity and atmosphere to get in the way, they will manufacture optical glass, new metal alloys, and other products that can't be made on earth. The need for constant change in manufacturing is now called *re-engineering the workplace*.

The Future of Construction

What new methods of construction will exist in the future?

Some cities of the future will be built underground. Others will be anchored or floating islands in our oceans. Still others will be constructed in outer space and on other planets. New materials and techniques will be developed to meet the specific needs of construction projects in these alien environments.

The Confederation Bridge in Canada was completed in 1997. All parts of this bridge were built on land to ¼" tolerance using information supplied by GPS satellites. When assembled, everything fit together exactly as planned.

New construction will also take advantage of new building materials such as the wrapping of concrete with reinforced carbon fiber for earthquake strength, geothermal heating and cooling systems, and new systems of natural lighting. Builders will increase their productivity by using advanced systems of modular, prefabricated construction. The buildings of tomorrow will be built on foundations of knowledge that date back centuries.

Your Place in Tomorrow's Technology

Where do you see yourself in the future?

When you dream about the future, what role do you see yourself playing? Are you an explorer? A scientist? An engineer? A secret agent?

Tech Talk

A new process has been developed that changes the structure of ordinary sawdust. The treated sawdust repels water and absorbs oil.

The manufacturer of Sea Sweep claims that its product can be spread on oil spills. One ton of this product should cost about $1,800 and be able to absorb 1,500 gallons of crude oil. When the sawdust absorbs its fill of oil, it congeals into floating balls that can be collected from the surface of the water. These balls can then be burned as an industrial fuel.

Your future is linked to the future of technology, regardless of the role you will play. You could become the builder of tomorrow's newest devices or a consumer of the products that will be created by one of your classmates. So continue to dream, for tomorrow's technology will come from your dreams of our future world.

■ Putting Knowledge To Work ■

The Martian Rover

The *Sojourner* Mars rover was designed to withstand the Martian night temperature of -100°F. Its suspension system is stable even if the vehicle climbs a rocky surface that has bumps that are as big as the rover is high. See Figure 16.17. Each wheel has independent drive and steering control, and under certain conditions, the *Sojourner* has been given a mind of its own.

On Mars, the rover is so far away from earth that it takes minutes for messages to travel between the two planets. For this reason, some of *Sojourner's* physical movement must be controlled by artificial intelligence. When in danger, it must override previous directions and act like a common insect.

To keep costs down, *Sojourner* wasn't given an artificial intelligence system that needs a powerful computer in order to work. For one thing, its builders couldn't be sure what situations their rover might roll into. So instead of programming the rover to handle particular situations, they programmed it to act as if it was an earth insect.

At MIT's Artificial Intelligence Laboratory, Rodney Brooks created the artificial intelligence that was based on bug responses to danger. He built a robotic insect named *Attila* that laid the foundation for developing the program that controls *Sojourner. Sojourner* and *Attila* handle their world in an entirely new way for robots, but a very old way for insects. Using small motors, tiny computers, and sensors, they move and react to stimuli with the intelligence of a small insect.

Attila and its friends have automatic responses that take over when the robot's sensors indicate a need. These responses include moving forward, moving back, and tracking prey. These responses aren't controlled by a central brain. Each stimulus causes a reaction. The reaction listed as most important for survival has priority and gets control over the motors that control motion.

When *Attila* senses a clear area ahead, a message is sent to its legs, and it walks forward. If another sensor suddenly sensed danger, it would send a danger message to *Attila's* legs. Both messages would now be received by the legs. The danger message has priority, and the little robot would run out of harm's way.

Today, *Attila's* technology helps *Sojourner* explore the surface of Mars. It does this without having any idea of what it is doing. *Sojourner* performs its work with the same instincts found naturally in the insect world.

People will someday develop the knowledge needed to help these little creatures evolve into smarter thinking machines—in the same way that we have evolved from our less-intelligent ancestors.

In the future, *Sojourner* technology coupled with GPS information might help the visually impaired walk among us using a robotic seeing-eye dog. This robot will have the intelligence to take its master to voice-given locations. This technology might also allow a robot to clean your home, mow your lawn, and even assemble stations in outer space.

Figure 16-17 Although small in size, the Martian rover called *Sojourner* was built tough to withstand the Mars alien environment and to avoid dangerous obstacles.

— Learning Time —

Recalling the Facts

1. How can people be searched for weapons and illegal substances?
2. What powers the new experimental artificial heart? How is this power transmitted to the unit?
3. What is *Sojourner*? What kinds of things will it do in the future?

Thinking for Yourself

1. How will developments in technology affect your future occupation?
2. The most notable negative effect of twentieth-century technology has been its effect on our environment. What do you think will be the most notable negative effect of technology during the first quarter of the twenty-first century?

Applying What You've Learned

1. Break up into teams. Construct a Lego vehicle that generates motion using legs instead of wheels. Attach a line between two such vehicles for an old-fashioned tug-of-war to see which team has the most powerful insect robot.
2. You have now studied the past, present, and future of technology. Design and construct a project using the tools, materials, and processes of technology. This project can be anything you want it to be. For example, you can build a futuristic bird feeder, speaker cabinet, desk organizer, note holder, or computing device.

Summary

Directions

The following sentences contain blanks. For each numbered blank, pick the answer that makes the most sense in the entire passage. Write your answer on a separate sheet of paper.

People who use current trends to predict the future are called futurists. Predicting many of the ___1___ changes that will occur during the 1990s only requires a look at the drawing boards of the present. As we look further into the ___2___ , predictions are more likely to be altered by major new ___3___ .

In the future all our present communication systems will join into one ___4___ system. How will the printing press, telephone, television, computer, camera, video camera, VCR, and CD player work as one machine? Your personal video computer/phone will allow you to ___5___ and talk to the person or machine that is on the other end of the call. Information and personal calls will be sent to your ___6___ and stored for your later use.

When you want to watch the news, this machine will provide you with a two-way communication system. You will request information from a computer-generated ___7___ . A newspaper and any other document will be available to you by computer ___8___ .

In the future you can expect the following changes in our ___9___ system. Planes will fly under the direction of on-board and on-the-ground ___10___ . This system already partially exists and is called flying-by-wire. On the Boeing 777 airliner, the ___11___ can refuse to carry out the pilot's ___12___ .

Fly-by-wire systems for the automobile are called smart highways. In time your only command to your car will be your ___13___ .

Future trains will travel at very high speeds suspended in air by magnetic levitation. By the first quarter of the ___14___ century, magnetic levitation will carry these trains along at over 600 mph. This same technology will be used to launch payloads into outer space. An ___15___ magnetic repulsion super cannon has already been built.

1.
   a. past
   b. old
   c. strange
   d. technological

2.
   a. future
   b. past
   c. present
   d. classroom

3.
   a. problems
   b. wars
   c. famines
   d. discoveries

4.
   a. stereo
   b. super
   c. old-fashioned
   d. television

5.
   a. touch
   b. feel
   c. see
   d. sense

6.
   a. house
   b. job
   c. car
   d. machine

7.
   a. salesperson
   b. sportswriter
   c. newspaper
   d. magazine

8.
   a. analysis
   b. printout
   c. virus
   d. disk drive

9.
   a. manufacturing
   b. biomedical
   c. transportation
   d. communication

10.
   a. pilots
   b. simulators
   c. stewardesses
   d. computers

11.
   a. plane
   b. train
   c. bus
   d. cruiser

12.
   a. demands
   b. thoughts
   c. requests
   d. instructions

13.
   a. destination
   b. address
   c. telephone number
   d. name

14.
   a. twentieth
   b. twenty-first
   c. nineteenth
   d. twenty-second

(Continued on next page)

Continued

Future electric power will be beamed down to earth by giant solar collectors located _____16_____. The principle has already been tested by beaming electricity up to a radio-controlled helicopter.

Today's gas-powered cars will also become a thing of the past. Future cars will be powered by hydrogen or solar energy.

The world of the future will have artificially _____17_____ tiny machines that clean your home, mow your lawn, and even build space stations.

Medical advancement will continue to improve the quality of _____18_____ life. Bioengineers will build lifelike bionic parts to replace most body parts that are _____19_____ at birth or injured during life.

You and your friends are destined to be the builders of tomorrow. The world of the future will someday be in your hands.

15.
a. obsolete
b. underwater
c. experimental
d. old

16.
a. in earth's orbit
b. under the ocean
c. in the desert
d. in your backyard

17.
a. manufactured
b. intelligent
c. processed
d. imported

18.
a. plant
b. animal
c. human
d. artificial

19.
a. perfect
b. defective
c. absent
d. missing

Key Word Definitions

Directions

The column on the left contains the key words from this chapter. The column on the right contains a scrambled list of phrases that describe what these words mean. Match the correct meaning with each word. Write your answer on a separate sheet of paper.

Key Word	Description
1. **Futurist**	a) Computer-controlled
2. **Optical computer**	b) Predicts the future
3. **Photon**	c) Cars under computer control
4. **Fly-by-wire system**	d) Uses lasers and fiber-optic cables instead of electricity
5. **Object avoidance system**	e) Works because same poles of a magnet repel
6. **Magnetic levitation**	f) Tiny unit of light
7. **Maglev train**	g) Senses that objects are too close
8. **Smart highway**	h) Ride suspended in air

Building a Magnetically Levitated Racing Vehicle

■ Did You Know? ■

In 1991 a contract to build our nation's first supertrain was awarded by the Texas High Speed Rail Authority. This system's 590 miles of track will allow two-thirds of the citizens of Texas quick access to five major cities. It will cost an estimated $5 billion to design and build this system that will transport passengers at 200 mph. Once in operation, it will only take an hour and a half to move commuters between Houston and Dallas.

The fastest-running train in America at this time is the Amtrak Metroliner. It has a maximum speed of 125 mph. The Texas system is the first of many American Maglev transportation system proposals to be approved. It won't be ready to start carrying passengers until 1998.

■ Problems To Solve ■

Design and construct a magnetic levitation racer and 8 inches of levitation track. Your vehicle must float over its track just like a Maglev train.

Your racer will be judged in a class competition. In this race your racer will be powered by a downward sloping track, just like a roller coaster at an amusement park. The slope of this track will be raised 10° per run. Each car will be marked for distance, speed, stability, and appearance. Other races can be run using powered vehicles and looped racing tracks.

Resources You Will Need

- **Minimum of 8 feet of Maglev Track (Kelvin Electronics no. 840433 or equal)**
- **Foam board**
- **Wood**
- **Racer hardboard platform ⅛-inch thick**
- **Four magnets**
- **Hot glue gun**
- **Hot glue**
- **Double-sided tape**
- **Markers**
- **Lucite**
- **Eight magnets to make a 10-inch display track**
- **Strip heater**

A.

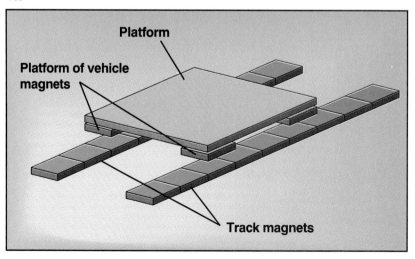

A Procedure To Follow

1. Use the Maglev track to determine which side of your magnets should be glued to the base of your car's platform. Mark that side with a marker.
2. Attach these magnets to the platform's four corners using the double-sided tape.
3. Test to see if your platform floats above the track. If any of your magnets are attached incorrectly, reverse them at this time.
4. Use the foam and other materials to build and decorate your racer. Your racer design must be light, aerodynamic, and balanced. Designers and engineers face these same problems when they design and build any transportation vehicle.
5. Test the operation of your vehicle.
6. Redesign your racer if necessary.
7. Construct a display track out of lucite for your racer.
8. Bend the lucite into a U-shaped channel using the strip heater.
9. Determine the correct polarity for the track's magnets so that the car can levitate.
10. Attach the magnets using double-sided tape.
11. Attach stops or more magnets so that your vehicle will continue to float above your display track.
12. Brainstorm the rules that your group will use to race your MagLev vehicles. Do you and your classmates want to add a power system to your racers? Does your class want to add a 360° loop to your racetrack?
13. Race and judge your vehicle.

What Did You Learn?

1. What advantages do Maglev trains have over nonmagnetic levitation trains?
2. How does your MagLev racer's system of magnetic levitation differ from a commercial MagLev system?
3. Your MagLev racer shows the principle of magnetic levitation. But what makes commercial MagLev trains move?

B.

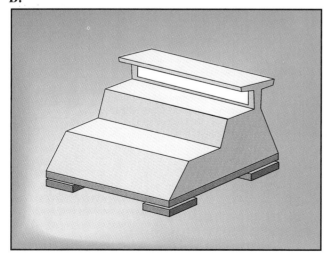

C.

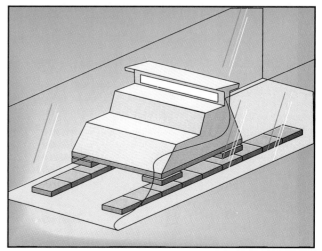

Connections

—**Key Words**—

In this chapter you will learn and study the meanings of the following important technology terms.

LIFE SCIENCE
PHYSICAL SCIENCE
EARTH AND SPACE SCIENCE
HYPOTHESIS
SCIENTIFIC THEORIES
SCIENTIFIC LAWS
END EFFECTOR
HYDRAULICS
PNEUMATICS
PSI
DENSITY
BUOYANCY
INFORMATION AGE
SHADOWING PROGRAM

In 1912 on its first voyage, the *Titanic,* a ship that everyone thought was unsinkable, hit an iceberg and sank. In the 1997 movie *Titanic,* the special effects were so fantastic that you felt you were watching the sinking of the real ship.

In a very dramatic moment in the movie, the builder of the *Titanic* informs the captain that the ship will sink. The man's exact words are, "It is mathematically inevitable that this ship will sink." How could math determine the fate of such a large vessel?

What role do mathematics, science, language arts, and social studies play in the development of past, present, and future technology? What role do these school subjects play in today's world of work and in your daily use of technological devices?

In this chapter you will begin to learn the answers to these questions.

—— Science ——

What is the scientific method, and what is the relationship between science and technology?

life science
the study of living organisms. Under the heading of life science, you study biology and environmental science.

physical science
the study of matter and energy. Under the heading of physical science, you study chemistry and physics.

earth and space science
the study of the earth's resources, structure, atmosphere, and oceans. It is also the study of our moon, solar system, and universe. Under the heading of earth and space science, you study mineral and material science, geology, oceanography, meteorology, and astronomy.

hypothesis
an explanation that can be tested

scientific theories
hypotheses that can stand up to rigorous scientific experimentation and investigation

scientific laws
scientific theories that can be expressed mathematically and have stood up to repeated experimentation and investigation over many years

Scientists working in the fields of **life science, physical science,** and **earth and space science** all apply the same scientific method to solve problems.

1. The scientists use observations to form questions.
2. Next, they gather information about what was observed.
3. They use this information to help them form a **hypothesis**.
4. Scientists then develop an experiment to test their hypothesis.
5. After completing their experiment, they carefully analyze the results.
6. They repeat the experiment to see whether they get the same results. Finally, they present their conclusions.

As they seek solutions to problems, scientists use technology as a tool for scientific discovery. Their discoveries often lead to the development of new technology. See Figure 17.1.

Scientific conclusions that have been carefully developed through experimentation are called **scientific theories**. Over time these theories will be tested many times. Eventually they may be accepted as **scientific laws**. All new technology is based on a clear understanding of the principles, theories, and laws of science.

In time scientific law can be challenged by new knowledge that was gained through the use of new technology. An example is Isaac Newton's third law of motion.

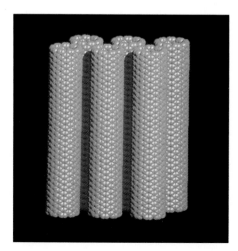

Figure 17-1 Researchers at the National Renewable Energy Laboratory in Golden, Colorado, have found a way to use tiny tubes of carbon molecules to store hydrogen gas. One day, this discovery may lead to the development of fuel-storage tanks for hydrogen-powered cars. Shown here is a computer model of the tubes.

Newton's third law of motion defines gravity as an attraction force between objects. The greater the object's mass, the greater its attraction force. See Figure 17.2.

Newton's law explains why apples fall to earth and also why the planets in our solar system revolve around the sun. Newton's law of gravitation doesn't explain why light, which has no mass, bends in a strong gravitational field. However, Albert Einstein's general theory of relativity accounts for the warping of light and time by the massive objects that make up our universe.

Can new superconductor technology change the laws of gravity? Most scientists feel that Newton's law and Einstein's theory can't be reversed. Perhaps they are wrong. A group of NASA researchers is currently using the scientific method to build an anti-gravity machine. See Figure 17.3 and read "Putting Knowledge to Work" on page 516 to learn more about this exciting research.

Tech Talk

Newton's first law states that an object will continue at rest or in motion unless acted on by a force. His second law is a formula: Force = Mass × Acceleration.

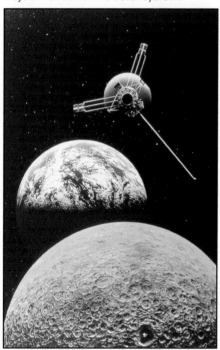

Figure 17-2 Gravity keeps the moon in orbit around the earth. The Pioneer 10 spacecraft, pictured here, escaped the pull of earth's gravity and became the first human-made object to leave the solar system.

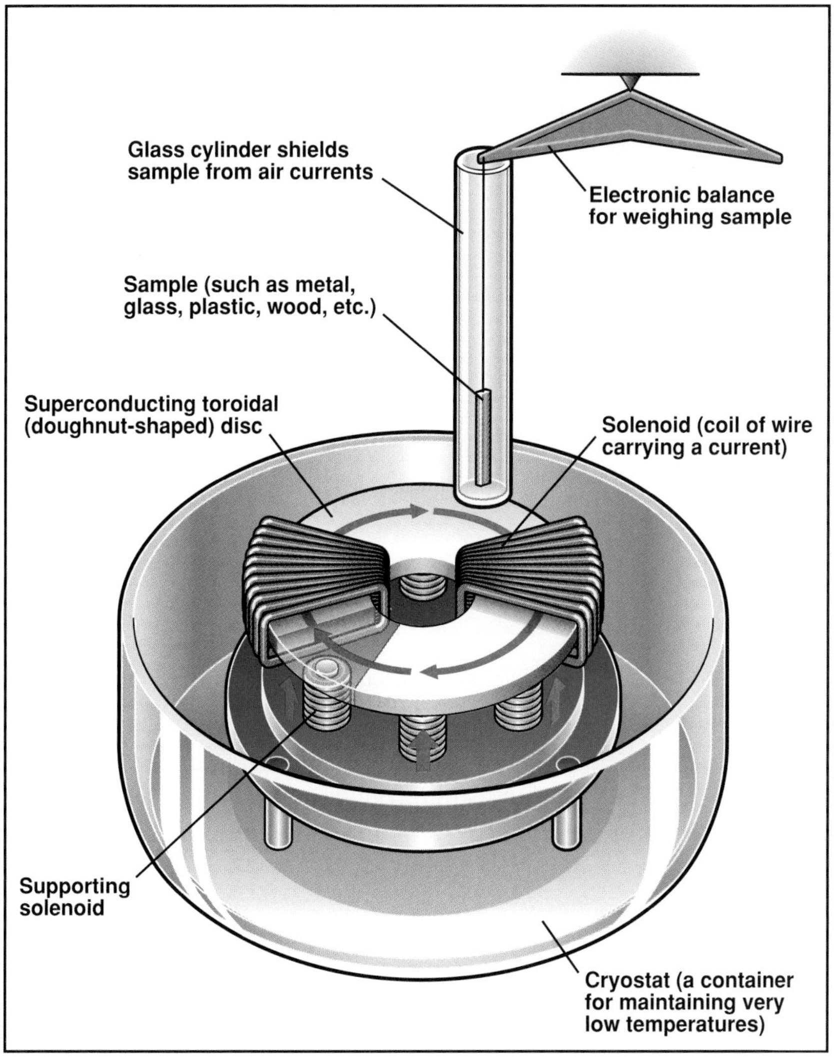

Glass cylinder shields sample from air currents
Electronic balance for weighing sample
Sample (such as metal, glass, plastic, wood, etc.)
Superconducting toroidal (doughnut-shaped) disc
Solenoid (coil of wire carrying a current)
Supporting solenoid
Cryostat (a container for maintaining very low temperatures)

Figure 17-3 This is a simplified drawing of the antigravity experiment. While the disc is spinning, objects suspended above it weigh less.

Building an Anti-Gravity Machine

The earth, moon, and stars all exert an invisible force called gravity. It is part of the glue that holds the universe together. If we could build a spaceship that would control the pull of gravity, we could explore our solar system without rocket power. We could tap the gravitational forces of the sun, moon, and neighboring planets to help us reach our destination.

To understand how our star engine would work, let's name our spaceship after Newton's famous apple. We turn on the *Apple's* engine and cancel the effect of earth's gravity on our spaceship. The *Apple* falls away from earth as it is attracted by the gravitational pull of our moon. Thus, we use the moon's gravity to launch our vehicle.

When the moon's pull can no longer help us speed toward our destination, we use the star engine to cancel the moon's gravity. Now our spaceship is attracted by the gravitational pull of another celestial body.

As impossible as this engine sounds to most scientists, researchers at NASA's Marshall Manned Space Flight Center in Huntsville, Alabama, are working to determine whether it can be built. In 1989 Dr. Ning Li, a scientist at the

Center, developed a theory that a superconductor that was rotating very fast, in a very strong magnetic field, could alter the force of gravity in its surrounding area.

At Tampere University in Finland, Dr. Eugene Podkletnov, while on leave from the Moscow Chemistry Science Research Center, performed experiments that paralleled Dr. Li's research. For four years Dr. Podkletnov conducted experiments in which he suspended objects over a supercooled superconductor disc. The disc was spinning at a very high speed while magnetically levitated (raised) over a very strong magnetic field. His data showed a variable but measurable weight loss of up to two percent of the weight of the suspended object.

Today Dr. Li is the senior research scientist of the University of Alabama's gravitational physics research team. She is helping NASA replicate Dr. Podkletnov's anti-gravity machine on a scale that could give larger results. In January 1998 the machine was more than 90 percent complete. Soon you should learn if an anti-gravity star engine will be feasible during the twenty-first century.

—— Mathematics ——

How can mathematics help us design technical products?

Few subjects are more important to technology than mathematics. Math is used to determine the flight characteristics of military airplanes, the manufacturing production rates for cameras, the operating speed of computers, and many other technical attributes. Math is often the reason products look the way they do.

For example, a narrow river might require only a simple girder bridge. A larger span, such as the Golden Gate Bridge in San Francisco, might require suspension cables. In both cases, mathematical calculations tell technologists which design is better.

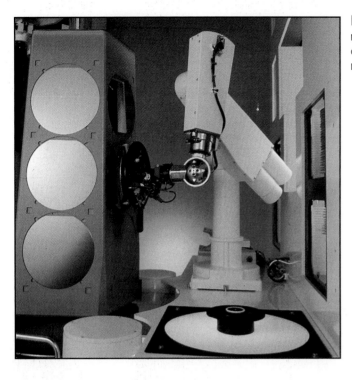

Figure 17-4 This industrial robot picks up silicon wafers and alters their electrical properties to make semiconductors.

Robots

What can mathematical calculations tell us about robots?

When the *RMS Titanic* was completed in Belfast, Ireland, in 1912, it was one of the first ships made of steel. The outer hull consisted of 1-inch-thick plates that were riveted in place. Today manufacturers rarely rivet heavy parts. They are more commonly welded together, often by industrial robots.

Industrial robots are important machine tools in manufacturing. Each robot has one arm that can be fitted with a different **end effector,** or hand. End effectors can be electric welders, paint sprayers, grippers, or other devices. Industrial robots are powered by electric motors, hydraulics, or pneumatics. **Hydraulic** robots use pressurized oil, and **pneumatic** ones use pressurized air. See Figure 17.4.

Hydraulic robots are used more often than the others. One reason is that they can lift heavier loads. Pneumatic robots operate at an air pressure of about 100 **psi**. Hydraulic robots operate at about 3,000 psi. Suppose a pneumatic robot is designed to lift up to 15 pounds. All other things being equal, a hydraulic robot could lift about 30 times as much because the pressure is 30 times higher (3,000 psi/100 psi). This hydraulic robot could lift up to 450 pounds (30 × 15 pounds).

An industrial robot can move quickly and precisely from one location to another. An arm extension speed of 40 inches per second is not unusual. How long would it take such a robot to move a part 4.5 feet? The robot could do it in 1.35 seconds! (Convert feet to inches: 4.5 feet × 12 = 54 inches. Divide this result by 40.)

Tech Talk

RMS stands for Royal Mail Steamship. The *Titanic* carried mail and cargo as well as passengers.

end effector
the part of an industrial robot that actually does the physical work for which the robot was designed

hydraulics
the branch of fluid power dealing with liquids under pressure

pneumatics
the branch of fluid power dealing with gases under pressure

psi
liquid or gas pressure measurement, in units of pounds per square inch

Figure 17-5 When it was launched in 1843, the *Great Britain* was the largest ship in the world. During its working life, it traveled 1.25 million sea miles. (Inset) The ship's designer, Isambard Kingdom Brunel, was a brilliant civil and mechanical engineer who also designed bridges and railways.

density
the ratio of the weight of an object to its volume

Construction

How can mathematics show that "impossible" structures can be built?

The world's first all-metal steamship made its first ocean voyage in 1845. The iron vessel was designed and built by Isambard Kingdom Brunel, England's greatest nineteenth-century technologist. His 252-passenger *SS Great Britain* is now on public display in Bristol, England. See Figure 17.5.

Many people scoffed at Brunel because they knew that metal sinks. Actually, it's easy to make metal float if it's properly shaped. As long as a rectangular solid has a density less than water, it will float. **Density** is weight divided by volume. The density of water is 0.0361 pounds per cubic inch (pci). Anything less dense than water will float. For example, the density of pine wood (0.0139 pci) is less than that of water. Therefore the wood floats.

Suppose you had a ¹⁄₁₆-inch-thick piece of copper cut as shown in Figure 17.6. If you crumpled it and threw it in a bucket of water, it would probably sink. But what would happen if you folded it on the dotted lines and sealed the edges?

You can solve this problem by using math. First find out how much the copper sheet weighs. You do that by multiplying the volume by the density.

- The area of the sheet is Area A plus Area B plus the other Area A.
  Area = (6 in × 2 in) + (10 in × 4 in) + (6 in × 2 in)
  Area = 12 sq in + 40 sq in + 12 sq in
  Area = 64 square inches
- The volume of the sheet is the area times the thickness.
  Volume = 64 square inches × ¹⁄₁₆ inch
  Volume = 4 cubic inches
- The density of copper is 0.324 pci. The weight of the sheet is its volume times its density.
  Weight = 4 cubic inches × 0.324 pounds/cubic inch
  Weight = 1.30 pounds (rounded up)

Now imagine that you fold the copper into a boxy boat. Determine the density of that rectangular solid.

- The box's volume is length times width times height.
  Volume = 6 in × 4 in × 2 in
  Volume = 48 cubic inches
- The density of the boxy copper boat is its weight divided by its volume.
  Density = 1.30 pounds/48 cubic inches
  Density = 0.0271 pci

The density of the copper boat (0.0271 pci) is less than the density of water (0.0361 pci). That means the small boat will float. Brunel used the same logic in constructing the *Great Britain*. It weighed 6.89 million pounds, but it was large enough to be **buoyant.**

The *Titanic* weighed 92 million pounds but had no trouble floating. Furthermore, it had a double-bottomed hull with sixteen watertight compartments. Four of these could have been flooded with no danger to the ship. When the *Titanic* hit the iceberg, six long, narrow slits opened up along the side. They totaled about twelve square feet in area and allowed sea water to rush in. The water quickly filled five of the front compartments, adding to the ship's weight and thus increasing its density. It became mathematically certain that the ship would sink.

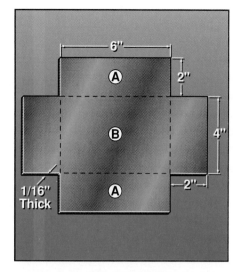

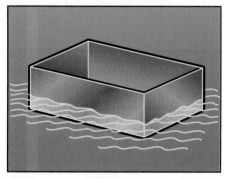

Figure 17-6 If you cut a sheet of copper and folded it (see dotted lines in drawing), you could make a small boat. Would it float?

buoyancy
the tendency of an object to float in water

— Learning Time —

Recalling the Facts

1. What characteristic is defined by Isaac Newton's third law of motion? What does it say about an object's mass?
2. What is a typical operating pressure for a pneumatic industrial robot? For a hydraulic industrial robot? Which one can lift more weight? Why?

Thinking for Yourself

1. Each of Newton's three laws of motion is at work during a Space Shuttle launch. Explain how for each law.
2. Repeat the copper boat example but change the 6-inch dimension to 10 inches and the copper thickness to $1/8$ inch. Will the boat float?

Applying What You've Learned

1. Choose five natural events and investigate how they affect technological systems in communication, production, transportation, and biotechnology. Use desktop publishing to publish your findings.
2. Investigate the details of one historic ship. Record its length, width, displacement, speed, power, etc. Some ships to consider are *Great Eastern, Mauretania, Savannah,* and *Queen Elizabeth 2.*

——— Language Arts ———

Tech Talk

If you have a very fast personal computer, you can now purchase word recognition software that will turn your spoken word into typed pages. With software for checking spelling, grammar, and literary form, you will soon be able to prepare your language arts assignments without lifting a finger.

What is the relationship between language arts and technology?

Verbal and written communication are in many ways the foundation of technology. We need communication in order to develop technology and teach it to others. In turn, technology provides the means for better communication.

The control of fire was one of the most significant developments in the history of technology. Archeological evidence indicates that our cave-dwelling ancestors improved their tool making and communication skills by firelight. Cave paintings (Figure 17.7), which are a form of written communication, were used in the same way your teacher uses an overhead projector to teach young people. The young cave dwellers were taught the art of hunting and the correct use of their hunting tools.

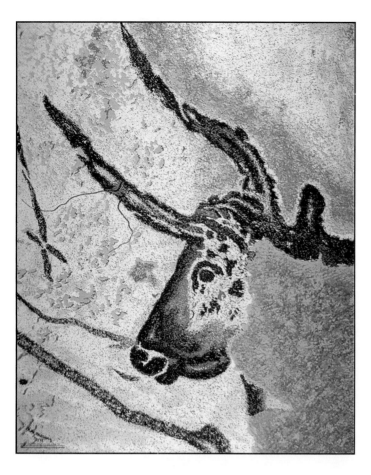

Tech Talk

Gutenberg invented a reusable two-part mold. He poured in molten metal to make a casting of a letter. He took the mold apart and used it to cast more letters.

It is also believed that our ancient ancestors told stories around their cave fires. Storytelling, of course, is verbal communication. It helped cave dwellers develop vocabulary and the part of grammar that language experts call syntax. Syntax is the way we organize the words in our sentences to convey our ideas to our listeners. All human languages have syntax. Language experts believe it is syntax that separates true language from animal forms of communication.

The use of the plow changed our early ancestors from hunters and gatherers into farmers. With farming came property ownership and the need to keep records. Written and spoken language grew in importance.

In the middle of the fifteenth century, Johannes Gutenberg's development of movable metal type allowed language arts to spur the development of all kinds of new technology. Gutenberg's printing press made book production faster and cheaper. In many ways the manufacturing of inexpensive books paved the way for universal education. Before Gutenberg, few people were schooled in the reading and writing arts. As more people became educated, and as communication increased, so did inventions.

Figure 17-8 The creation and distribution of information are key parts of most jobs today. This woman is creating a Web page.

Information Age
the modern era. Commonly used to describe a far-reaching system of computer networks capable of carrying vast amounts of information to homes, schools, work places, and government institutions.

—— Social Studies ——

How have technology and social studies worked together?

The study of technology deals with more than things. It deals also with how people use technology and how it changes their lives. Most American technical museums are not simply displays of historical items. They show how past inventions have improved people's lives. Social studies and technology have always had strong links.

The Industrial Revolution is a good example. It began in about 1760 and resulted in worldwide social changes. Society had been based on farming. Now it became based on factory production. People are still researching the effects of the Industrial Revolution, and many modern books cover the topic in great detail.

Today we are in another revolution. Our society has moved beyond factory production into the **Information Age**. We still need factories, of course, but more and more people make their living by creating, processing, and distributing information. See Figure 17.8.

The Information Age

≡ *What is the "Information Age"?*

It is difficult to say precisely when the Information Age began. Some say it started in 1844 when Samuel Morse tapped out the first long distance telegraph message between Baltimore and Washington DC. It may have been in 1901 when Guglielmo Marconi sent the first transatlantic wireless radio signal from England to Newfoundland. Perhaps it began in 1944 when Howard Aiken operated the world's first large-scale digital computer at Harvard University. The Information Age could have started in 1967 when a government group named the Advanced Research Projects Agency (ARPA) established a computer network they called ARPANET. It evolved into the modern Internet, which connects the worlds' citizens, schools, businesses, and countries. Whenever the Information Age began, it is here to stay.

Linking the world by computers has allowed everyone to share in the specialized knowledge of a few. A medical group in Europe might have found a cure for a particular illness. Their information can be immediately read throughout the world to help save lives. NASA works on projects in medicine, business operations, agriculture, all types of transport vehicles, alternative energy programs, and many others. Much of the information is available on NASA's Internet Web site.

Global Positioning System (GPS) satellites help map the world with great accuracy. The system was developed by the Pentagon in the 1970s, and its information is now universally available. A series of 24 satellites orbits the earth. Twenty-one are active, and three others are kept in reserve. From any given spot on the earth, five to eight of the satellites are visible. A small hand-held receiver can pick up data from several satellites. The information is used to precisely determine the receiver's position with an accuracy of ±25 meters.

The GPS information permits ship captains to plan a long ocean voyage so as to use the least amount of fuel. It allows highway engineers to determine the most practical route for constructing roads in undeveloped regions. Precise satellite mapping helps oil companies locate likely sources of underground petroleum. Another application is in the field of agriculture. Farmers can use satellite data to measure crop yields in sections of fields as small as $\frac{1}{100}$ of an acre. See Figure 15.6 on page 463.

Tech Talk

NASA developed a very low voltage cold light for plant experiments in space. This light is now being tested on the human body as a way of activating new tumor-killing drugs. In the future, surgeons will be able to use photon surgery to destroy tumors without affecting healthy tissue.

Tech Talk

NASA's Internet Web site is at **www.nasa.gov**. Another useful technical site is sponsored by the National Institute of Standards and Technology. The address is **www.nist.gov**.

The rapid exchange of information has altered the way people visualize the world. Television, computers, fax machines, telephones, and other products of the Information Age have changed the way schools educate students. Unlike their ancestors, today's young people have almost immediate access to large amounts of information from all over the world.

Besides advancing society in general, the Information Age also helps people conduct personal business. In retail stores, bar code readers rapidly sense the price of food, clothing, and other items. People can get through the check-out lines faster. With a credit card and access to a telephone (or to an online computer service), people can make travel reservations and buy products. Accurate banking information is instantly available over the computer. People can also do their banking on ATMs (automated teller machines).

Expanding and improving these systems will require workers with technology-based abilities. Many new jobs are available in the Information Age for people with education and training. The number increases every year.

Tech Talk

Four popular search engines are:
www.altavista.digital.com
www.inference.com/ifind/
www.yahoo.com
www.lycos.com

─── Your Future Career ───

Will your future career require language arts skills and knowledge of science, mathematics, and social studies?

Can you think of one career that doesn't require spoken or written communication or a basic understanding of the principles of science and mathematics? Social studies and foreign language are also very important because transportation and communication technologies have changed our world into a global village. Today we can fly to countries on other continents in just a little more time than it took our ancestors to walk to a distant neighbor. It used to take months to send a letter to the other side of the world. Today it takes a few seconds to connect by telephone or computer.

Your future career might require you to communicate and work with people from all over the world. This mobility makes your social studies and foreign language learning experiences very important to your future career.

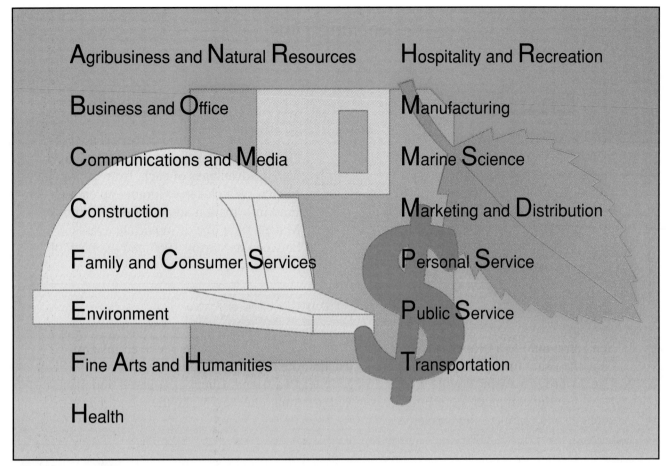

Agribusiness and Natural Resources	Hospitality and Recreation
Business and Office	Manufacturing
Communications and Media	Marine Science
Construction	Marketing and Distribution
Family and Consumer Services	Personal Service
Environment	Public Service
Fine Arts and Humanities	Transportation
Health	

■ **Figure 17-9** ■ All occupations can be divided into these job clusters, which have been established by the U.S. Office of Education. The first order of business for your research group is to determine the job cluster of your chosen occupation.

Try this activity challenge. With your teacher's guidance, divide the class into groups that have the same occupational interests. See Figure 17.9. Each group should research its chosen field to determine what technological devices are used and how communication, science, mathematics, social studies, and knowledge of a foreign language might be needed.

Research for this project should involve contacting people who have careers in your group's chosen field. You might invite some people to come to your class and discuss their profession. With your teacher's help, you could establish a **shadowing program** in which you and your classmates spend a day in the work environment of your chosen field. You might want to develop a survey and send it out to individuals or professional organizations that represent members of the chosen profession. You can find most of these groups on the Internet by typing the occupation into one of the search engines.

shadowing program
a program in which students observe employees and employers at companies, factories, and government agencies. During the observation period, students stay as close as the person's real shadow, hence the name shadowing program.

— Learning Time —

Recalling the Facts

1. Why was Johannes Gutenberg's development of movable type so important to language arts (two reasons)?
2. For each of these persons, state why he was important to the Information Age and give a significant date: Samuel Morse, Guglielmo Marconi, and Howard Aiken.

Thinking for Yourself

1. Research the scientific careers named in this chapter. Make a list of the advantages and disadvantages of each, from your point of view. Which career might you enjoy?
2. Find information about the ARPANET. Make a time line of important events between its start in 1967 and the birth of the Internet.

Applying What You've Learned

1. Send an e-mail to a nearby friend. Electronic messages are not always received instantly. Note the exact time you sent it. Have your friend note the exact time it arrived.
2. Use a word processor to write a 1- or 2-page letter to a friend. Check it yourself and print it. Next, use the computer's spelling checker and grammar checker. Make the recommended changes and compare the two letters.

Summary

Directions

The following sentences contain blanks. For each numbered blank, pick the answer that makes the most sense in the entire passage. Write your answer on a separate sheet of paper.

There is a connection between the development of technology and the knowledge gained by studying science, mathematics, language arts, and social studies. Let us review the connection between these ___1___ and technology.

As they seek solutions to problems, scientists use technology as a ___2___ for scientific discovery. In turn, their discoveries often lead to the development of new ___3___.

Few subjects are more important to technology than mathematics. Math is often the reason products look the way they do. ___4___ can reveal whether a proposed design will work as intended.

Verbal and written communication are in

1.
   a. people
   b. subjects
   c. buildings
   d. technologies

2.
   a. tool
   b. checklist
   c. subject
   d. goal

3.
   a. products
   b. designs
   c. technology
   d. formulas

4.
   a. Mathematical calculations
   b. Trial and error
   c. Scientific theories
   d. Technology

(Continued on next page)

many ways the foundation of technology. We need ___5___ in order to develop technology and teach it to others. In turn, technology provides the means for better communication.

The study of technology deals with how people ___6___ technology and how it changes their lives. Social studies and technology have always had strong links.

Technology depends on the knowledge that you gain by studying other subject areas. Your future as a productive member of the 21st century will depend upon your ___7___ to form connections between all of your school subject learning experiences. Only then will you be able to use and build the technological systems of tomorrow.

5.
a. foreign languages
b. technology
c. pictures
d. communication

6.
a. learn
b. use
c. calculate
d. theorize

7.
a. ability
b. desire
c. wish
d. gullibility

Key Word Definitions

Directions
The column on the left contains the key words from this chapter. The column on the right contains a scrambled list of phrases that describe what these words mean. Match the correct meaning with each word. Write your answer on a separate sheet of paper.

Key Word	Description
1. Life science	a) Branch of fluid power; deals with liquids under pressure
2. Physical science	b) Study of the earth's resources, structure, air, and oceans
3. Earth and space science	c) The tendency of an object to float in water
4. Hypothesis	d) The modern era
5. Scientific theories	e) Scientific theories that have stood up to repeated experimentation and investigation
6. Scientific laws	f) The study of living organisms
7. End effector	g) Unit of liquid or gas pressure; pounds per square inch
8. Hydraulics	h) The part of a robot that actually does the physical work
9. Pneumatics	i) An explanation that can be tested
10. psi	j) Students observe employees and employers at work
11. Density	k) Hypotheses that can stand up to rigorous scientific experimentation and investigation
12. Buoyancy	l) The ratio of the weight of an object to its volume
13. Information Age	m) The study of matter and energy
14. Shadowing program	n) Branch of fluid power; deals with gases under pressure

Producing a Movie

Resources You Will Need

- Outside consultants
- Video camcorder and accessories
- Large tub or other container of water
- Motion picture props
- Plastic boat, about 6 to 8 inches long
- Small screwdriver
- Metal weights, such as washers, nuts, or bolts
- Several Styrofoam packing peanuts
- Electric drill
- Ear syringe
- Double sticky tape
- Glue
- 3 feet or more of flexible plastic tubing, small diameter

Did You Know?

Movie production involves the skills and talents of many people. The stars and the director receive most of the publicity, but they are only a small part of the team. Writers, musicians, set designers, and camera operators are just some of the many people needed to produce a movie.

Think of a movie you've enjoyed recently. What did you like about it? How did the dialogue, the music, the setting, and the cinematography add to your enjoyment?

Problems To Solve

In this activity, your technology class will write, edit, produce, and direct a video and/or computer simulation of *Titanic* significance.

The theme of your production will determine your need for outside consultants. For the writing part of this project you will of course want to ask your English teacher to serve as a consultant. Perhaps your teacher will allow your class to substitute this writing project for some of the assignments that you would have been doing this term in English.

If your class decides to develop a historical theme, you will also want to bring in your social studies teacher as a consultant. You might choose a theme that revolves around the location of sunken treasure ships, the building of *Titanic*-size vessels in foreign or American ports, the history of navigation, the development of water transportation, or some other theme that deals with social studies issues.

Your class will have to build the movie props that will serve as the co-stars of your produc-

A.

tion. To film the *Titanic* movie, it was necessary for the studio to build a large number of movie props that could be sunk and then quickly restored to seaworthiness for the next "take" (the shooting of the next scene). The vessels you build in your technology laboratory might combine the buoyancy of an ocean liner with the ability to dive under the sea like a submarine. Your technology, science, mathematics, and art teachers can help you design the props that you will use.

To make this production you will need to re-create the management organization of a movie studio and pick producers, directors, writers, special effects coordinators, and camera crews.

You and your class will move through all the phases of pre-production, which will include everything from writing the story to preparing the story boards that show how your video will unfold. See Figures A and B.

During the production phase of this project you will build and test the props, and you will shoot your video. See Figures C and D.

The final editing of your video—which might include cutting or rearranging scenes for better effect, dubbing in music, and taping the titles and ending credits—will take place during post production. See Figure E.

As you can see, you have a lot to do before you get to say, "Scene 1, take 1. Silence on the set, camera, and action!"

A Procedure To Follow

1. Building a realistic model boat that floats and submerges can be a challenging project. You can save time by buying a ready-made plastic boat. The boat will give you something with which to experiment, and you can decide later if you want to use a different one. Try to locate an inexpensive toy boat that is about 6 to 8 inches long, or even a bit

B.

C.

longer. You may want the boat to sink by the bow (front), and longer models do it more realistically.

2. Make sure you can remove the top from your model boat to expose the inside of the hull. If you're careful, you can usually remove the top with a small screwdriver and careful prying.

3. Plastic doesn't sink well, and neither does an all-plastic boat. You will need to add weights. Use double sticky tape on metal washers, bolts, nuts, or fishing sinkers and position them along the inside of the hull. For the ship to sink bow first, tape a Styrofoam packing peanut or small piece of balsa wood at the back of the hull for flotation. Either will keep water away from that area. As water enters the hull, the front will be more heavily loaded, and the model will sink head first. See Figure F.

4. Experiment by floating the exposed hull in a container of water. Try different combinations and placements of weights and flotation devices to get the desired effect of a ship sinking. You don't need to duplicate exactly what you may have seen in some motion pictures. Those effects were created by people with years of experience who were supported by expensive model-making shops. Just do the best you can with the materials you have.

5. Drill a small hole in the bottom of the hull near the front. Glue the flexible plastic tube into the hole. You will use this tube and an ear syringe to pump water into the boat. Or, if you want to make the boat refloat automatically, drill two holes. Put a balloon on the end of the plastic tube, insert the balloon and tube in the first hole, and glue the tube in place. The second hole is for water to enter. You can con-

D.

E.

trol the amount of water that enters the boat by letting air out of the balloon. You can refloat the boat by filling the balloon with air, forcing water out of the boat.

6. Replace the top of the boat but do not permanently attach it. Put water in the ear syringe and slowly force it into the boat's hull through the plastic tube. See Figure G. (Note: In the auto refloat design, the tube will carry air, for the balloon, instead of water.) Evaluate the boat's sinking technique. It may have changed after you put the top back in place. Make any necessary adjustments to the weights and flotation devices inside the hull.

7. Use the video camera to record a "rough draft" of the more dynamic aspects of your story. Those scenes may include the ship plowing through heavy seas,

moving at high speeds, or sinking after hitting an object in the water. You may want two or more boats to appear at the same time. Choreography will be an important part of the video. *Choreography* means mapping out the activity.

8. Try shooting from different angles and with different lighting effects. Critically evaluate your results and make adjustments so that the production will be as good as you can make it. Don't expect to get everything right the first time. It is not unusual for motion picture directors to make many shots of the same sequence. The old saying "practice makes perfect" certainly holds true in the motion picture industry.

9. After practicing time and again, you will at last be able to say, "Scene 1, take 1. Silence on the set, camera, and action!"

What Did You Learn?

1. How long did it take to plan, write, produce, and edit your video? Compare that to the actual length of your finished video. What percentage of your total time investment actually ended up as viewable tape?

2. What editing features does your video camera have? How were you able to use these to improve your movie?

F.

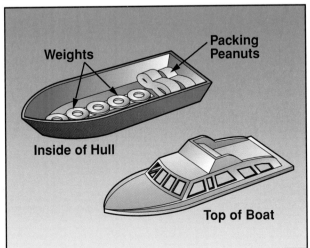

Weights

Packing Peanuts

Inside of Hull

Top of Boat

G.

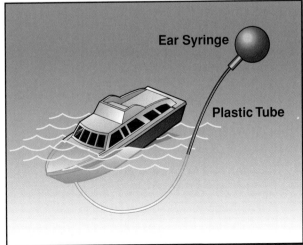

Ear Syringe

Plastic Tube

Important Formulas

with Representative US Customary Units

1. AREA
   A = Length × Width {square feet}
2. VOLUME
   V = Length × Width × Height {cubic feet}
3. DENSITY
   D = Weight/Volume {pounds/cubic feet}
4. SPEED (or VELOCITY)
   V = Distance/Time {miles/hour}
5. ACCELERATION
   A = (Final Velocity – Initial Velocity)/Time {feet/second/second}
6. PRESSURE
   P = Force/Area {pounds/square inch}
7. WORK
   W = Force × Distance {foot-pounds}
8. EFFICIENCY
   E = (Work Output/Work Input) × 100 {percent}
9. POWER
   P = Work/Time {foot-pounds/second}
   P = Energy/Time {British Thermal Units/hour}
10. MECHANICAL ADVANTAGE
    MA = Resistance Force/Effort Force {no units}
11. MECHANICAL ADVANTAGE OF A LEVER
    MA = Effort Arm Length/Resistance Arm Length {no units}
12. MECHANICAL ADVANTAGE OF AN INCLINED PLANE
    MA = Inclined Plane Length/Inclined Plane Height {no units}
13. GEAR RATIO
    GR = Number of Teeth in Driving Gear/Number of Teeth in Driven Gear
    {no units}
14. VOLTAGE
    V = Amperage × Resistance {volts}
15. AMPERAGE (or ELECTRICAL INTENSITY)
    I = Voltage/Resistance {amperes}
16. RESISTANCE
    R = Voltage/Amperage {ohms}
17. ELECTRICAL POWER
    P = Voltage × Amperage {watts}

Approximate Customary-Metric Conversions

	When you know:	You can find:	If you multiply by:
Length	inches	millimeters	25.4
	feet	millimeters	304.8
	yards	meters	0.9
	miles	kilometers (km)	1.6
	millimeters	inches	0.04
	meters	yards	1.1
	kilometers	miles	0.6
Area	square inches	square centimeters (cm²)	6.5
	square feet	square meters	0.09
	square yards	square meters	0.8
	square miles	square kilometers (km²)	2.6
	acres	square hectometers (hectares)	0.4
	square centimeters	square inches	0.16
	square meters	square yards	1.2
	square kilometers	square miles	0.4
	hectares (ha)	acres	2.5
Mass	ounces	grams	28.4
	pounds	kilograms	0.45
	tons	metric tons (t)	0.9
	grams	ounces	0.04
	kilograms	pounds	2.2
	metric tons	tons	1.1
Liquid Volume	ounces	milliliters	29.6
	pints	liters	0.47
	quarts	liters	0.95
	gallons	liters	3.8
	milliliters	ounces	0.03
	liters	pints	2.1
	liters	quarts	1.06
	liters	gallons	0.26
Temperature	degrees Fahrenheit	degrees Celsius	0.6 (after subtracting 32)
	degrees Celsius	degrees Fahrenheit	1.8 (then add 32)
Power	horsepower	kilowatts (kW)	0.75
	kilowatts	horsepower	1.34
Pressure	pounds per square inch (psi)	kilopascals (kPa)	6.9
	kPa	psi	0.15
Velocity (Speed)	miles per hour (mph)	kilometers per hour (km/h)	1.6
	km/h	mph	0.6

Suggested Student Readings

- *The New Book of Popular Science.* Danbury, CT: Grolier, 1979.
- *Science Now.* New York: Arco Publishing, 1984.
- *Walt Disney's How It Works in the Country.* Chicago: J.G. Ferguson, 1982.
- *Walt Disney's How It Works in the Home.* Chicago: J.G. Ferguson, 1982.
- Altman, Linda Jacobs. *Women Inventors.* New York: Facts on File, 1997.
- Asimov, Janet and Issac. *Norby Through Time and Space.* New York: Ace Books, 1987.
- Barron, T.A. *Heartlight.* New York: Philomel Books, 1990.
- Bender, Lionel. *Invention.* Eyewitness Books. New York: Alfred A. Knopf, 1991.
- Duane, Diane. *Spock's World.* Pocket Books, 1989.
- Earnshaw, Brian. *Dragonfall 5 and the Empty Planet.* New York: Lothrop, Lee and Shepard Co., 1973.
- Feldman, Anthony. *Technology at Work.* New York: Facts on File, 1980.
- Hann, Judith. *How Science Works: 100 Ways Parents and Kids Can Share the Secrets of Science.* Pleasantville, New York: Reader's Digest Association, 1991.
- Hilton, Suzanne. *How Do They Cope with It?* Westminster Press, 1970.
- Humberstone, A. Eliot. *Finding Out About Everyday Things.* Usborne Publishing Ltd., 1981.
- Jeffrey, Laura. *American Inventors of the Twentieth Century.* Springfield, NJ: Enslow Publishers, Inc., 1996.
- Kaufman, Joe. *Joe Kaufman's What Makes It Go? What Makes It Work? What Makes It Fly? What Makes It Float?* Golden Press, 1971.
- Lambert, David. *Great Discoveries and Inventions.* Facts on File Publications, 1985.
- Lodewijk, T. *The Way Things Work: An Illustrated Encyclopedia of Technology.* Simon and Schuster, 1973.
- Mahy, Margaret. *Aliens in the Family.* New York: Scholastic, Inc., 1985.
- Maynard, Christopher. *I Wonder Why Planes Have Wings and Other Questions About Transportation.* New York: Kingfisher Books, 1993.
- Morgan, Sally and Adrian. *Technology in Action.* Facts on File Designs in Science Series, 1994.
- Noble, Holcomb B., (Ed.) *Next: The Coming Era in Science.* Boston: Little, Brown, 1988.
- Platt, Richard. *Smithsonian Visual Timeline of Inventions.* New York: DK Publishing, Inc., 1994.
- Poling, James. *Story of Tools: How They Built Our World and Shaped Man's Life.* Norton, 1969.
- Preuss, Paul. *Hide and Seek.* Avon, 1989.
- Reader's Digest. *How in the World: A Fascinating Journey through Human Ingenuity.* Random House, 1990.
- Sleator, William. *Strange Attractions.* New York: E.P. Dutton, 1990.
- Sutton, Caroline. *How Do They Do That? Wonders of the Modern World Explained.* New York: Morrow, 1981.
- Taylor, Paula. *The Kid's Whole Future Catalog.* New York: Random House, 1982.
- Wilson, Anthony. *Visual Timeline of Transportation.* London: Dorling Kindersley Publishers, Inc., 1995.
- Zubrowski, Bernie. *Structures: The Way Things Are Built.* Cuisenaire, 1993.

Glossary

acid rain polluted rain caused by the use of fossil fuels such as coal, oil, and gasoline. It causes much damage to the environment.

adapting improving or changing to make fit for a new use or situation

added value the increase in how much a piece of material is worth after it becomes part of a finished product

agricultural technology the branch of technology that specializes in improving techniques and products for raising livestock, cultivating soil, and producing crops

alternating current electrical flow that constantly changes direction. The electrical flow that powers your home is Alternating Current (AC).

amperage the strength or power of the electrical energy through a circuit. It is also called the *current*.

anthropologist a scientist who studies human beings, especially in the areas of origin, physical characteristics, relationship of races, culture, and development

aperture the device on a camera that controls how much light the film is exposed to when the shutter is open. Also known as a *diaphragm*, it works much like the iris of your eye.

artificial intelligence a type of specially designed computer program that gives the impression a computer can think for itself

asphalt a brownish-black, flexible material made from tar or pitch found in nature and used to make and repair some road surfaces

assembly line an arrangement of workers or machines where each person or machine completes a certain part of the product and then passes it on to the next person or machine in the line

autopilot a device that mechanically controls the steering in aircraft, spacecraft, and ships

axle a rod or shaft on which one or more wheels turn

binary code an electronic Morse code based on the binary number system that the computer can understand

bioengineering the branch of technology that designs products with the comforts and needs of people in mind. Also known as *ergonomics*.

biological communication communication that uses natural body parts, such as the brain, voice box, ears, arms, and hands to transmit and receive messages

biometrics the branch of technology that uses voice prints, finger prints, and retina eye scans as positive forms of identification. Used for security and other purposes.

bionics the branch of bioengineering that develops replacement parts for the human body

biotechnology the branch of technology that applies or uses the information found in biology to improve such things as health care, the environment, and agriculture

blood plasma the fluid part of blood that can be given to anyone, regardless of blood type

blueprints drawings showing the location and size of rooms in a building. When the drawings are reproduced, they turn out with white lines on blue paper.

boiler an enclosed vessel, such as a pot or kettle, in which water is boiled and converted to steam

brainstorming a group problem-solving technique in which group members call out possible solutions

building code the construction regulations, set by a community, that builders in that area must follow

building permit the certificate given by local governments that gives a builder permission to begin construction

buoyancy the tendency of an object to float in water

CAD (computer-aided design) a method of planning or creating a product using a computer

CAM (computer-aided manufacture) a system that uses computers to operate the machinery in a factory

CIM (computer-integrated manufacture) the use of one computer system to control the design, manufacturing, and business functions of a company

CNC (computerized numerical control) machine tool operation controlled by numerical commands from a computer

CPU (central processing unit) the part of a computer that processes data into useful information. It is the "brain" of the computer.

calories the measure of food energy in a person's body. Also, the energy required to raise the temperature of one milliliter of water one degree Celsius.

camera obscura a darkened chamber or room in which a small hole in one wall causes an upside down image to be projected on the opposite wall

cantilever bridge a bridge made of two self-supporting beams, each of which is fastened to the ground at one end. The beams meet in the middle of the bridge.

capital the money, goods, and possessions used in a business in order to make more money

carbon monoxide an odorless, colorless, poisonous gas formed when gasoline burns

cargo ship a large ship used to transport products such as oil, grain, iron ore, and automobiles

cartridge a plastic case that film comes in

caulking a toothpaste-like material used to seal cracks in a house

celluloid roll film a thin plastic material that has emulsion already on it; used to take photographs

city mpg the miles per gallon of fuel that a vehicle gets while being driven in a city or town

closed-loop system a system that has feedback and quality control for an improved end product

coal reserve the amount of coal which has not been mined and is still underground

cold-type composition all typesetting methods that do not use letters cast in metal. Computer typesetting and photographic typesetting are examples.

combustion burning

commercial airplane an airplane that carries passengers or freight in order to earn money

commercially prepared provided and marketed by a company in order to make a profit

commission a payment made to a salesperson or agent for business they have done

communication the process of exchanging information between individuals. This can be done through behavior or by use of symbols or signs.

communication device an instrument, such as a telephone, telegraph, radio, or television, that enables a person to send or receive messages

communication system a plan or route that takes a message from the sender to the receiver. Includes the steps of input, process, output, and feedback.

communication technology the use of knowledge, skill, tools, machines, and materials to enhance communication through the development and use of communication devices

composing stick a tray that type is set into by a typesetter

compression stroke the second part of an engine cycle that follows the intake stroke in which the piston moves to the top of the cylinder and squeezes the gasoline and air mixture in the cylinder to about one-eighth its original volume

computer a programmable, electronic device that calculates, stores, and processes information

computer-aided design (CAD) a computer system that is used, instead of a drawing board and drawing tools, to design an object

computer-aided design and drafting (CADD) a combination of computer-aided design and computer-aided drafting in which the computer is designing and drawing an object at the same time

computer-aided drafting the process of using a computer instead of regular drafting tools (T-square, paper, triangles, etc.) to make the drawings used by engineers, architects, and craftspeople

computer bit the smallest piece of information that a computer can use

computer byte an information unit made up of eight bits

computer-generated made or produced on a computer

computer hardware the physical parts of a computer system. Included are such parts as the CPU, monitor, keyboard, disk drives, mouse, CD-ROM drives, and joystick.

computer axial tomography (CAT) a type of computerized X ray that gives a three-dimensional image of the interior of the object scanned

computer program software which contains a set of instructions which the computer follows

computer software the programs the computer follows to execute a specific function

computer virus a destructive computer program that "infects" the computer system and can cause damage to data in the system

concrete a mixture of cement, gravel, sand, and water used in the foundation of houses. It's also used to make stairs, sidewalks, birdbaths, and many other items.

condenser an instrument that changes steam back into water

conductor material, such as copper or silver, that electricity flows through easily

construction site the location where a house or other type of structure is being built

consumers the people who buy and use products

corrosive having the power to eat away or wear away

craft a skilled occupation or job, usually done with the hands, such as carpentry or sewing

craftsperson a person who specializes in a particular job or craft

crankshaft the part of an engine that changes the reciprocating piston motion to rotary motion to turn the wheels

creativity the ability to use knowledge and imagination to make, develop, invent, or produce something new

current the flow of electricity along a wire

cycle an event, or series of events, that repeats itself on a regular basis

cylinder in an engine, the tube in which a piston slides

data facts, such as numbers and symbols, that are put into the computer using an input device such as a computer keyboard

density the ratio of the weight of an object to its volume

design department the part of a company that has engineers who decide the size, color, shape, and material of a new product

designing the process of planning and drawing an idea

desktop publishing the computer process that allows people to create their own reports, booklets, and other items in their own school, office, or home

developed nation a country that has a high economic level of manufacturing, industrial production, and standard of living. Most manufacturing is done by machines in factories.

developer the name given to all chemicals that are used to bring out the latent image in photographic processes

developing nation a country that has a fairly low economic level in areas such as manufacturing, and standard of living. Much of the manufacturing in developing nations is done by hand.

developing tank a container that contains chemicals in which photographs are developed

diagram a drawing that shows how parts are arranged and the relationship of those parts

diaphragm another name for the aperture or the part of a camera that controls how much light passes through the lens when the shutter is open

dimensions the measurements that give the size of an object, size and location of holes, cut-outs, and other features of the object

direct current the one-directional flow of electrons in an electrical circuit. All batteries have Direct Current (DC) flow.

disk drive the device used to read and record data on a floppy disk

displacement the weight of the water that is moved out of the way by a floating ship

division of labor the idea of dividing work into many steps. A different person is responsible for each different step. This allows the job to be completed quickly and accurately.

drafter the person who prepares technical drawings

drafting the drawing techniques that accurately represent the size, shape, and structure of objects. *Technical drawing* and *mechanical drawing* are other terms used to describe drafting.

drive-by-wire car system a future electronic navigation system that will direct cars to their destinations. Also called *smart highways.*

drywall the inside covering of walls and ceilings, made from plaster and sturdy paper

dual dimensioning a measurement label that uses both U.S. and metric units of measurement

earth and space science the study of the earth's resources, structure, atmosphere, and oceans. It is also the study of our moon, solar system, and universe.

electrical circuit an electrical pathway that begins and ends at the same power source. Includes parts such as a wire for a path to conduct the flow and a device the electricity is being delivered to.

electricity a form of energy that comes from the movement of invisible particles called electrons through an electrical conductor. Electric current is used as a source of power.

electric switch a device that opens and closes an electric circuit and allows electrical items to be turned on and off

electron theory the study of how electrons flow through an electric circuit from the negative terminal toward the positive terminal

electrostatic printing the process in which negatively charged toner is transferred to positively charged paper and then melted into place. This is also known as xerography.

emulsion a light-sensitive chemical coating used on glass, paper, or photographic film

end effector the part of an industrial robot that actually does the physical work for which the robot was designed

energy the capacity or ability to do work. There are six types of energy: mechanical, heat, electrical, chemical, nuclear, and light.

energy conservation the management and efficient use of all energy sources

engine a device that uses heat to convert energy into motion

engineering the application of science and mathematics that makes matter and energy found in nature useful to people in such forms as machines, products, structures, processes, and systems

ergonomics the branch of technology that designs products with the comfort and needs of people in mind. Also known as *bioengineering*.

exhaust stroke the fourth part of an engine cycle that follows the power stroke in which the piston moves up and pushes out the exhaust gases

external combustion engine an engine that converts the fuel to energy outside the engine itself

FRAM (ferroelectric random access memory) flash memory that is saved indefinitely without electricity

face-to-face communication sending and receiving a message within sight of, or in the presence of, the person the message is going to or coming from

factory building, or group of buildings, where products are manufactured

feedback the part of a system that measures and controls the outcome of that system

fertilization the process of applying different substances, such as manure or chemicals, to the soil as food for plants

fiberglass brittle fibers of glass used to make items such as insulation, textiles, and structures

fiber optic a type of system that uses thin, flexible glass strands or cables to transmit light over great distances

fixer a chemical that makes the photographic material no longer sensitive to light

flexography a relief printing process that uses a raised, rubber printing plate which is inked and pressed against paper. A good example is a rubber stamp.

floor plan sketches that show the size and location of rooms in a building

floppy disk a low cost magnetic recording device made of plastic on which computer programs and data are stored

fly-by-wire system a computer controlled method of flying aircraft

focus range the area in a picture that will be sharp

footing the bottom part of a foundation, made of hardened concrete, under the foundation wall

fossil fuels fuels such as coal, oil, and natural gas that are found in the earth and come from the remains of plants and animals

foundation the bottom support of a house, or other building, that rests on the ground

foundry type metal-cast letters used in hot-type composition

four-color process printing a printing method that uses the colors cyan, magenta, yellow, and black (referred to as CMYK) to create all the colors in nature and different tones of those colors

four-stroke cycle an engine in which the pistons move up or down four times before they repeat themselves

four-wheel drive a power system in which the power from the engine is transferred to all four wheels of the vehicle

freehand sketch accurate pencil and paper drawings done without mechanical devices

frequency the number of cycles or changes in direction of the Alternating Current. Frequency is measured in hertz, or cycles per second.

front-wheel drive a power system in which the power from the engine is transferred to the front wheels of the vehicle

futurist a person who deals with what might happen in the future, based on what is currently happening

gable roof a roof with two sloping sides that meet at the ridge and form a triangular shape at either end

genetic engineering the process of changing genes that contain information for developing certain characteristics or traits in plants and animals

geodesic dome a rounded structure made of many small, pyramid-shaped frames

geosynchronous orbit a particular path of a satellite that keeps it directly above the same place on earth all the time

geothermal describes thermal, or heat, energy in the ground that is caused by the radioactive decay of certain elements

girder bridge a bridge made of beams that rest on the ground on either side of the span

global village the idea that, because of advanced technology in communication and transportation, each country is a neighbor rather than a distant place

going into production manufacturing a product

graphic communication the methods of sending and receiving messages using visual images, printed words or symbols. Includes books, magazines, newspapers, photographs, etc.

graphic reproduction all of the processes necessary to change artwork, photographs, and text into printed material

gravure printing a printing process that is achieved by etching or scratching letters and designs onto a metal plate. Ink fills these grooves, and is then transferred to paper.

greenhouse effect the gradual, steady increase in the temperature of the earth's atmosphere caused by the rise in the amount of industrial gases in the earth's atmosphere

halftone a picture produced by using a series of dots. The picture is rephotographed through a special screen. A continuous tone picture is produced when the halftone is printed.

halftone contact screen a screen through which a photograph is rephotographed and caused to break up into dots for printing purposes

hand tool instrument such as a hammer, pliers, or screwdriver that is powered by the user's hand or arm

helicopter an aircraft that is lifted straight up off the ground by a rotor, or horizontal propeller-like device

high technology advanced or highly specialized technology

highway mpg the miles per gallon of fuel that a vehicle gets while being driven on highway or open road

horse power a unit of power equal to a horse lifting 550 foot-pounds one foot in one second, or 550 foot-pounds of work per second

hot-type composition typesetting letters made out of metal

hydraulics the branch of fluid power dealing with liquids under pressure

hydroelectric describes the electricity generated by turbines propelled by flowing water

hydroponic farming the process of growing plants in a mixture of water and fertilizer without the use of soil

hypothesis an explanation that can be tested

Industrial Revolution a period during the late 1700s when machines were used to complete many of the tasks previously done by hand. Factory production replaced much of the home manufacturing.

industrial technology an area of education formerly known as industrial arts in which students develop manual skills and familiarity with machines and tools

Information Age the modern era. Commonly used to describe a far-reaching system of computer networks capable of carrying vast amounts of information to homes, schools, work places, and government institutions.

ink-jet printing the process that uses tiny ink spray guns to shoot ink onto the printing surface

input something that is put in, such as data or conversation

input device any control device, such as a keyboard, mouse, game controller, or joystick, that is used to get information into a computer's CPU

insulator material, such as ceramic, rubber, or plastic, that electricity will *not* easily flow through

intake stroke the first part of an engine cycle in which the piston moves down and a mixture of air and gasoline is brought into the cylinder

integrated circuit a tiny chip that contains dozens of electronic components. It replaced the transistor.

internal combustion engine an engine in which the fuel is burned inside the engine

irrigation the process of supplying water to fields by use of pipes, canals, or sprinklers

isometric drawing a pictorial drawing that shows three, equally distorted sides of the object

jet engine a type of gas turbine engine that is moved forward because of the hot air and exhaust that are shot out of the back part of the engine

joist one of many parallel beams used to hold up a floor or ceiling

jumbo jet a very large airplane that carries several hundred passengers at one time. It can make very long flights before needing to refuel.

just in time a method of scheduling the arrival of materials at the time they are needed, so that it is not necessary to store those materials

laser an instrument that produces a very powerful, narrow beam of light. Used to cut and process material, carry communication signals, build and guide weapons, and produce 3D pictures.

latent image the image, or picture, on the film that is invisible until after the film is developed

lens the part of a camera that magnifies light and the size of an object

letterpress printing a printing process that uses raised metal letters, symbols, or designs that are inked and then pressed against paper. It is also known as *relief printing*.

life science the study of living organisms

lighter-than-air craft (LTA) an airship that is lifted into the air by helium or hot air. Many, such as dirigibles, zeppelins, and blimps, have propellers to move them forward.

lithography a printing process that is based on the principle that oil and water don't mix. The image is created from oil-based ink. The printing surface is covered with water to prevent ink from sticking to it.

livestock animals raised or kept for pleasure or for use and profit

machine tool with a power system that takes advantage of certain scientific laws and makes the tool work better

machine tool a large instrument, such as a drill press or grinder, that is bolted to the floor and powered by an electric motor

maglev train a rail system in which the vehicle rides suspended on a cushion of magnetism

magnetic levitation a system based on the principle that same poles of a magnet will repel, or push, the other away. This repulsion is used to lift objects so that they rest on a cushion of magnetism.

magnetic resonance imaging (MRI) a system that uses magnetic waves to create an image of the interior of the object being imaged

mainframe computer a large, often room-sized, computer that is very fast and can handle many tasks at the same time

manufactured material raw materials that have been so altered by processes that they are unrecognizable

manufactured products items that are made by people in a factory or workshop

market to sell

market research the process of getting people's opinions about a product so that a company knows what changes to make or whether to sell the product

mass communication the systems that make it possible for many people to hear the same message at the same time. Radio and television are examples of mass communication systems.

mass production the system of making many of the same items very quickly, usually by machine

mass transportation public passenger service used to move large numbers of passengers from one place to another

mechanical drawing the drawing techniques that accurately represent the size, shape, and structure of objects. *Drafting* and *technical drawing* are other terms used to describe the same techniques.

medical technology the branch of technology that specializes in developing products to aid in the improvement of health care

metric measurement system a decimal system of measures and weights based on the kilogram and meter

monitor a computer output device that resembles a television screen

monoculture the cultivation, or growing, of only one crop

monolithic integrated circuit computer chips that contain integrated circuits that equal millions of transistors

mortar a mixture similar to concrete, made of sand and cement, used to glue or fasten concrete blocks together and sometimes used in plastering

motor a device that converts energy into motion

moving assembly line a system that moves parts to the people who then complete their specific jobs with those parts

mpg requirements the government regulations that set the minimum number of miles per gallon a vehicle may produce during city driving and highway driving

multicultural made up of many groups that have different customs, traditions, religions, and national origins

multiview drawing a drawing that includes the top, front, and right side views of an object. The three views are drawn on the same piece of paper.

NIOSH (National Institute of Occupational Safety and Health) the agency that approves for use protective equipment, such as safety glasses, hard hats, and steel-toed shoes

National Aeronautics and Space Administration (NASA) the government agency that regulates and directs the space program for the United States

negative photographic film that contains an image in which the light parts of the image appear dark, and the dark parts of the image appear clear

nonrenewable energy source fossil fuels, such as coal, oil, and natural gas, that cannot be replaced when used

nuclear reaction the process by which atomic particles hit each other and split apart, or react, thus creating nuclear energy

OSHA (Occupational Safety and Health Administration) the government agency that sets safety rules and checks to make sure the rules are being followed by companies

object avoidance system an advanced automobile system that enables a car to avoid hitting other objects or cars

oblique drawing a pictorial drawing that shows one side of the object without any distortion

offset lithography a printing process that begins with an image photographed onto a thin metal plate. The plate is then mounted on a press and the image is transferred to a rubber blanket which then transfers the image to the paper.

ohm the unit used to measure the resistance in an electrical circuit

Ohm's Law the relationship between amperage, voltage, and resistance

open-loop system a system that has no self-correcting action, or way of measuring or controlling the product

operating system program computer programs that tell the computer how to run its own systems

optical computer an experimental computer that uses fiber optic cable and laser signals instead of electricity

organization chart a diagram of the jobs needed to make a product

orthographic projection a set of drawings that shows a straight-on view of the six individual sides of an object

output what the system produces, the end product

output device equipment such as a printer or monitor that receives information from the CPU

parallel circuits multiple pathways that electrical current flows through to individual electrical devices. If one device stops working, the others are not affected. Homes are wired with parallel circuits.

particulates very small particles

passenger vessels ships and boats specially built to carry people

perspective drawing a pictorial drawing that makes the object look realistic. The parts of the object that are farther away look smaller.

petroleum a natural liquid that comes from the ground and can burn; a fossil fuel known as "crude oil"

photography the process that uses the energy of light to create pictures of objects

photon a very tiny unit of light

photovoltaic cells devices that transform solar energy into electricity. Also called solar cells or photo cells.

physical science the study of matter and energy

pick-and-place robots industrial robots with the function of picking up an item at one spot and placing it somewhere else

pictorial drawing a drawing that shows a three-dimensional object so accurately that it resembles a photograph

pinhole camera a small camera obscura that was the first photographic camera

piston a cylinder-shaped object that slides inside a cylinder of an engine. A downward stroke creates power.

pneumatics the branch of fluid power dealing with gases under pressure

power the measurement of the work done by using energy; energy under control

power stroke the third part of an engine cycle that follows the compression stroke in which the gasoline and air mixture burns very rapidly and forces the piston down

power tool a hand-held instrument powered by a small motor

precast describes concrete parts that have been made ahead of time and shipped to the construction site

prefabricated describes parts of houses that are built in factories and then delivered to the construction site

printing all of the tasks necessary to the process of transferring images to a printed page. Also called *graphic reproduction.*

problem solving finding solutions to problems

process changing ideas or activities into useful products by the use of machines, resources, and labor

processed material raw material changed by technology into a more useful form that still resembles the raw material

production making products available for use

production department the part of a company responsible for actually making the company's product

production materials everything necessary to make a product

propellant the fuel and oxygen that causes a rocket engine to run

prototype a handmade test model of a product by which later stages or designs are judged

psi liquid or gas pressure measurement, in units of pounds per square inch

quality control the process of inspecting products to make sure they meet with all standards that have been set

RAM (random access memory) one of two types of memory contained in the CPU. This memory is lost when the computer is turned off.

ROM (read only memory) one of the two types of memory contained in the CPU. It is permanent memory that cannot be deleted or changed by a computer's instructions.

radioactive describes certain materials that give off invisible, high-energy rays or radiation

radiology the branch of medicine that uses radiation, as in X rays, to diagnose and treat diseases

rafter one of the wooden supports for a roof

raw material natural materials that can be converted into new and useful products

real McCoy a phrase that is used to describe something that is not an imitation, fake, or substitute

rear-wheel drive a power system in which the power from the engine is transferred to the rear wheels of the vehicle

reciprocating motion up-and-down or back-and-forth motion that occurs in a straight line. The pistons in an engine have this type of motion.

recycle to reuse all or part of substances, such as metal, glass, paper, plastics, etc.

relief printing a kind of printing in which the printing surface is raised above the rest of the plate. Also called letterpress printing.

renewable energy source energy that comes from plants and animals and so can be replaced when more is needed

research and development department the part of a company that looks at new ideas and creates a new product from those ideas

resistance the force that opposes or slows the flow of the electrical current

resource something that supplies help or aid; could be a source of information or expertise, wealth or revenue, supply or support

retailer a merchant who buys products from a wholesaler and sells them to final consumers

robotics technology that deals with the design, manufacture, and use of robots in industrial and automated situations

rotary motion circular motion. In an engine, the reciprocating motion of the pistons is changed to rotary motion by a crankshaft.

scale objects or drawings made where the size is smaller or larger than the real thing. For example, one inch on the drawing might equal one foot of real distance.

schedule a timed plan that includes the work to be done and when it should be done

schematic diagrams drawings prepared for the electric/electronic industry that show the circuits and components of electrical appliances

science knowledge covering general truths or laws that explain why something happens

scientific laws scientific theories that can be expressed mathematically and have stood up to repeated experimentation and investigation over many years

scientific management the system of developing standard ways of doing particular jobs. One goal of the system is to cut out all wasted time and motion.

scientific theories hypotheses that can stand up to rigorous scientific experimentation and investigation

sectional drawing a drawing that shows the interior of an object

self-propelled a vehicle that contains within itself an engine or other device that provides power for movement

semiconductors materials, such as the transistor, that can be used as either conductors or insulators

series circuit a single pathway that current flows through to more than one electrical device. If one device in the path stops working, they all stop. Old Christmas tree lights are a good example.

serigraphy a printing process that uses a printing plate made of an open screen of silk, nylon, or metal mesh. The process is also known as *screen printing* or *silk screening.*

shadowing program a program in which students observe employees and employers at companies, factories, and government agencies. During the observation period, students stay as close as the person's real shadow, hence the name shadowing program.

shutter the mechanical part of a camera that opens and closes quickly to determine if objects will be sharp or blurred. Also determines the amount of light that reaches the film.

sisal a strong white fiber usually made into cord or twine

site the specific location where something, such as construction, takes place

skill the ability to use knowledge and practice to perform activities well

skyscraper a tall building with an inside frame that is usually made of strong steel

smart highway a system in which cars are under control of a computer that navigates and directs the cars to their destinations

smog a heavy, dark fog caused by smoke and chemical pollutants mixing with fog

solar cells devices that convert sunlight into electrical energy. Solar cells are most commonly used in space vehicles, but commercial, industrial and residential use is increasing.

solar collectors a device used to collect solar energy for heating water or houses. These devices are often large, flat panels found on the roofs of houses.

solar heating a process by which water is heated by the sun in a system of tubes (usually on the roof of a house) and then pumped into the house for use

solar-powered having solar cells that convert the sun's energy into usable energy. Energized by the sun.

sound carrier wave communication signals carried by air

specialty craft ships, boats, and other water vessels made for special purposes, such as barges for transporting coal and other goods, and tugboats for pulling large ships into dock

spectrographic analysis the process that uses a special camera to record light waves onto film in order to determine what things are

standard a guideline set up as a rule for comparison of such things as quantity, value, weight, or quality

standard of living the necessities and comforts people enjoy and desire to make life more comfortable

static electricity electricity that is stationary, or unmoving

stop bath a chemical that is used to stop the developing process and remove excess developer from the film or paper

stroke a one-direction movement of a body or machine part

stud one of the upright, or vertical, wall supports of a building. Usually placed 16 inches apart.

subfloor the first layer of flooring, usually made of plywood, that takes much abuse during the construction of the building

subgrade the soil or rock that forms the first layer of a road

subsystem a minisystem that cannot exist outside of the larger system it is a part of

sulfur a compound found in coal that, when burned, gives off sulfur dioxide, an ingredient in acid rain

superconductors certain materials that, when cooled to very low temperatures, allow an electrical flow without resistance

supertanker a very large ship that is fitted with tanks for carrying oil across oceans

Supplier Certification Award the symbol that a company has passed a test and received approval for its quality control system

suppliers people or companies that provide one or more of the parts for a manufactured product

suspension bridge a roadway that is hung from large cables and goes across a wide span, or space

synthetic a product that is not found naturally and is made chemically

synthetic material materials created artificially by scientifically combining chemicals and elements

system an organized or established way of doing something through objects or ideas that work together to complete a task

Taylor system of manufacture standard ways of completing factory operations that make it easier for the worker. Also called *scientific management*.

technical drawing the drawing techniques that accurately represent the size, shape, and structure of objects. *Drafting* and *mechanical drawing* are other terms with the same meaning.

technician a person whose occupation requires specialized training in a certain subject area or who is known for having skill in a particular area

technologist a person who has special training and skill in technology

technology a scientific method that makes practical use of human knowledge. It involves using knowledge of scientific principles to change resources into products needed by a society.

telecommunication the methods of communicating over long distances; includes telephone, television, radio, etc.

test model the handmade full-scale form of a new design. Also called a *prototype*.

textile knit or woven cloth, or the fiber used to make cloth

thermography a printing process that involves the use of heat

third-angle projection a multiview drawing that shows drawings of the front, top, and right side of an object on the same paper

thrust the high pressure exhaust that pushes a jet engine forward

time and motion study an investigation into the best ways of doing a job. Things such as working conditions, wasted time, and unnecessary movement are observed

tolerance the allowable difference in size (smaller or larger) that a part can have from the design size and still be usable

tool something such as an instrument or an apparatus that increases one's ability to do work

toxic poisonous

tractor trailer a two-part truck that includes a tractor, or engine and cab, and a trailer that holds the load to be hauled

transformer an electrical device that changes electricity from one voltage to another

transistor an electronic device that performs the same functions as the vacuum tube, which it replaced. The transistor runs cooler, uses less power, and works faster than the vacuum tube.

transmission the parts and gears that cause the power to be sent from the engine to the axle

transmitter the part of a communication system, such as a radio, that changes the sender's message into electrical impulses and sends it through the channel to the receiving end

transportation moving people and products from one place to another

transportation system the parts and processes necessary to move people or things from one place to another in an organized way

transportation terminal a building or location used for entering and leaving a transportation system; a place where people and products are loaded and unloaded

tripod a three-legged stand on which a camera rests

truss bridge a bridge that is held together by steel beams that are fastened together in triangular shapes

turbine a disk or wheel that is made to turn continuously by air, water, or steam currents

typesetting putting together, or setting up, pages of type to be printed

ultrasound an image created by ultrahigh-frequency sound waves striking against the body part to be viewed

unlimited energy source an energy source, such as sunlight and wind, that will never run out

U.S. Customary measurement system a system of measures and weights based on measurements such as ounces and pounds, and feet and inches

utilities the electric, gas, water, and sewage services provided to homes and other buildings

vacuum a space that contains less pressure than is in the atmosphere

vehicle an object used to move people or things from one place to another

videophone a phone that incorporates a video camera, telephone, and computer so that people are able to both see and hear the person with whom they are communicating

volt the unit used to measure voltage or electrical pressure

voltage the force or pressure needed to push electricity through a circuit

vulcanization the process of treating rubber to give it strength, hardness, and elasticity

wattage the measure of electrical power needed to operate an electrical appliance. Wattage can be determined by using the following formula: watts = amperage × voltage.

wave communication the communication methods that have their messages made up of sound, electrical, or light waves that move through air, water, outer space, or solids

wholesaler the merchant who buys large quantities from a manufacturer and sells smaller quantities to retailers

wind farm a large number of windmills located in an area that has a fairly constant wind speed

working drawings scale drawings of an object that contain all of the information needed by the engineers, architects, and builders to construct the object

xerography the printing process that transfers negatively charged toner to positively charged paper. The toner is then melted into place. Also known as *electrostatic printing.*

X ray a type of radiation that penetrates different thicknesses of solid materials and creates a photographic picture of the interior of the solid penetrated making it possible to see inside the solid without cutting it open

Index